AF400551

# Communications and Control Engineering Series

Editors: A. Fettweis · J. L. Massey · J. W. Modestino · M. Thoma

Scientific Fundamentals of Robotics 7

M. Vukobratović · B. Borovac
D. Surla · D. Stokić

# Biped Locomotion

## Dynamics, Stability, Control and Application

With 136 Figures

Springer-Verlag
Berlin Heidelberg NewYork
London Paris Tokyo Hong Kong

Professor MIOMIR VUKOBRATOVIĆ, D. Sc., Ph. D.
Corr. member of Serbian Academy of Sciences and Arts
Foreign member of Soviet Academy of Scienes
Institute »Mihailo Pupin«, Beograd,
Volgina 15, POB 15, Yugoslavia


Assoc. professor BRANISLAV BOROVAC, Ph. D.
Faculty of Technical Scienes
University of Novi Sad, Yugoslavia


Assoc. professor DUŠAN SURLA, Ph. D.
Mathematical Institute
University of Novi Sad, Yugoslavia


Assoc. professor DRAGAN STOKIĆ, Ph. D.
Institute »Mihailo Pupin«, Beograd,
Volgina 15, POB 15, Yugoslavia


ISBN-13:978-3-642-83008-2     e-ISBN-13:978-3-642-83006-8
DOI: 10.1007/978-3-642-83006-8


Library of Congress Cataloging-in-Publication Data
Biped locomotion : dynamics, stability, control and application / M. Vukobratović . . . [et al.].
(Scientific fundamental of robotics: 7)
(Communications and control engineering series)
Translated from the Serbo-Croatian (Cyrillic).
Includes bibliographical references.
 ISBN-13:978-3-642-83008-2
1. Robots--Dynamics. 2. Human locomotion.
I. Vukobratović, Miomir. II. Series. III. Series: Communications and control engineering series.
TJ211.4.B57 1990
629.8'92--dc20                                                            89-26195



2161/3020 543210 – Printed on acid-free paper

# Preface

Fourteen years ago, the "Mihajlo Pupin" Institute published in English
a comprehensive research monograph by M. Vukobratović under the title
LEGGED LOCOMOTION ROBOTS AND ANTHROPOMORPHIC MECHANISMS. In this monograph were presented results of the seven-year work of a small research group in the Biocybernetics Department of the Institute "Mihajlo
Pupin", as well as results of researchers from several other centres
in the world. In 1976, the monograph was published in Japanese and Russian, and in 1983 in the Chinese language.

The period of vigorous research in the domain of mathematical modelling
and motion synthesis of the legged locomotion mechanisms and machines which
may be considered as the pre-manipulation era, was primarily concerned
with the applications in the rehabilitation of disabled people and with
the problems of specific (legged) transport. It is true, however, that
in the "locomotion era" of robotics various manipulation systems had
been developed, but only those aimed at the rehabilitation of severely
disabled people. It should be noticed, for example, that the first journal
articles published in English were devoted to the locomotion control,
especially of anthropomorphic mechanisms. We shall mention here only
two of them; one, "Contribution to the Synthesis of Biped Gait" by Vukobratović M., and Juričić D., published in IEEE Trans. on Bio-Medical
Engineering, Vol. 16, No. 1, 1969, the first paper on mathematical modelling of biped gait, and the other, "Some Considerations Relating to
the Design of Autopilots for Legged Vehicles", by Frank A.A., and McGhee
R.B., published in Journal of Terramechanics, Vol. 6, No. 1, 1969.

In these and several other articles, cited in particular chapters of
this book, a number of questions have been raised and some of them solved that were related to the essence of the activity which has later
been established as the scientific-technical discipline under the general name ROBOTICS. Thus, in the course of studies of locomotion mechanisms, synthesis and stabilization of artificial gait, and the realization of rehabilitation devices, the appropriate mathematical procedures have been developed for computer-forming of mathematical models

of the dynamics of complex active spatial mechanisms which are usually
encountered in the systems of anthropomorphic structure. The application of computer methods for automatic forming of locomotion mechanisms
models have coincided with the beginning of the development of these
methods for the needs of the open-chain manipulation mechanisms.

In parallel with the computer-oriented procedures for forming dynamic
models of active spatial mechanisms, the locomotion mechanisms, and the
anthropomorphic ones in particular, have induced genuine needs for dynamic control, and this has also been extended on the domain of control of manipulation robots for the industrial and other purposes. Thus,
with the aim of solving the problem of dynamic equilibrium of the gait
in a perturbed regime of the initial condition type and of small parameters variation, the force feedback mechanisms based on dynamic reactions at the points of contact of the mechanism's foot and ground have
been for the first time indicated (Vukobratović M., Juričić D., Frank
A.A., "On the Control and Stability of One Class of Biped Locomotion
Systems, Trans. of the ASME, June, 1970) and introduced (Vukobratović
M., Stepanenko Yu., "On the Stability of Anthropomorphic Systems",
Math. Biosciences, Vol. 15, Oct. 1972). Several years later, the load
and force feedbacks have been introduced as a means of stabilization
of manipulation robots, especially in the tasks of automatic assembly
and constrained motion of the manipulator gripper in the tasks of mechanical metal working. Even today, the problem of the so-called hybrid control (positional and force control) attracts great attention of
researchers in many research centres in the world.

The following questions may be asked: Why has this book been written in
the monograph series devoted to manipulation robots? What is new in this
monograph in comparison to the first research monograph mentioned at
the beginning of this preface? The answer to the first question is based on the fact that general problems of mathematical modelling, simulation, and stability, are common to all robotic systems, and that there are certain specificites in view of which the class of anthropomorphic mechanisms should be studied separately, what is actually the subject of the present monograph.

As for the novelties in comparison to the previous monograph on anthropomorphic mechanisms, this monograph represents a step forward in dealing with the programming support for generating mathematical models
of dynamics of an arbitrary anthropomorphic mechanism; the support is
having now the form and properties of the customer's software package. In

contrast to the previous one, this monograph considers the problems of stability and stabilization of the biped gait in their fully resolved forms using the methods of large-scale systems stability that have been developed in the meantime, as well as relying upon the adopted strategy of the decentralized control structure, and including all the necessary feedback aiming at controlling the dynamics of anthropomorphic mechanisms in a disturbed regime, taking into account that the system possesses the uncontrolable degrees of freedom. If compared to the 1975 monograph, this monograph presents also the results on the new concept of modular active orthoses which have, to a great extent, replaced the first realizations of complete exoskeletons, and which at the time of their appearance, represented a technological and medico-biological challenge in the rehabilitation of the most severely disabled people with the insufficient or nonexistent motor activity of the lower extremities. These were the reasons why we, after a longer period of time, decided to prepare a substantially innovated text. The subject matter of this monograph represents thus a sound background for mathematical modelling of the anthropomorphic bipedal gait, the investigation of its stability, as well as the synthesis of its dynamic control while taking into account the complete dynamic information of the system. The complexity attained in the modelling and synthesis of control of bipedal mechanical structure offers the possibility of its application in the synthesis of artificial biped gait of high anthropomorphic fidelity in its exoskeletal version of realization.

If the results obtained in the meantime in several research centres abroad are concerned, apart from some of them that have been repeated, this monograph encompasses new results from the USA (Hemami et al.) and in the USSR (Beletskii, Formal'skii, etc.) which are related to modelling and posture control of biped systems. It should be noticed that the contributions of the Soviet authors, and especially of Beletskii, have been covered in more detail if compared to the English-speaking authors, as we are of the opinion that because of the language barrier, these results are less accessible.

The only text that has been taken from the previous monograph is the section entitled "Method based on Euler's angles". The reason for doing so is the fact that this method, developed by Juričić and Vukobratović already in 1972, because of its high functionality in regard of modelling of anthropomorphic locomotion mechanisms, was very close to the current computer-oriented procedure for forming a dynamic model in its symbolic form. For this reason the authors wanted to draw attention of the concerned researchers to adapt this form of the model for its automatic generation in symbolic form, which would be of great importance for the microprocessor implementation of the model in real time.

This book consists of four chapters and two appendices.

Chapter 1 considers the dynamics of biped gait. First, a survey is given of the results on mathematical modelling of biped locomotion mechanisms in regard of their complexity. Further, the synthesis of anthropomorphic gait is described that is based on the original method of the prescribed synergy, known in the literature as the semi-inverse method, and which has been adopted by practically all the researchers that, after the appearance of the fundamental works of the Belgrade school of locomotion robotics, have been systematically concerned with the dynamics of anthropomorphic gait.

Appendix provides a detailed account of the method for modelling dynamics of the biped gait using Euler's angles.

Chapter 2 is entirely devoted to the problems of synthesis of the nominal dynamics of biped mechanisms. A special programming package is used which has been developed on the basis of the mathematical model using the general theorems of mechanics; the locomotion mechanism is modelled as a set of branched chains of the open configuration in the single-support gait phase, while in the double-support phase the chain representing the mechanism's legs is closed.

At the end of this chapter are given five examples of the nominal dynamics synthesis. In Example 1 we consider a mechanism having fourteen links, eight of them being powered, and involving the compensation by the trunk. Example 2 considers the same mechanism but with an additional degree of freedom at both the ankle joint and hip. The compensation in the frontal plane is achieved by the ankle joint, and in the sagittal plane by the trunk. In both examples the mechanism's "arms" are fixed to the trunk, whereas in Example 3 they are considered as free passive pendulums. Mechanical configuration and the choice of joints for the synthesis of compensating movements are the same as in Example 2. In contrast to the previous three examples which are concerned with the nominal dynamics synthesis for a single-support phase, Example 4 includes also the double-support phase. Example 5 considers modelling of a two-link foot.

Chapter 3 is devoted to the control of biped motion and to the problems of its stability analysis. Its first section reviews the results in the field, including also some results in the domain of realization of the artificial gait. The concept of two-stage control synthesis is used which has been already announced in the earliest period of the biped gait synthesis (Vukobratović M., Juričić D., "Contribution to the Synthesis of Biped Gait", IEEE Trans. on Biomedical Eng. Vol. 16, No. 1, 1969). In the first stage the control is synthesized which should ensure realization of the nominal (programmed) motion, while in the second stage (the stage of perturbed regimes) - the problems considered concerned with the gait realization under the conditions deviating from the nominal ones. A special method of control synthesis at the stage of perturbed regimes is developed in which accelerations of correctional movements of the mechanism are constrained in order to prevent significant disturbances of dynamic equilibrium of the system. In addition, the problem of global feedback with respect to the position of the mechanism's foot reaction is considered, whose purpose is to prevent overturning about the foot edge. All these considerations have been illustrated by the appropriate simulation results.

The remaining part of the chapter is concerned with the stability analysis of a biped gait mechanism possesing also the unpowered degrees of of freedom (the angles between the foot and ground). The aggregation--decomposition method is applied which is based on the Lyapunov vector functions in the bounded regions of the state space. In view of the fact that this method has been developed for the mechanisms with powered joints, it has to be extended onto the specific case involving both the powered and unpowered mechanism degrees of freedom. The models of unpowered degrees of freedom are associated with the mathematical model of the one (chosen in advance) powered degree of mechanism freedom and their common stability is analyzed. In addition, the results are presented concerning the stability analysis for the case when the mathematical model of an unpowered degree of freedom is associated with mathematical model of the hip, or of the ankle joint and for different control structures employed.

Chapter 4 presents a brief account on the fundamentals of biodynamics of locomotion and of biped gait in particular. The attention is paid to reviewing previous results in the field of realization of active devices in the form of exoskeleton for generating basic locomotor activity of the disabled with a special emphasis on the new results of modular design of active orthoses for a special class of insufficient motor activity of human extremities. Also, the most recent results are reviwed concerning the concept of hybrid actuator which uses the residual motor activity of the muscle system and which is additionally supplied with external energy from artificial actuators.

This new volume of the monographic series is intended for researchers interested in the dynamics of biped gait and its stabilization. Having in mind that this is a rather narrow field of robotics, the book could not expect a wider readership. However, taking into account that the problems of modelling of complex kinematic chains, synthesis of anthropomorphic gait, stability analysis and dynamic control synthesis, have been tackled from the point of view of a general approach, the book might attract attention of a much wider audience outside the limited number of people involved in studying strictly the biped gait biodynamics, its synthesis, and exoskeletal realization of the artificial gait.

In view of the above, the authors believe the book may be useful to students of general courses in robotics and biomechanics at both graduate and postgraduate level and of some special courses at technical faculties and faculties for physical culture and sport. The authors strongly believe in this statement because they also believe in the value of the results presented in the book, as far as their application is concerned, even in some new applications of mathematical modelling, simulation and dynamic control, for example in the biomechanics of athletics and other sports.

At the end of the many-year period of studies in the field of biped locomotion and its application we feel obliged to mention with great pleasure some people who gave highly valuable contribution to the results in this research.

First of all, it is our great pleasure to name Professor Davor Juričić, who has not been involved in this research field for a number of years, but who with the first author of this book, twenty years ago, laid the foundations of mathematical modelling of biped gait and traced the way of its further development. The section devoted to the mathematical models of biped system via Euler's angles, written by D. Juričić in the previous research monograph is included as the appendix of Chapter 1 in this monograph.

Also, we would like to express our gratitude to Dragan Hristić Ph.D.,

the unavoidable man in the field of rehabilitation robotics who, as
the closest co-worker of Professor Vukobratović for many years, has given substantial contribution both in the ideas and realization of artificial gait synthesis. He participated in both conceiving and writing of Chapter 4 of the present monograph. We are also indebted to
Zora Konjović, M.Sc., for writing the appendix on the programming package for the gait dynamics modelling. Finally our thanks are due to
Professor Luka Bjelica for translating the book into English, and to
Miss Vera Ćosić for her highly professional preparation of the type-
script for publication.


September 1989                              A u t h o r s
Beograd

# Contents

# Chapter 1:
# Dynamics of Biped Locomotion

## 1.1. Introduction

The motion of living organisms by means of legs, especially the locomotion of bipeds, has always been a challenging problem to scientists
of different vocations: biologists, physiologists, medicine specialists,
mathematicians, and engineers. In spite of their efforts, however, this
problem has not been solved yet in a satisfactory way.

From the viewpoint of mechanics the motion of living organisms can be
interpreted as a result of changes in equilibrium conditions within the
fields of forces in which the system finds itself. The spontaneous motion due to the redistribution of tension in muscle groups modifies the
relations between forces, bringing these relations to equilibrium or
taking them away from the equilibrium position. Study of these systems
and their motion requires certain simplifications because the legged
locomotion systems, and particularly the anthropomorphic mechanisms,
represent extremely complex dynamic systems both from the aspect of
mechanical-structural and control system complexity [1]. Nearly 350
muscle pairs are available to man for his complete skeletal activity
[2]. Such a system involves great dynamical complexity, even if it is
idealized to a system of rigid levers with simple torque generators
acting at each joint. In fact, mathematical analysis of the relationship between force and movement indicates that this form of interaction
does not have a unique dependence. This absence of uniqueness stems
from the fact that the relationship between force and movement is generated in a biomechanical sense based on a second-order differential
equation whose solution requires two initial values. These constants
of integration (initial position, initial velocity, ...) can lead to
quite different effects during the same initial innervation. This complexity (mechanical and any other) is the reason why any attempt of
practical realization of biped, based on a simple copy of the human
pattern, may be considered absurd.

Mechanical complexity of locomotion systems is only one of characteristics that makes the study very complicated. There are some other features to be mentioned, which also determine basic characteristics of the
behaviour of locomotion systems.

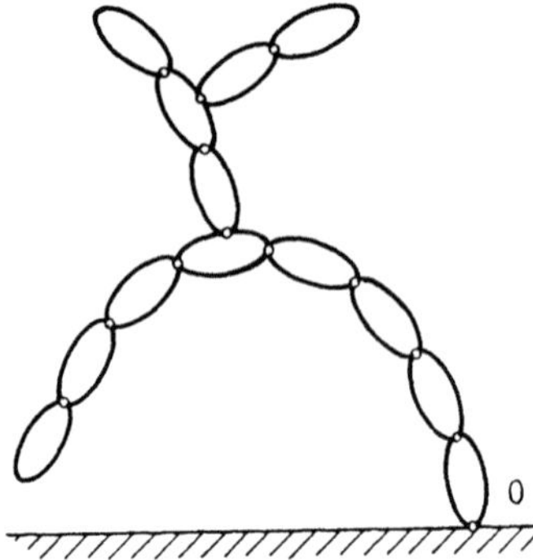

Fig. 1.1. System consisting of a set of kinematic chains

Let us consider a system consisting of links connected in such a way to constitute one or more kinematic chains (Fig. 1.1). At each of joints is applied an actuator whose function is to control the motion of the corresponding link. The only exception is the joint 0 which is, in fact, a degree of freedom (d.o.f.) formed in contact of the foot and the ground surface. It is not possible to effect it in a direct way, i.e., by the action of corresponding actuator. The effect of the ground surface can be replaced by the total reaction force (it is supposed the friction is large enough to prevent slippage), whose intensity, on the other hand, depends on the whole mechanism dynamics. Hence, the motion of an unpowered d.o.f. can be effected (or controlled) only by an appropriate motion of the rest of the system, i.e., by the appropriate actions of the powered d.o.f. If an unwanted situation happened, the system as a whole would rotate around the foot edge and collapse, even if the exact law of change of the internal angles was realized. In this case, the stability of the overall system is not preserved. However, it is obvious that for a locomotion system it is more important to preserve the realization of a walking process than to achieve certain level of tracking quality of the desired trajectories of link. In other words, we may allow that the system walks with certain deviations in the links position, but it is absolutely unacceptable that the system realizes the exact change of internal angles, and, at the same time, falls down by rotating around the foot edge.

Presence of an unpowered d.o.f. is the most important characteristic of locomotion mechanisms, especially of bipeds, because of its crucial influence on the system stability.

Another fundamental characteristic of every legged motion is a certain repeatability of movements which generate it. For a two-leg locomotion, the period on which movements are repeated over and over again is one step. That means that the positions and velocities at the beginning and at the end of each step are the same, and only the motion satisfying these conditions is acceptable for bipeds. They are known as the repeatability conditions [1]. These conditions result from the nature of legged motion and impose additional constraints on the possible solution in the process of motion synthesis. Also, it may happen that a particular gait type is required, what is of special importance in designing active orthoses used for reestablishing the locomotion function of the handicapped. These problems of artificial gait synthesis will be discussed in more detail later.

A third characteristic of two-leg locomotion is a permanent change of situations when the mechanism is supported on one foot and when the both feet are in contact with the ground. When the kinematic chain playing the role of legs is in contact with the ground only at one end, while the other is in the swing phase, the situation corresponds to a single-support phase. If however both ends are in contact with the supporting surface we speak of double-support phase. Each of these two cases is characterized by quite different dynamic situations and is to be studied separately.

The presence of a closed kinematic chain lowers the system order, and the mathematical equations describing it are much more complicated than those for the open kinematic chain used to describe the single-support phase. Additionally, when the system is in the double-support phase, there is no a unique set of driving torques and reaction forces which can be associated with the motion performed. In fact, the closed chain can support the internal torques which act in opposition but do not contribute to the mechanism motion, and which are only useless load to joints. In order to overcome this, it is necessary either to make some assumption about the way in which the ground reaction forces are divided between the feet during a double-support phase, or to directly measure the forces acting on the feet.

These problems will be discussed in more detail in the paragraphs to follow.

There are two approaches to studying locomotion activity: experimental, by investigating the motion of living organisms, and theoretical, by mathematical modelling. We shall focus our attention on the latter approach.

As already mentioned, the motion of a complex mechanism such as the human skeleton involves great dynamic complexity. Consequently, its mathematical description results in a high order system of nonlinear differential equations. However, the writing of such a large set of differential equations is always associated with the possibility of making mistakes. Another, more serious, problem is the impossibility to solve them in analytic way. This problem can be overcome in two ways. One of them is the simplification of the mechanical model used to represent the locomotion system, up to the level when only the characteristics of interest are preserved. For example, the human stance behaviour is well approximated by a single inverted pendulum model, controlled by torques applied at the pendulum base. Then, the whole body is approximated by one single massive link and the controlling torque corresponds to the ankle torque of the human body. Of course, more complex motions require more complex models. A further simplification is concerned with the system of differential equations. This is usually achieved by linearization, though some other methods have been employed, too. However, some very important features of locomotion systems may be lost by simplification, and for this reason it should be used with additional care. Another approach to mathematical modelling (apart from simplification of either the mechanism, or the mathematical model) is transferring the task of forming and solving the model to a computer. Then, the mathematical complexity is no more a limiting factor, and it can be chosen to match in the best way all requirements of the motion under investigation. If an appropriate software package is available, all changes, including the structural ones, can be realized by changing simply the input data [3]. Of course, the solution is obtained in numerical form, which is not convenient for further analysis. However, some recent software packages have offered the solution in analytic form [4], even in the case of closed kinematic chains [5]. The analytic models are very convenient if real time computation of dynamics is required.

In the following section we shall review the mathematical models used for studying locomotion systems.

## 1.2. A Brief Survey of Mathematical Models Used to Study the Effects Arrising with Locomotion Systems

The earliest systematic study of principles of human and animal locomotion is apparently due to Muybridge [6, 7] who invented a type of
motion picture camera which was successfully used in 1877 to obtain
the first photographic record of a quadruped gait. In his earliest
work, Muybridge was interested primarily in the sequence in which feet
are lifted and placed during the steady forward motion of a quadruped.
His investigations ultimately revealed a total of eight patterns of
gait employed by various animals, some of which were previously unknown to either horsemen or zoologists.

The first mathematical model of a tree-structure mechanism with rotational joints was introduced by Fischer [8] who wrote a set of Lagrange's equations which describes the motion of each link. The equations
were written "by hand" and a set of Euler's angles was used to define
the mechanism positions. However, because of their complexity, he wrote
in an explicit (open) form only the equations for two bodies connected
by a spherical joint.

The simplest model that can represent some locomotion activities is a
single massive link modelled as an inverted pendulum [9-14]. Two situations can be distinguished as the base joint is concerned: it can be
fixed to the supporting ground [9-11], or move in space [12-14]. Hemami et al. [9] used such a model to study the behaviour of a body in
standing position when no muscle dynamics is involved. Then, the torque applied at the base joint is equivalent to the ankle joint in human
body and it should maintain the upright vertical position. The same
model can be used when the torso motion is studied. Hemami et al. [10]
and Bavarian et al. [11] investigated the case when the base joint connected to the ground surface is powered with no locomotion involved.

In [12-14] is supposed that the torso base is not fixed, but follows a
prescribed trajectory by an appropriate activity of legs. Chow and Jacobson [12] assumed that two legs (each of them consisting of three
links) are attached to the torso and derived the law of the legs angle
change using optimization techniques. In the most of his examples Beletskii [13] studied the model of massive body supported on massless
legs. The body trajectories are determined by prescribing the position
of the hip joint and a sequence of legs supporting points on the gro-
und. Then, for each time instant the legs angles are uniquely defined.

Gubina [14] considered a spatially placed massive body supported on massless legs of variable length. The variable leg length should replace the knee function. The body positioning in space is achieved by extension of the supporting leg (it is supposed the "knee" can generate force) while the orientation has to be ensured by the three torques applied at the hip joint. The spherical coordinate frame was used as the closest to that operating in humans. The system behaviour was investigated by simulation on digital computer.

Various types of constraints can be imposed onto the torso base joint. It is usually supposed that the torso joint is a simple hinge which allows only rotation. In [10], for the purpose of a more realistic modelling, some soft constraints are taken into consideration. That means, the system is able to penetrate the surface of constraints, or, in other words, the base of the pendulum is set free with additional displacement to represent compliance in natural joints. In [11], the hinge is replaced by abstract constraints which attempt to model the cartilage and ligament behaviour.

Single inverted pendulum, in spite of the fact that it is a very rough approximation of a real locomotion system, can be considered very useful in studying certain activities of locomotion systems. There are a lot of situations when the entire body sways with no muscle dynamics involved; in fact, this is a class of problems of postural stabilization. The task imposed to the control system is the maintenance of the vertical upright position if a deviation occurred.

Owing to their greater complexity, multi-link planar models offer studying the locomotion effects that are not possible to investigate with the single-mass model. The application of internal torques at mechanical joints, what corresponds to the muscle dynamics of human body enables changing of the relative positions of all mechanism links. In this way, the gait, a basic feature of all locomotion systems, can be performed.

Planar models of different complexity have been used by Hemami and his co-workers, mainly to study the problem of posture control in the presence of various types of constraints. For this purpose, simple movements are simulated such as the side-to-side sway, sway of the body, return to vertical stance from certain initial displacement, and similar. No gait in any case was involved. A two-link model representing a massive body and a massive leg is used in [15]. It is supposed that each connection constraint is maintained by a single ligament which consists of one viscoelastic element, a spring-dashpot combination. In addition, each joint is driven by a group of active muscles. Simulation with the initial deviation in positions of links is carried out, and the system stability thus proved. A three-link model is used in [16-19]. The mechanism consists of two legs and a torso. Every part is supposed to be a rigid link. In [16-18], various constraint forces are imposed on the system with the purpose of simulating system's behaviour under different conditions. In [15], the constant forces of constraints and the forces being functions of the state vector are introduced. In addition in [17] and [18], some on-off constraints and the constraint forces being explicit functions of the state and the input respectively, are considered. In addition, in [18], the postural stability is investigated by simulation of a side-to-side biped sway in the frontal plane. In [20], a five-link planar model is derived. Each link is supposed to be massive and its returning from eight initial poses to the vertical stance is simulated. In [21], a five-link planar model represents the torso, thigh, shank, and two-link complex foot. The links of heel and toe are connected with a spring-dashpot pair to represent the plantar fascia. The system is subjected to an external impact force. Its behaviour is studied by simulation, and the results show that the spring-dashpot pair prevents the arch structure from collapsing and makes the foot work like a rigid body. The most complex planar model used by the same author is a nine-link biped [22]. The model has two legs (each consisting of the thigh, shank, and a two-linked foot) and a torso. It is supposed that at the joints act both ligamentous forces and forces produced by muscles. The simulation is carried out for initiation of walking and for the system going on tiptoe.

A further improvement of joint modelling is done in [23, 24], and, because of its great importance in locomotion processes, special attention is paid to the knee function. A model of planar motion of the human knee joint, developed in [23], involves a relative motion of the geometry of the contacting surface between the tibia and the femur. The pure gliding motion and the pure rolling motion are formulated including the holonomic and non-holonomic constraints that must be satisfied. To the model in [24] the ligaments are added and their function as local controllers, independent of the central nervous system in maintaining the integrity of the joint, is shown. Three kinds of surface motion of the knee: gliding, rolling, and combined gliding and rolling are considered, while the holonomic and non-holonomic constraint equations are used to

8

describe these modes of motion. Computer simulation of rolling movement
of the knee is presented.

Another five-link planar model, for simulation of locomotion processes,
is developed by Formal'skii [25]. The mechanism considered consists of
the heavy body and two two-link legs (without feet). All data about
links are similar to those of the human body. The differential equations of the mechanism motion are written and solved then with additional constraints of repeatability conditions.

The mathematical models described in all previous papers were used for
a complete investigation. In some cases, the results were compared with
those measured for the human body. However, only the mathematical model
given in [26, 27] served as the basis for a built-up active mechanism
for realization of the artificial locomotion. Miura and Shimoyama [26]
used the model of a three-link mechanism (two one-link legs and the
pelvis) to develop the construction BIPER-3. The mechanism behaviour
was modelled and studied in the frontal and sagittal plane, and the results served as a basis for choosing the appropriate feedback gains. The
validity of the adopted control structure was proved by the experiment
involving the real mechanism BIPER-3. Furusho and Masubuchi [27] used
a planar five-link model as a basis for developing a spatial mechanism.
The model had two two-link legs with no feet, and the torso. Since a
planar model was used to investigate the behaviour of a spatial mechanism, a steel pipe was attached to the lowest end of the leg, in order
to maintain the lateral balance, and constrain robot's motion in the
sagittal plane. The mathematical model was used for both simulation of
the locomotion process and stability analysis. Validity of these considerations was then checked by experiment.

Moreynis and Gritsenko [28] developed and investigated a mathematical
and physical model of gait with the aim of acquiring information about
the dynamic characteristics of both healthy persons and those using
prosthesis. They modelled a nine-link mechanism with eleven d.o.f.
whose motion was described by eleven second-order differential equations, written in the form of Lagrange's equations. All geometrical and
physical data about the mechanism are supposed to be known. The accelerations due to the movements of the hip joint with respect to the
orthogonal coordinate frame were obtained experimentally, and the
ground reaction forces recorded using a force plate. The angles were
measured by means of suitable potentiometers and the second derivatives

were obtained by numerical differentiation. Then, from the system of differential equations they were able to compute the driving torques which can be further used for computing the power consumption at each joint. The procedure enables an estimation of the energy demands during the gait with prostheses of various design, and the information acquired can be used for designing the anthropomorphic mechanisms. A similar method for forming mathematical models of the simplified locomotion mechanisms was developed by some others authors (for example, Gurfinkel et al. by [29]).

The model of a spatial mechanism with all links massive was developed first by Vukobratović and Juričić [30, 31]. The mechanism consisted of six links: two two-link legs, massive pelvis and the torso approximated by a concentrated mass attached to a rigid massless cane. The torso possesses two d.o.f. which enables it to move in both the frontal and sagittal plane; the position is defined by two angles in these two planes measured from the vertical. The system dynamics is maximally simplified by introducing a specific gait pattern. It is supposed that in such gait shanks are parallel during the whole cycle. Both leg ends are permanently at the level of the ground surface, but the end of the front leg is just above it. Thus, the contact between the system and the ground is realized at one point - the rear leg end. The rear leg, fixed to the pelvis, is permanenlty streched. When the half-cycle is finished and the rear leg becomes the front one, the mechanism instantaneously changes the supporting leg and continues the walk. In this way, the double-support phase is avoided. Because of these simplifications, related to the kinematic connections between the links of the locomotion mechanism, it is possible to define the gait by one angle only. The equations of the mechanism motion are written with respect to the supporting point on the ground. Owing to the fixed kinematic programme of the lower extremities motion, the angles of the torso position can be computed for each time instant. In this way, the compensating motion of mechanism's upper part is defined under conditions of preserving the overall system stability.

By the method of prescribing the law of legs motion, the problem of obtaining the solutions with repeatability conditions (i.e., the problem of the desired gait type) is practically avoided. The repeatability conditions are automatically satisfied with the chosen gait pattern, and the compensating movements of the rest of the mechanism are then computed to fulfil  the stability requirements of the system as a whole.

To each law of legs motion corresponds a different law of the torso motion. A whole range of compensating movements, computed in this way, was presented in [1, 30, 31].

A model much closer to the actual anthropomorphic system is known as the "system with fixed arms", also presented by Vukobratović [1]. The system consists of twelve heavy links, spatially placed. Each leg has the foot, shank, and thigh. Both legs are connected to the pelvis which is an intermediate link between the legs and the upper part of the body. To the torso, represented as a heavy inverted pendulum, the two--link arms are attached. The arms are fixed on the chest, and they do not change their position during the walk. All the links are connected by simple rotational joints.

The law of legs motion is adopted from man, and is characterized by a very smooth behaviour of the pelvic link, which is of importance for practical use of the active exoskeleton. The pressure under the supporting foot can be replaced by the appropriate reaction force acting at a certain point of the mechanism's foot. Since the sum of all moments of active forces with resepect to this point is equal to zero, it is termed the zero-moment point (ZMP). During the gait the pressure diagram changes, what corresponds to a change of the direction of the reaction force, its intensity and position of the acting point. It is supposed that for the considered "system with fixed arms" the reaction force changes three times during one half-step. The first phase is characterized by the foot striking the ground ("heel strike"), the second one corresponds to the situation when the system is supported on the full foot, and in the third phase the heel deploys, and the contact point is under the toes. Then the mechanism changes its supporting leg and the ZMP passes over under the other foot, presently being in contact with the ground. It should be noted that this transfer of the ZMP causes the gait to be smoother to some extent. However, even a more natural gait can be realized by prescribing the ZMP trajectory corresponding to the double-support phase.

If the kinematic algorithm (chosen gait type) and the ZMP trajectory are known, it is possible to compute the upper part dynamic algorithm (compensating synergy). An example of compensating movements thus obtained has been presented, together with the diagrams for the prescribed synergy (legs trajectories) in three cases: walk upon level ground, climbing stairs, and descending stairs, for all three angles (at the ankle, knee, and hip).

The mathematical models in all above papers were formed for particular structures, and no general method in mechanism modelling has been used. One of the first general mathematical methods for modelling anthropomorphic mechanisms, known as the "method based on Euler's angles", was developed by Juričić and Vukobratović [31]. The method will be presented in detail in Appendix, and only its main characteristics will be outlined here. The approach used considers the structure as composed of a set of heavy links interconnected by "ball-and-socket" type of joints. Due to this type of connections, the number of d.o.f. for each link is reduced to three. These are described by three Euler's angles making them appear as the variables of dynamic equations of motion. The equilibrium of the moments is considered for each joint separately. That leads up to a system of equations on which the prescribed synergy method is applied: the motion of the legs is assumed to be known, while the moments about the main supporting point, as well as about the joints of passive links, have to vanish for the equilibrium condition. Then, the set of equations is split up into an algebraic and differential part. The differential part, the part having Euler's angles of compensating links as unknowns, can be used to complete their law of change during one step.

All data necessary to define the mechanism structure and the task imposed have to be defined as input data. This means that the inertial and geometrical data of all the links, together with the dimensions of the system and its structure defined by structural matrices, are all the data necessary to reduce the general mathematical model to the equations of motion of a specific structure. In this way, modelling of any biped structure is made possible by using the same general mathematical model, changing merely the structural matrix. Although some limitations to the model are imposed, they lead only to the usual restrictions, as for example, the exclusion of a leaping gait and of a walk with hands held together. The differential equations obtained are used to solve the boundary condition problem stated by the repeatability conditions of the gait. The solution of the non-linear boundary condition problem is implemented by using the sensitivity matrix approach.

From this brief review of the results concerning mathematical modelling of locomotion mechanisms it can be seen how serious problems arise with the greater mechanism complexity. In some cases, it is quite acceptable to use simple models which can be easily formed and solved. However, in the most cases, especially if it is not only a posture problem, but the gait is involved, the model to be used is much more complex. To handle such complex models specific computer programmes have been developed [1]. Thus the problem of forming and solving the mathematical model is transferred to a computer, and any change, even in mechanism's structure, can be made simply by changing the input data. Much more about it will be said in Chapter 2.

## 1.3. Artificial Gait Synthesis – A Method Based on Prescribed Synergy

The synthesis of artificial motion of a biped is a very complex problem. The goal is to achieve a human-like gait, as the best example of such type of motion. An important practical reason for studying such systems is the application of the results in restoring the locomotion function of disabled persons.

Studies of human locomotion by measuring all its characteristics of interest can answer the question how man walks. However, the question why he walks in such a way remains still unanswered. As stated by Bernstein [32], the human walk is a most automated motion, and no central nervous system is involved in its steady regime. Thus, the locomotion process appears to be similar to a certain algorithm which is executed over and over again when no disturbances occur.

Some authors have tried to define the motion of an artificial biped mechanism, but in all cases they had to introduce certain simplifications. Most often they assume massless legs. As a consequence, the motion of legs does not influence the body motion, which is of great importance in preserving a low order of model complexity. Such a simplification, introduced by Beletskii [13] and Gubina [14], implies the reduction of the system of differential equations describing the body motion to one vector equation. If the motion is restricted to one plane [13], this is a differential equation of second order, but if the torso moves in space [14] then three such differential equations are necessary to define it. If the hip trajectory is prescribed, and the positions of the supporting points of legs on the ground are known, then the legs' positions are uniquely defined [13]. However, only the legs angles at the end of each half-step are thus defined, while the legs trajectories between these two terminal positions remain unknown. Therefore, the assumption of massless legs can hardly be accepted as adequate for the actual system.

A quite different approach to the gait synthesis, based on optimal programming, has been proposed by Chow and Jacobson [33]. They have considered a seven-link planar mechanism consisting of two three-link legs (the shank and thigh are massive, whereas the foot is massless) and the torso modelled as a heavy inverted pendulum. All geometrical and inertial data are supposed to be known.

The mechanism motion is described by five nonlinear differential second-order Lagrange's equations. Then, the problem of gait synthesis can be defined as: how to determine generalized forces such to perform a periodical walk. Because of great complexity of the problem, the authors introduced the following simplifications. The first simplification is to prescribe the hip trajectory, which allows splitting the system into three independent parts: the torso and two legs. The second assumption is that all partial derivatives of the hip trajectory are zero. This enables the hip joint to be considered as the origin of a local coordinate frame which performs the desired motion. In this way, the so-called "simplified dynamics", is defined. By additional studies of the foot trajectory in the support and deploy phase, and after some simplifications related to the forces which act on the foot, the model is reduced to two d.o.f. with kinematic constraints and the force and torque at the ankle joint. In this way the model is prepared for the optimization procedure. The authors prescribed the boundary conditions ensuring repeatability, so that the problem of gait synthesis can be restated as: how to define the torques at the hip and knee joints to perform the system's periodical trajectories, and, at the same time, to minimize the desired criterion. After numerous simplifications, the system of differential equations reduces to four equations of the first order. To obtain the solution, the maximum principle was employed. As a result, the authors arrived at a two-point boundary value problem which they solved numerically.

This work can serve as proof of how difficult the optimization task is by itself. On the basis of the criterion of minimum energy consumption, the authors tried to obtain the driving torques at the hip and knee, and the appropriate "optimal" trajectories of the corresponding joints. But, to make this optimization problem solvable, they made some assumptions that has considerably simplified the initial locomotion problem. Thus, the motion problem was reduced to the sagittal plane. Here, the problem of dynamic equilibrium was not solved, and the problem of the unknown dynamic reactions was avoided by using experimental data. Given these simplifications, it would be rather difficult to expect that result of the optimization of such a model could present some realistic and truly optimal results.

Another possible approach is to obtain a direct solution of the model with some additional constraints imposed on the system. A very good example of this, is the work of Formal'skii [25]. He considers a five--link planar model with all heavy links in the single-support phase only. The problem is how to define driving torques at the mechanism's joints such to ensure the mechanism performs a periodical walk. In addition, Formal'skii assume the driving torques act for a very short time in the beginning of each half-step (the so-called impulse control). The constraints of repeatability conditions are also imposed. The solution is obtained in two ways. The first approach is to linearize the model around the vertical position and then solve it. In the second approach, a nonlinear  model is solved by using an iterative procedure. Two solutions are obtained: the symmetric - which is very close to the linearized case, and non-symmetric - resembling the way in which man walks. All solutions were illustrated by stick diagrams.

In this way, by a direct solving of the whole model, only the repeatability conditions are satisfied. However, the very important problem of the overall mechanism stability is not taken into account, and neither is given the possibility to choose a desired gait type. If these results are supposed to be applied in rehabilitation, this would be a serious drawback.

These problems can be overcome by using the prescribed synergy method. Simply speaking, to the one part of the system dynamics is prescribed, whereas from the rest of the system such "compensating" dynamics is required to bring the system to equilibrium with respect to the task specifications and corresponding conditions of dynamic connections. It should be pointed out that in this way we come to a specific procedure of dimensionality reduction which is neither system's simplification, nor its mathematical linearization. By prescribing various synergies for the one part of the system we can obtain an adequate set of compensating synergies, and thus replace all dynamic possibilities of the system by those dynamic forms being of interest for the example considered. Let us denote the desired system dynamics (both the prescribed and compensating part) as the nominal dynamics, whose mathematical formulation can be carried out in the following way.

Let $P \in R^n$ be a vector of generalized forces of an arbitrary dynamic process, and let $q \in R^n$ be its dynamic state vector. Then, this process can be represented by the following set of nonlinear dynamic equations

$$P = A \cdot \ddot{q} + B \cdot [\dot{q}^2] + C \cdot [\dot{q}\,\dot{q}] + G \tag{1.3.1}$$

where the elements of the matrices A, B, and C, and the column G depend on the adopted generalized coordinates, while $[\dot{q}^2]$ and $[\dot{q}\,\dot{q}]$ denote vectors whose each element correspond to square and product of generalized velocities, respectively. That means, equation (1.3.1) represents a very large class of dynamic systems.

By analogy with the classical tasks in mechanics, the following three cases can be defined, corresponding to the first, second, and combined type of the problem of dynamics.

a) In the simplest case, the motion (in a mechanical sense), i.e., the dynamics of the complete system (in a general sense) is known. Then, the resulting set of equations with respect to the unknown generalized forces becomes a trivial set of algebraic equations from which the control actions of the dynamic process can be computed.

b) In the second case, the dynamics (motion) should be determined if the prescribed control actions (generalized forces of the dynamic system) are applied. For this purpose, equation (1.3.1) should be transformed into

$$\ddot{q} = A^{-1}(P - B[\dot{q}^2] - C[\dot{q}\,\dot{q}] - G) \tag{1.3.2}$$

The resulting system of second-order differential equations can be solved by numerical methods, assuming the initial conditions $q(0) = q_0$ and $\dot{q}(0) = \dot{q}_0$ are known.

c) In the combined type of problem, the generalized forces (moments) for one part of dynamic system are known and for the rest of the system they have to be determined in such a way to satisfy certain dynamic connections.

16

Let us denote the known generalized forces and dynamic states with $P_o \in R^k$, $(k<n)$ and $q^o \in R^{n-k}$, $(k<n)$, respectively. Accordingly, $P_x \in R^{n-k}$ and $q^x \in R^k$ are the unknown forces and states. Let the transformation matrices T and R be

$$\left[\begin{array}{c} T_o \\ \hline T_x \end{array}\right] P = \left[\begin{array}{c} P_o \\ \hline P_x \end{array}\right] , \quad [R_o \mid R_x] \left[\begin{array}{c} q^o \\ \hline q^x \end{array}\right] = q \qquad (1.3.3)$$

After applying such transformations to (1.3.1), it follows

$$\left[\begin{array}{c} P_o \\ \hline P_x \end{array}\right] = \left[\begin{array}{c} T_o \\ \hline T_x \end{array}\right] A [R_o \mid R_x] \left[\begin{array}{c} \ddot{q}^o \\ \hline \ddot{q}^x \end{array}\right] + \left[\begin{array}{c} T_o \\ \hline T_x \end{array}\right] (B[\dot{q}^2] + C[\dot{q}\,\dot{q}] + G) \qquad (1.3.4)$$

where

$$\left[\begin{array}{c} T_o \\ \hline T_x \end{array}\right] A [R_o \mid R_x] = \left[\begin{array}{c|c} A_{oo} & A_{ox} \\ \hline A_{xo} & A_{xx} \end{array}\right] . \qquad (1.3.5)$$

and where the submatrices $A_{ox}$ and $A_{xo}$ are symmetrical since the sets of the known generalized forces and motions (i.e. dynamics) are complementary. Equation (1.3.4) can be rearranged in such a way that $P_x$ and $\ddot{q}^x$ appear explicitly, that is

$$\left[\begin{array}{c} P_x \\ \hline \ddot{q}^x \end{array}\right] = - \left[\begin{array}{c|c} 0 & A_{ox} \\ \hline I & A_{xx} \end{array}\right]^{-1} \left( \left[\begin{array}{c|c} -I & A_{oo} \\ \hline 0 & A_{xo} \end{array}\right] \left[\begin{array}{c} P_o \\ \hline \ddot{q}_o \end{array}\right] + \right.$$

$$\left. + \left[\begin{array}{c} T_o \\ \hline T_x \end{array}\right] (B[\dot{q}^2] + C[\dot{q}\,\dot{q}] + G) \right) \qquad (1.3.6)$$

System (1.3.6) consists of both algebraic and differential equations, and can be divided in two subsystems with respect to the unknown dynamics (the system of generalized coordinates) and unknown generalized forces:

$$\ddot{q}^x = A_{ox}^{-1} [P_o - A_{oo} \cdot \ddot{q}^o - T_o (B[\dot{q}^2] + C[\dot{q}\,\dot{q}] + G)] \qquad (1.3.7)$$

$$P_x = A_{xx} \cdot \ddot{q}^x + A_{xo} \cdot \ddot{q}^o + T_x (B[\dot{q}^2] + C[\dot{q}\,\dot{q}] + G) \qquad (1.3.8)$$

Equations (1.3.7) and (1.3.8), or system (1.3.6) is a mathematical basis for the nominal regime synthesis based on the prescribed synergy concept.

As we are dealing with mechanical systems for artificial locomotion, further considerations of applicability of the prescribed synergy will be restricted only to the problem of nominal regime synthesis for this class of mechanisms.

### 1.3.1. Single-support phase

The locomotion activity of a biped has certain important characteristics which have to be taken into account when an artificial motion is to be synthesized. They have already been mentioned in Section 1.1., but we are going to review them briefly and then, concentrate on the problem of artificial gait synthesis based on the prescribed synergy method.

The most important characteristic is the presence of the unpowered d.o.f. which occur only with the locomotion mechanisms, and are of basic importance for their stability as a whole. Namely, in the contact of the foot and the ground, an additional d.o.f. is formed. The mechanism would rotate around the foot edge if an undesirable situation happened, and the task of control system is to prevent this. However, each "regular" d.o.f., except the unpowered one, is powered and controlled by its own actuator and any deviation from its nominal state can be corrected only by an appropriate dynamics of the rest of the system. Because the "rest of the system" in case of locomotion mechanisms is, in fact, the whole mechanism (the unpowered d.o.f. always appears at the lowest end of the system), the dynamics of the whole locomotion mechanism is "responsible" for the behaviour of the unpowered d.o.f. If a complex mechanical structure is involved (the structure with a greater number of d.o.f.), the problem appears to be extremly inconvenient both for the nominal motion synthesis and the control of locomotion processes in the presence of disturbances.

The second characteristic of such mechanisms which is of importance for the gait synthesis, is the variable mechanism structure. In the walking process, the system is alternatively supported on the one and both feet, and in this way, the configuration of the legs' kinematic chain changes from the open to a closed one. When the mechanism is supported on one leg, the situation is recognized as single-support phase, when both feet are on the ground, as double-support phase. Each of these cases gives a quite different dynamic portrait and has to be modelled separately.

A third characteristic is related to the periodical character of the mechanism motion in the walking process. The positions and velocities at the beginning and at the end of each step have to be the same, so that the walk can be performed continuously.

It is obvious that the motion of biped systems is very complex either from kinematic or dynamic point of view. The dynamics has a dominant role because of the existence of unpowered d.o.f. A further understanding of the walking process require a detailed description of the behaviour of those d.o.f. related to the motion of the whole system. The information about the forces, i.e., about the torques at mechanism's joints and the ground reaction forces under the foot (concerning both intensities and directions), are very useful in determining the system's behaviour. The following experiment can serve as proof that the human control system regulates locomotion with respect to forces. A healthy man with a temporary artificially paralysed vestibular system of natural gyroscopes can walk in a stable gait without great difficulties under the condition that a visual feedback is preserved. In such case, he is simply not able to keep a permanently steady attitude during the gait. However, the same person, with his vestibular system functioning properly but with the locomotor - muscle system of his lower extremities artificially paralysed, moves in the same way as a paralysed paraplegic person. It is quite clear that a healthy man "feels" forces, i.e., dynamic reactions, and distributes them in some manner in the form of driving torques along his skeleton activated by numerous muscle groups. Force sensing at the feet of a healthy person is a global feedback which takes care of the overall system behaviour and can be applied to artificial mechanisms, too.

Let us suppose the system is in the single-support phase and the contact with the ground is realized by the full foot (Fig. 1.2). Then, it is possible to replace all vertical elementary reaction forces by the resultant $R_V$. If we reduce it to the centre of supporting area, the reaction force N and moment M will be obtained. As the pressure diagram is of the same sign, the reaction force $R_V$ can be always computed. Obviously, the ZMP in single-support phase cannot be out of the supporting area (area covered by one foot), while in double-support phase it

can be anywhere inside the dashed area (Fig. 1.3). Within these areas
the ZMP can move in accordance with different laws, continuously or
not, depending on which gait type is performed.

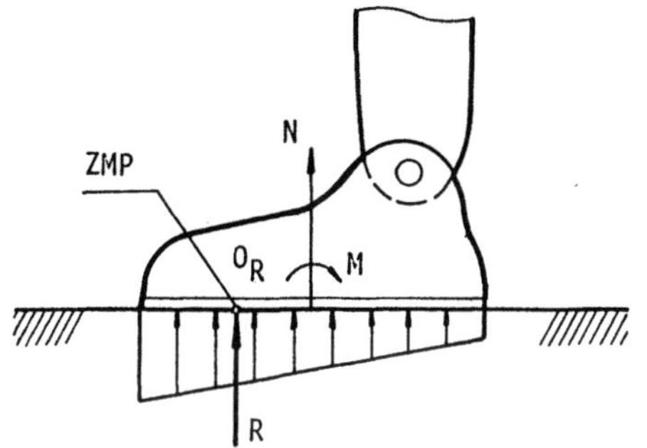


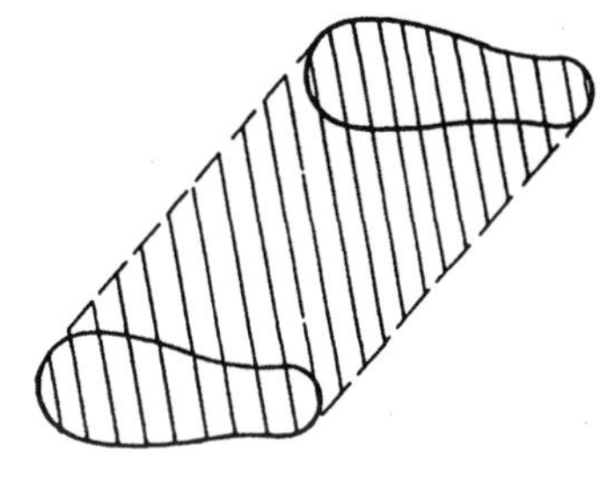

Fig. 1.2. Longitudinal distribution    Fig. 1.3. Area of allowable
of pressure on the foot                    positions of ZMP in
and ZMP position                           double-support phase

The basic idea used in artificial synergy synthesis is that the law of
change of vertical reaction and friction forces (i.e., total reaction
force) under the foot is known in advance, or prescribed. The prescribed part of dynamic characteristics which, in a dynamic sense, additionally restricts the system, is named "dynamic connections". Thus, if a
certain point represents the ZMP, and if the ground reaction forces $\vec{R}_V$
are reduced to it, then the moment $\vec{M}$ should be equal to zero. The vector $\vec{M}$ has always a horizontal direction and, hence, two dynamic conditions have to be satisfied: the projection of the moment on two mutually orthogonal axes X and Y in the horizontal plane should be equal
to zero.

$$M_X = 0, \quad M_Y = 0 \tag{1.3.9}$$

As for the friction forces, it can be stated that their moment with
respect to a vertical axis V is equal to zero:

$$M_V = 0 \tag{1.3.10}$$

The axis V can be chosen to be in any place, but if it passes through
the ZMP, then the axes X, Y, and V constitute an orthogonal coordinate
frame, and V will be denoted by Z.

The external forces acting on the locomotion system are the gravity,
friction, and ground reaction forces. Let reduce the inertial forces
and moments of inertial forces of all links to the ZMP and denote

them by $\vec{F}$ and $\vec{M}_F$, respectively. The system equilibrium conditions can be derived using D'Alambert's principle, and conditions (1.3.9) can be rewritten as

$$(\vec{M}_G + \vec{M}_F) \cdot \vec{e}_X = 0, \qquad (\vec{M}_G + \vec{M}_F) \cdot \vec{e}_Y = 0 \qquad (1.3.11)$$

where $\vec{M}_G$ is the total moment of gravity forces with respect to ZMP, while $\vec{e}_X$ and $\vec{e}_Y$ are unit vectors of X and Y axes of the absolute coordinate frame. Since the gravity forces are parallel to the V axis, the third equation of dynamic connections (1.3.10) becomes

$$(\vec{M}_F + \vec{\rho} \times \vec{F}) \cdot \vec{e}_V = 0 \qquad (1.3.12)$$

where $\vec{\rho}$ is a vector from the ZMP to the piercing point of the axis V through the ground surface; $\vec{e}_V$ is a unit vector of axis V.

Let us adopt the relative angles between two links to be the generalized coordinates and denote them by $q^i$. The relative (internal) angles do not depend on the choice of the absolute coordinate frame, and they are very convenient for defining the mechanism's position. Additionally, let suppose the mechanism foot rests completely on the ground, so the angle between them is zero, $q_o \equiv 0$. The inertial force $\vec{F}$ and moment $\vec{M}_F$, in a general case, can be represented in linear form of the generalized accelerations, and in quadratic form of generalized velocities

$$F^k = \sum_{i=1}^{n} a_i^k \cdot \ddot{q}^i + \sum_{i=1}^{n} \sum_{j=1}^{n} b_{ij}^k \cdot \dot{q}^i \dot{q}^j, \qquad k=1,2,3$$

$$\qquad (1.3.13)$$

$$M_F^k = \sum_{i=1}^{n} c_i^k \cdot \ddot{q}^i + \sum_{i=1}^{n} \sum_{j=1}^{n} d_{ij}^k \cdot \dot{q}^i \dot{q}^j, \qquad k=1,2,3$$

where the coefficients $a_i^k$, $b_{ij}^k$, $c_i^k$, $d_{ij}^k$ ($k=1,2,3$; $i=1,\ldots,n$; $j=1,\ldots,n$) are the functions of generalized coordinates, and $F^k$ and $M_F^k$ ($k=1,2,3$) denotes projections of vectors $\vec{F}$ and $\vec{M}_F$. By introducing these expressions into (1.3.11) and (1.3.12), one obtains [1, 3]:

$$\vec{M}_G \cdot \vec{e}_X + \sum_{i=1}^{n} c_i^1 \cdot \ddot{q}^i + \sum_{i=1}^{n} \sum_{j=1}^{n} d_{ij}^1 \cdot \dot{q}^i \cdot \dot{q}^j = 0$$

$$\vec{M}_G \cdot \vec{e}_Y + \sum_{i=1}^{n} c_i^2 \cdot \ddot{q}^i + \sum_{i=1}^{n} \sum_{j=1}^{n} d_{ij}^2 \cdot \dot{q}^i \dot{q}^j = 0$$

$$\sum_{i=1}^{n} c_i^3 \ddot{q}^i + \sum_{i=1}^{n} \sum_{j=1}^{n} d_{ij}^3 \cdot \dot{q}^i \dot{q}^j + \rho^X ( \sum_{i=1}^{n} a_i^2 \ddot{q}^i + \sum_{i=1}^{n} \sum_{j=1}^{n} b_{ij}^2 \dot{q}^i \dot{q}^j ) -$$

$$- \rho^Y ( \sum_{i=1}^{n} a_i^1 \ddot{q}^i + \sum_{i=1}^{n} \sum_{j=1}^{n} b_{ij}^1 \dot{q}^i \dot{q}^j ) = 0 \qquad (1.3.14)$$

where the superscripts X, Y denote the components in the direction of the corresponding axis.

If the locomotion system has only three d.o.f., the trajectories for all angles $q^i$ can be computed from (1.3.14).

If, however, the mechanism is designed to perform a gait, it usually possesses more than three d.o.f. Because (1.3.14) does not contain the information about the gait type, the trajectories for the rest (n-3) coordinates should be prescribed in such a way to ensure the desired legs' trajectories. It is the easiest and most suitable way to adopt this information from measurements of human gait parameters. In this manner, the problem of legs' trajectories synthesis is avoided, since the criteria which have to be satisfied in the synthesis of a desired gait type are unknown. The trajectories for this part of the system trajectories are prescribed, while the dynamics of the rest of the system is determined in such a way to preserve the overall mechanism stability. This method, apart from the advantage of attaining a specific reduction of the system order which is neither simplification nor linearization of the mechanism, offers the possibility of realization of any desired gait type, which is of special interest in designing exoskeletons for rehabilitation of disabled persons.

Now, the set of coordinates $q^i$ can be divided in two subsets: the first one containing all coordinates whose motion is prescribed, denoted as $q^{oi}$, and the second subset comprising all coordinates whose motion is to be defined using the prescribed synergy method, denoted as $q^{xi}$. Accordingly, the condition (1.3.14) becomes

$$\sum_{i=1}^{n} c_i^k \ddot{q}^{xi} + \sum_{i=1}^{n} \sum_{j=1}^{n} d_{ij}^k \dot{q}^{xi} \dot{q}^{xj} + g^k = 0, \qquad k=1,2,3 \qquad (1.3.15)$$

where $c_i^k$ and $d_{ij}^k$, (k=1,2,3) are the vector coefficients dependent on $q^o$, and $q^x$, whereas vector $g^k$, (k=1,2,3) is a function of $q^o$, $\dot{q}^o$, $\ddot{q}^o$, and $q^x$.

In regard to the nature of legged motion of living organisms, certain repeatability conditions reflecting the feature of the gait in a stationary regime, have to be added to equations (1.3.15). Since the gait is symmetric, the repeatability conditions can be written in the form

$$q^i(0) = \pm q^i\left(\frac{T}{2}\right), \quad \dot{q}^i(0) = \pm \dot{q}^i\left(\frac{T}{2}\right)$$

where the sign depends on the physical nature of the appropriate coordinates and their derivatives; $\left(\frac{T}{2}\right)$ is the duration of the one half-step period. As the motion of the prescribed part of the mechanism has been already defined (the repeatability conditions are implicitely satisfied), only for the rest of the mechanism, i.e., for the part of the mechanism whose motion has to be determined, the repeatability conditions

$$q^{xi}(0) = \pm q^{xi}\left(\frac{T}{2}\right), \quad \dot{q}^{xi}(0) = \pm \dot{q}^{xi}\left(\frac{T}{2}\right) \tag{1.3.16}$$

are to be added to the original set of equations describing the mechanism motion.

System (1.3.15), together with conditions (1.3.16), enables one to obtain the necessary trajectories of the coordinates $q^{xi}$, i.e., to carry out the compensation synergy synthesis. Accordingly, the synergy synthesis for $q^{xi}$ coordinates is reduced to the solving of system (1.3.15) with conditions (1.3.16). For this purpose, various iterative methods for solving the boundary value problem can be applied [1, 31].

It has to be emphasized that though the equations of dynamic relations (1.3.15) have to be written only for three coordinates $q^{xi}$, the dynamic properties of all of the adopted anthropomorphic models are taken into account. This is a consequence of the fact that the coefficients in (1.3.15) depend on the motion of the whole mechanism. So, the mathematical model includes the inertial terms of links $q^{oi}$ on which the prescribed synergy is imposed. In this way, the dynamics of the adopted synergy is also considered.

After the synergy synthesis is completed, the driving torques, which have to force the system to follow nominal trajectories, have to be computed. For this reason, the conditions of kinetostatic equilibrium around all joints' axes should be written and then the driving torques computed.

Dynamic connections can be set not only for the point where the ground reaction force acts on the system, but also for other joints of the mechanism. A good example is the arms motion in a walking process. Let us suppose each arm is approximated by one heavy link. Then, if there is no any specific task to be fulfilled (as, for example, the active participation of arms in the artificial gait) and if there is no any actuator at the shoulder joint, but with viscous friction involved, the behaviour of the arms is similar to that of free pendulums. It practically means that the equilibrium conditions can be written for the shoulder joints. The sum of moments of all gravity and inertial forces with respect to the shoulders joints axes, forms, along with relations (1.3.14), the additional relations of dynamic connections. During the motion, the repeatability conditions (1.3.16) have to be satisfied, too. That means the arms behaviour has to be periodic with the same period as for the other mechanism links, namely, the one-step period T. In such a case, the set $q^x$ will contain as many $q^{xi}$ coordinates as there are possible conditions of dynamic connections. The matrices and vectors in (1.3.15) will be of the corresponding order.

Therefore, the problem of artificial locomotion synthesis, by its nature, belongs to combined type of problem, as, for the one part of the system - motion is known, and for the rest of it - forces. The biped locomotion thus defined has certain specificities which deserve some additional explanations.

In spite of the fact that the motion of the one part of the system is prescribed (in the case we are dealing with, this is the legs' trajectories), the driving torques remain unknown. Hence, the dynamic reactions are the function of both the prescribed and compensating synergy, and they are acting on the system as unknown external forces. Because of that, it is not possible to compute the driving torques, even at the joints of prescribed trajectories. This is the reason why in the synergy synthesis a special attention has to be paid to solving these dynamical indefinitness.

The solution has been obtained by prescribing the position of the point where the ground reaction force acts on the mechanism (ZMP), to which a coordinate frame has been attached. Then, the equations of dynamic equilibrium are formed with respect to the ZMP. That means, if the legs trajectories are prescribed and the position of ZMP is chosen in advance, the sum of all moments with respect to the ZMP and axes of passive joints (shoulder joint) are equal to zero. Then, the system of

differential equations with respect to the unknown part of synergy
(1.3.7) becomes

$$\ddot{q}^X = -A_{OX}^{-1}(A_{OO} \cdot \ddot{q}^O + T_O(B[\dot{q}^2] + C[\dot{q} \; \dot{q}] + G)$$

and, the corresponding driving torques can be computed from (1.3.1).

In this case, the unknown synergy describes not only the motion of the
compensating links for maintenance of the equilibrium with respect to
the supporting point, or used for establishing the gait repeatability,
but also the free motion of passive links.


### 1.3.2. Double-support phase

We shall consider now the double-support phase which is characterized
by simultaneous contact of both mechanism's feet with the ground. In
this case the kinematic chain playing the role of legs is closed, i.e.,
the unknown reaction forces to be determined act on its both ends. We
shall describe only an approximate procedure described in monograph
[1], though some other methods have been reported in the literature
[5, 34].

The procedure for synergy synthesis is in the most part analogous to
that for a single-support phase. Let the position of axis V be selected within the dashed area in Fig. 1.4. Then, by writing the equilibrium equations with respect to three orthogonal axes (two horizontal and
one vertical) passing through the ZMP, and by setting the sum of all
moments of external forces to zero, the compensating movements for the
corresponding part of the body can be computed.

The next problem is how to chose the position of the axis V with respect to the ZMP. The information on ZMP and axis V is insufficient for
computation of the driving torques. For this reason, it is necessary
to provide some additional relations concerning the ground reaction
force. These relations are concerned with the characteristics of the
friction between the feet and the ground surface. The total reaction
under one foot can be expressed as a sum of three reaction forces and
moment components in the direction of coordinate axes. The components
$M_X$ and $M_Y$ can be equal to zero since the diagram of the vertical
forces is of the same sign. The third component, $M_V$, should also be

equal to zero, according to the following considerations. Generally speaking, the friction forces can produce moments, but in the synergy synthesis the moment $M_V$ should also be equal to zero. As a consequence, if moments of friction forces are generated, they should be of the opposite sign under each foot. However, in such a case these moments do not affect the system motion but only load additionally the legs drives and joints. Because of this, it is reasonable to synthesize the gait in such a way as to reduce each of these moments to zero. Consequently, it can be assumed that total moments of reaction forces under each foot are equal to zero,

$$\vec{M}_a = \vec{M}_b = 0 \qquad\qquad (1.3.17)$$

where the subscripts a and b denote the left and right foot, respectively.

Now we shall discuss in more detail the total reaction force $\vec{R}$. This force has to fulfil certain relations between its horizontal and vertical component. Characteristics of the friction between the foot and the ground can be represented by a friction cone (Fig. 1.4). If the total ground reaction force is within the cone of angle $2\gamma$, its horizontal component, opposing the sliding (i.e., the friction force) will

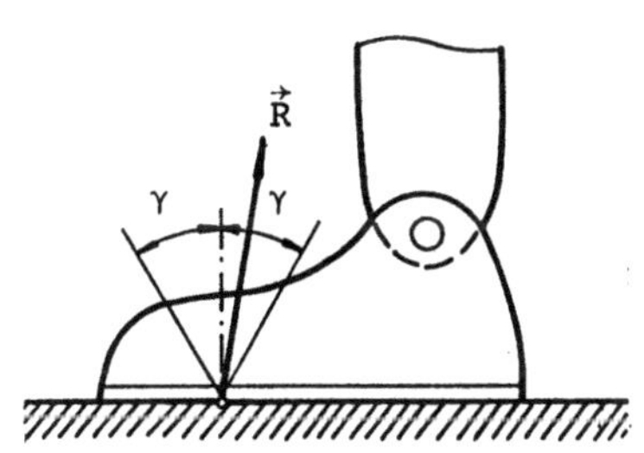


Fig. 1.4. Friction cone

be of sufficient intensity to prevent an unwanted horizontal motion of the supporting foot over the ground surface. This can be expressed as

$$\frac{|\vec{R}_X + \vec{R}_Y|}{|\vec{R}_V|} \leq tg\gamma = \mu \qquad\qquad (1.3.18)$$

where $\mu$ is the friction coefficient of the surfaces in contact. Thus we come to an important conclusion that it is reasonable to distribute the horizontal components of ground reactions per foot proportionally to the normal pressure. The vertical components are inversely proportional to the distances between the ZMP and corresponding foot. So,

$$\frac{|\vec{R}_{Va}|}{|\vec{R}_{Vb}|} = \frac{\ell_b}{\ell_a} \qquad\qquad (1.3.19)$$

Then, from (1.3.18), the relation [1, 3]:

$$\frac{|\vec{T}_a|}{|\vec{T}_b|} = \frac{\ell_b}{\ell_a} \qquad (1.3.20)$$

holds for the horizontal components, where $\vec{T}_a$ and $\vec{T}_b$ are the friction forces under the corresponding foot (Fig. 1.5). On the basis of similarity of the triangles $\triangle OAD$ and $\triangle OBC$, it can be concluded that relation (1.3.20) does not depend on the direction of the force $\vec{T}$ (i.e.,

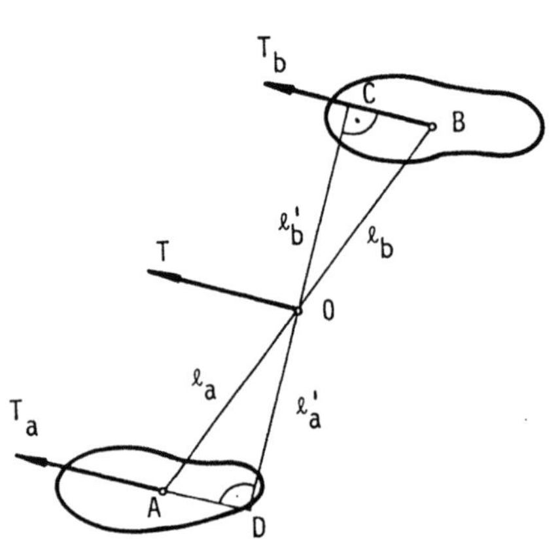

the distances $\ell_a'$ and $\ell_b'$) but only on the distances between the feet, $\ell_a$ and $\ell_b$. Thus we come to an important conclusion that, in order to have the friction forces divided in proportion to the vertical pressures, a necessary and sufficient condition is that the axis V passes through the ZMP. Then, for the synergy synthesis in double-support phase the following vector equation holds

Fig. 1.5. Determination of total friction force

$$\sum_{i=1}^{n} (\vec{r}_i \times (\vec{G}_i + \vec{F}_i) + \vec{M}_i) = 0 \qquad (1.3.21)$$

where $\vec{r}_i$ is a radius vector from the ZMP to the gravity centre of the i-th link; $\vec{F}_i$ and $\vec{M}_i$ are the inertial force and the corresponding moment of the i-th link reduced to its centre of gravity.

When the synthesis of compensating laws of motion is completed, it is possible to determine total horizontal and vertical reactions

$$R_Z = -\sum_{i=1}^{n} (F_{iZ} + G_i), \qquad \vec{T} = -\sum_{i=1}^{n} (F_{iX} \cdot \vec{e}_X + F_{iY} \cdot \vec{e}_Y) \qquad (1.3.22)$$

where $F_{iZ}$ is the projection of $\vec{F}_i$ to the vertical axis and $F_{iX}$ and $F_{iY}$ to axes X and Y, respectively. Here, the axis Z corresponds to a vertical axis (previously denoted by V) which passes through the ZMP. The axes X, Y, and Z with unit vectors $\vec{i}$, $j$ and $b$, respectively, constitute, the absolute orthogonal coordinate frame. Furthermore the relations $\vec{T} = \vec{T}_a + \vec{T}_b$ and $\vec{R} = \vec{R}_a + \vec{R}_b$ are obvious, and they, together with the relations

$$\vec{\ell}_a \times \vec{T}_a + \vec{\ell}_b \times \vec{T}_b = 0$$

$$\vec{\ell}_a \times \vec{R}_a + \vec{\ell}_b \times \vec{R}_b = 0 \qquad (1.3.23)$$

extend the possibility of defining the vertical reactions $\vec{R}_a$ and $\vec{R}_b$, as well as the friction forces $\vec{T}_a$ and $\vec{T}_b$. The $\vec{\ell}_a$ and $\vec{\ell}_b$ are vectors from the ZMP (denoted by 0) to the centres of the corresponding supporting surfaces A and B, respectively.

When the reactions are defined according to (1.3.22) and (1.3.23), it is necessary to check out whether the inequality (1.3.18) still holds. If not, another ZMP ought to be selected and the synergy synthesis performed again as described above. When, however, (1.3.18) is satisfied, then the determination of driving torques at mechanism's joints can be worked out. For this purpose, the closed kinematic chain of legs should be broken at one end, and the corresponding reactions (for example, $\vec{R}_a$ and $\vec{T}_a$) applied. Such a situation corresponds to the case when all kinematic chains are open, and driving torques can be computed in the way already described.

# References


[1] Vukobratović M., <u>Legged Locomotion Robots and Anthropomorphic Mechanisms</u>, Monograph, Institute "Mihailo Pupin", Beograd, 1975.

[2] Vukobratović M., "How to Control the Artificial Anthropomorphic Systems", IEEE Trans. on Systems, Man and Cybernetics, SMC-3, pp. 497-507, 1973.

[3] Stepanenko Yu., Vukobratović M., "Dynamics of Articulated Open-Chain Active Mechanisms", Mathematical Biosciences, Vol. 28, No. 1/2, 1976.

[4] Vukobratović M., Kirćanski N., <u>Real-Time Dynamics of Manipulation Robots</u>, Series: Scientific Fundamentals of Robotics 4, Springer-Verlag, 1985.

[5] Wittemburg J., "Analytic Methods in Mechanical System Dynamics", Computer Aided Analysis and Optimization of System Dynamics, NATO ASI Series, Vol. F9, 1984.

[6] Muybridge E., "Animals in Motion", Dower Publications, New York, 1957 (first published in 1899.

[7] Muybridge E., "Animals in Motion", Dower Publications, New York, 1955 (first published in 1901).

[8] Fischer O., "Theoretische Grundlagen für eine Mechanik der lebenden Körper", B.G. Teubuer, Leipzig, Berlin, 1906.

[9] Hemami H., Wall C.III, Black F., Golliday G., "Single Inverted Pendulum Biped Experiments", Journal of Interdisc. Model. and Simul., Vol. 2, No. 3, 1979.

[10] Hemami H., Katbab A., "Constrained Inverted Pendulum Model for Evaluating Upright Postural Stability", Journal of Dynamic Systems, Measurement and Control, Vol. 104, 1982.

[11] Bavarian B., Wyman B., Hemami H., "Control of Planar Simple Inverted Pendulum", Int. J. of Control, Vol. 37, No. 4, 1983.

[12] Chow C.K., Jacobson D.H., "Postural Stability of Human Locomotion", Mathematical Biosciences, Vol. 15, No. 3/4, 1972.

[13] Beletskii V.V., "Biped Gait - Modelling Problems in Dynamics and Control (in Russian), Nauka, Moscow 1984.

[14] Gubina F., "Stability and Dynamic Control of Certain Types of Biped Locomotion", IV Symposium on External Control of Human External Control of Human Extremities, Dubrovnik, 1972.

[15] Hemami H., Chen B.R., "Stability Analysis and Input Design of a Two-Link Planar Biped", The International Journal of Robotics Research, Vol. 3, No. 2, 1984.

[16] Hemami H., Wyman B.F., "Indirect Control of the Forces of Constraints in Dynamic Systems", Journal of Dynamic Systems, Measurement and Control, Vol. 101, December 1979.

[17] Goddard R.Jr., Hemami H., Weimer F., "Biped Side Step in the Frontal Plane", IEEE Trans. on Automatic Control, Vol. 28, No. 2, 1983.

[18] Hemami H., Wyman B., "Modelling and Control of Constrained Dynamic Systems with Application to Biped Locomotion in the Frontal Plane", IEEE Trans. on Automatic Control, Vol. 24, No. 4, 1979.

[19] Hemami H., Hines M., Goddard R., Friedman B., "Biped Sway in the Frontal Plane with Locked Knees", IEEE Trans. on Systems, Man and Cybernetics, Vol. 12, No. 4, 1982.

[20] Hemami H., Robinson C., Ceranowicz A., "Stability of Planar Biped Models by Simultaneous Pole Assignment and Decoupling", Int. J. Systems Sci., Vol. 11, No. 1, 1980.

[21] Zheng Yuan-Fang, Hemami H., "Impact Effects of Biped Contact with the Environment", IEEE Trans. on Systems, Man and Cybernetics, Vol. 14, No. 3, 1984.

[22] Hemami H., Zheng Yuan-Fang, Hines M., "Initiation of Walk and Tiptoe of Planar Nine-Link Biped", Mathematical Biosciences, Vol. 61, pp. 163-189, 1982.

[23] Wongchaisuwat C., Hemami H., Buchner H., "Control of Sliding and Rolling at Natural Joints", Trans. of ASME, Journal of Biomechanical Engineering, Vol. 106, Nov. 1984.

[24] Wongchaisuwat C., Hemami H., Hines M., "Control Exerted by Ligaments", Journal of Biomechanics, Vol. 18, No. 7, 1984.

[25] Formal'skii A.M., Locomotion of Anthropomorphic Mechanisms (in Russian), Nauka Moscow, 1982.

[26] Miura H., Shimoyama I., "Dynamic Walk of Biped", The International Journal of Robotics Research, Vol. 3, No. 2, 1984.

[27] Furusho J., Masubuchi M., "Control of a Dynamical Biped Locomotion System for Steady Walking", Journal of Dynamic Systems, Measurement and Control, Vol. 108, June 1986.

[28] Moreynis J.Sh., Grytsenko G.P., "Physical and Mathematical Model
of Human Locomotor Apparatus", Protezirovanie i protezostroenie,
Vol. 33, Moscow, 1974.

[29] Gurfinkel V.S., Fomin S.V., Stilgind G.I., "Determination of Joint Moments During Locomotion" (in Russian), Biofizika, Vol. 40,
No. 2, 1970.

[30] Vukobratović M., Juričić D., "Contribution to the Synthesis of
Biped Gait", IEEE Trans. on Biomedical Engineering, Vol. 16,
January 1969.

[31] Juričić D., Vukobratović M., "Mathematical Modelling of Bipedal
Walking System", ASME Publication 72-WA/BHF-13.

[32] Bernstain N.A., On the Motion Synthesis, (in Russian), Medgiz,
Moscow, 1947.

[33] Chow C.K., Jacobson D.H., "Studies of Human Locomotion via Optimal Programming", Technical Report No 617, 1970, Division of Engineering and Applied Physics, Harvard University, Cambridge,
Mass.

[34] Vukobratović M., Potkonjak V., Applied Dynamics and CAD of Manipulation Robots, Monograph, Springer-Verlag, 1985.

# Appendix – Description of the Mechanical Model Using Euler's Angles

For the mathematical description of a biped system the following definitions and notation (Fig. A.1) will be used.

- There are n rigid links subscripted by i=1,2,...,n.

- The rigid links are interconnected by ball-and-socket type joints.

- The links are only simply interconnected.

- The origin of the coordinate system is at the mass centre of the
  rigid link, and its orientation in the body is chosen arbitrarily.

- Each rigid link i has its mass $m_i$, its inertia tensor $J_i$ and a distance $\vec{d}_i$ between its first joint and its mass centre. The first joint
  of a link is defined as the joint closest to the fixed point.

- Because the links are simply interconnected, there are as many joints
  as there are links. Here the fixed point is included as the joint
  between the first link and the fixed space. The joints used in the
  analysis are subscripted by p = (0,1,...n).

- Each rigid link has none, one, or more lengths $\vec{l}$ between the first
  and the other joints on the same link. There are m lengths in the
  whole system; they are subscripted by j=1,2,...,m. The position and
  attitude of a rigid link is described by the position of its mass
  centre relative to the fixed point and by three Euler's angles $\theta$, $\psi$,
  $\phi$, describing the orientation of the local x, y, z frame relative
  to the absolute frame, X, Y, Z. The position of the mass centre depends
  on the attitude of all the links connecting the considered link and
  the fixed point.

The structure of the mechanical model of anthropomorphic locomotion
mechanism was not defined in advance, but was left to be described by

the structural matrices which enable the mechanism structures definition. In this way, a broad class of mechanical anthropomorphic structures can be covered by the same mathematical model.

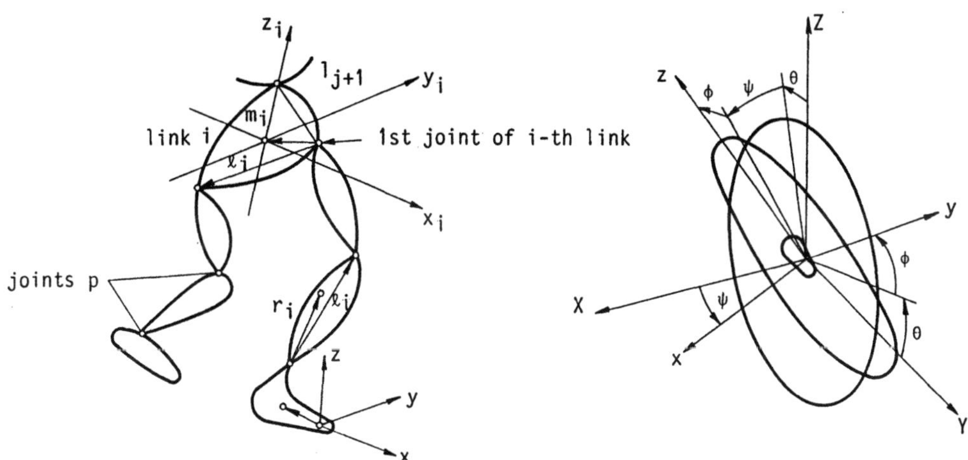


Fig. A.1. Notation of the anthropomorphic     Fig. A.2. Set of adopted
        chain                                                        Euler's angles

The structural matrices used are defined in the following way:

- The matrix $[\varepsilon]$ is defined as having the term $\varepsilon_{ip}$ equal to unity if
  the link i contributes to the moment about the joint p; otherwise
  it is equal to zero. Hence it describes the role of the links regarding the joint moments.

- The matrix $[\delta]$ is defined as having the terms $\delta_{ijp}$ equal to unity if
  the length $\ell_j$ lies on the joint p to the first joint of the link i;
  otherwise it is equal to zero. It describes the role of the length $\ell_j$
  in deriving the moment about the joint p due to link i.

- The matrix $[\gamma]$ is defined as having the terms $\gamma_{ij}$ equal to one if
  the length $\ell_j$ lies on the link i; otherwise it is equal to zero.
  Hence it connects a link and its lengths $\ell$'s.

The moments about the joint p are expressed relative to the arbitrary
axes described by the directional cosines relative to the neighbouring
links. Each joint p will have the subscript of the first and the second
link it connects, defined by the numbers (p, 1) and (p, 2). The directional cosines to the first axis relative to the first link will be

given by $(\ell, m, n)_{p,1}$ and the directional cosines of the second axis relative to the second link will be given by $(\ell, m, n)_{p,2}$. The choice of these two axes will result from the hardware used or from the natural structure of the joint. The third axis is defined as being perpendicular to the first two.

A mathematical model of described generalized biped structure will be derived in the following paragraphs. The inertial and geometric data of all the links, together with the dimensions of the system and the structural matrices, are all the data necessary to reduce the general mathematical model to the equations of motion of a specific structure. The equations of motion obtained are then used to study the behaviour of the system under the imposed gait and the repeatability conditions.

*Angular motion of links*

The angular motion of links is treated by using the coordinate frames given in Fig. A.2. The Euler's angles adopted ($\theta$, $\psi$ and $\phi$) are shown in the same figure. The transformation of coordinates X, Y, Z of the fixed coordinate frame to the coordinates x, y, z of a coordinate frame rotating with the link considered is given by

$$\begin{bmatrix} x \\ y \\ z \end{bmatrix} = [\phi]\ [\psi]\ [\theta] \begin{bmatrix} X \\ Y \\ Z \end{bmatrix}$$

where the transformation matrices are as follows:

$$\theta = \begin{bmatrix} 1 & 0 & 0 \\ 0 & \cos\theta & \sin\theta \\ 0 & -\sin\theta & \cos\theta \end{bmatrix}$$

$$\psi = \begin{bmatrix} \cos\psi & 0 & -\sin\psi \\ 0 & 1 & 0 \\ \sin\psi & 0 & \cos\psi \end{bmatrix} \qquad (A.1)$$

$$\phi = \begin{bmatrix} 1 & 0 & 0 \\ 0 & \cos\phi & \sin\phi \\ 0 & -\sin\phi & \cos\phi \end{bmatrix}$$

Because of the orthogonality properties, the inverse transformation can be expressed by the transposed transformation matrices, i.e.

$$\begin{bmatrix} X \\ Y \\ Z \end{bmatrix} = [\Theta]^T [\Psi]^T [\Phi]^T \begin{bmatrix} x \\ y \\ z \end{bmatrix}$$

The angular velocity of the link, expressed in terms of time deriatives of Euler's angles will be

$$\begin{bmatrix} \omega_x \\ \omega_y \\ \omega_z \end{bmatrix} = [\Phi][\Psi] \begin{bmatrix} \dot{\theta} \\ 0 \\ 0 \end{bmatrix} + [\Psi] \begin{bmatrix} 0 \\ \dot{\psi} \\ 0 \end{bmatrix} + \begin{bmatrix} \dot{\phi} \\ 0 \\ 0 \end{bmatrix}$$

that reduces to the following form:

$$\begin{bmatrix} \omega_x \\ \omega_y \\ \omega_z \end{bmatrix} = [\P] \begin{bmatrix} \dot{\theta} \\ \dot{\psi} \\ \dot{\phi} \end{bmatrix}$$

Here the transformation matrix $[\P]$ stands for

$$[\P] = \begin{bmatrix} \cos\psi & 0 & 1 \\ \sin\psi\sin\phi & \cos\phi & 0 \\ \sin\psi\cos\phi & -\sin\phi & 0 \end{bmatrix} \qquad (A.2)$$

The time rate of change of components of angular velocity will be

$$\begin{bmatrix} \dot{\omega}_x \\ \dot{\omega}_y \\ \dot{\omega}_z \end{bmatrix} = \frac{d}{dt} \begin{bmatrix} \omega_x \\ \omega_y \\ \omega_z \end{bmatrix} = \frac{d}{dt} [\P] \begin{bmatrix} \dot{\theta} \\ \dot{\psi} \\ \dot{\phi} \end{bmatrix}$$

that after expanding the right-hand side gives

$$\begin{bmatrix} \dot{\omega}_x \\ \dot{\omega}_y \\ \dot{\omega}_z \end{bmatrix} = [\pi] \begin{bmatrix} \ddot{\theta} \\ \ddot{\psi} \\ \ddot{\phi} \end{bmatrix} + [\P_1] \begin{bmatrix} \dot{\psi}\dot{\phi} \\ \dot{\phi}\dot{\theta} \\ \dot{\theta}\dot{\psi} \end{bmatrix}$$

Here the matrix $[\P_1]$ stands for

$$[\P_1] = \begin{bmatrix} 0 & 0 & -\sin\psi \\ -\sin\phi & \sin\psi\cos\phi & \cos\psi\sin\phi \\ -\cos\phi & -\sin\psi\sin\phi & \cos\psi\cos\phi \end{bmatrix} \qquad (A.3)$$

The external moment that causes the angular motion of the link as described above, will be expressed in terms of components along the x, y, z axes. For that purpose, the angular momentum of the link is expressed in matrix form as follows:

$$\begin{bmatrix} H_x \\ H_y \\ H_z \end{bmatrix} = [J] \begin{bmatrix} \omega_x \\ \omega_y \\ \omega_z \end{bmatrix}$$

where the matrix [J] represents the inertia tensor

$$[J] = \begin{bmatrix} J_{xx} & -J_{xy} & -J_{xz} \\ -J_{yx} & J_{yy} & -J_{yz} \\ -J_{zx} & -J_{zy} & J_{zz} \end{bmatrix} \qquad (A.4)$$

The moment is obtained now as the time rate of change of angular momentum:

$$\begin{bmatrix} M_x \\ M_y \\ M_z \end{bmatrix} = \frac{d}{dt} \begin{bmatrix} H_x \\ H_y \\ H_z \end{bmatrix} + [\omega] \begin{bmatrix} H_x \\ H_y \\ H_z \end{bmatrix} \qquad (A.5)$$

Here the matrix $[\omega]$ stands for

$$[\omega] = \begin{bmatrix} 0 & -\omega_z & \omega_y \\ \omega_z & 0 & -\omega_x \\ -\omega_y & \omega_x & 0 \end{bmatrix}$$

As the axes x, y, z are fixed to the body, the expression (A.5) after differentiation takes the form

$$\begin{bmatrix} M_x \\ M_y \\ M_z \end{bmatrix} = [J] \begin{bmatrix} \dot{\omega}_x \\ \dot{\omega}_y \\ \dot{\omega}_z \end{bmatrix} + [\omega][J] \begin{bmatrix} \omega_x \\ \omega_y \\ \omega_z \end{bmatrix}$$

This form can be expanded further to yield

$$\begin{bmatrix} M_x \\ M_y \\ M_z \end{bmatrix} = [J] \begin{bmatrix} \dot{\omega}_x \\ \dot{\omega}_y \\ \dot{\omega}_z \end{bmatrix} + [J_1] \begin{bmatrix} \omega_x^2 \\ \omega_y^2 \\ \omega_z^2 \end{bmatrix} + [J_2] \begin{bmatrix} \omega_y\omega_z \\ \omega_z\omega_x \\ \omega_x\omega_y \end{bmatrix} \qquad \text{(A.6)}$$

where the inertia matrices $[J_1]$ and $[J_2]$ stand for

$$[J_1] = \begin{bmatrix} 0 & -J_{zy} & J_{yz} \\ J_{zx} & 0 & -J_{xz} \\ -J_{yz} & J_{xy} & 0 \end{bmatrix}$$

$$\text{(A.7)}$$

$$[J_2] = \begin{bmatrix} J_{zz}-J_{yy} & J_{yx} & -J_{zx} \\ -J_{xy} & J_{xx}-J_{zz} & J_{zy} \\ J_{xz} & -J_{yz} & J_{yy}-J_{xx} \end{bmatrix}$$

The square terms of angular velocities can be expressed by using the time rate of change of Euler's angles. This can be accomplished by forming the following expression:

$$\begin{bmatrix} \omega_x \\ \omega_y \\ \omega_z \end{bmatrix} [\omega_x \; \omega_y \; \omega_z] = [\P] \begin{bmatrix} \dot{\theta} \\ \dot{\psi} \\ \dot{\phi} \end{bmatrix} [\dot{\theta} \; \dot{\psi} \; \dot{\phi}] [\P]^T$$

By comparing the corresponding terms the following equalities can be established:

$$\begin{bmatrix} \omega_x^2 \\ \omega_y^2 \\ \omega_z^2 \end{bmatrix} = [\P_2] \begin{bmatrix} \dot{\theta}^2 \\ \dot{\psi}^2 \\ \dot{\phi}^2 \end{bmatrix} + [\P_3] \begin{bmatrix} \dot{\psi}\dot{\phi} \\ \dot{\phi}\dot{\theta} \\ \dot{\theta}\dot{\psi} \end{bmatrix}$$

$$\text{(A.8)}$$

$$\begin{bmatrix} \omega_y\omega_z \\ \omega_z\omega_x \\ \omega_x\omega_y \end{bmatrix} = [\P_4] \begin{bmatrix} \dot{\theta}^2 \\ \dot{\psi}^2 \\ \dot{\phi}^2 \end{bmatrix} + [\P_5] \begin{bmatrix} \dot{\psi}\dot{\phi} \\ \dot{\phi}\dot{\theta} \\ \dot{\theta}\dot{\psi} \end{bmatrix}$$

where the transformation matrices $[\P_2]$ to $[\P_5]$ are given as follows:

$$[\P_2] = \begin{bmatrix} \cos^2\psi & 0 & 1 \\ \sin^2\psi\ \sin^2\phi & \cos^2\phi & 0 \\ \sin^2\psi\ \cos^2\phi & \sin^2\phi & 0 \end{bmatrix}$$

$$[\P_3] = \begin{bmatrix} 0 & 2\cos\psi & 0 \\ 0 & 0 & 2\sin\psi\ \sin\phi\ \cos\phi \\ 0 & 0 & -2\sin\psi\ \sin\phi\ \cos\phi \end{bmatrix}$$

$$[\P_4] = \begin{bmatrix} \sin^2\psi\ \sin\phi\ \cos\phi & -\sin\phi\ \cos\phi & 0 \\ \sin\psi\ \cos\psi\ \cos\phi & 0 & 0 \\ \sin\psi\ \cos\psi\ \sin\phi & 0 & 0 \end{bmatrix} \tag{A.9}$$

$$[\P_5] = \begin{bmatrix} 0 & 0 & \sin\psi\,(\cos^2\phi - \sin^2\phi) \\ -\sin\phi & \sin\psi\ \cos\phi & -\cos\psi\ \sin\phi \\ \cos\phi & \sin\psi\ \sin\phi & \cos\psi\ \cos\phi \end{bmatrix}$$

After substituting the expressions (A.8) into the moment equation (A.6) the following form of the moment equation results:

$$\begin{bmatrix} M_x \\ M_y \\ M_z \end{bmatrix} = [J][\P]\begin{bmatrix} \ddot\theta \\ \ddot\psi \\ \ddot\phi \end{bmatrix} + ([J_1][\P_2] + [J_2][\P_4])\begin{bmatrix} \dot\theta^2 \\ \dot\psi^2 \\ \dot\phi^2 \end{bmatrix} +$$

$$+ ([J][\P_1] + [J_1][\P_3] + [J_2][\P_5])\begin{bmatrix} \dot\psi\dot\phi \\ \dot\phi\dot\theta \\ \dot\theta\dot\psi \end{bmatrix} \tag{A.10}$$

The moment components obtained are in the directions of the moving co-ordinates axes. The same moment expressed in terms of the components along the fixed axes X, Y, Z can be obtained by the inverse transformation

$$\begin{bmatrix} M_X \\ M_Y \\ M_Z \end{bmatrix} = [\Theta]^T[\Psi]^T[\Phi]^T\begin{bmatrix} M_x \\ M_y \\ M_z \end{bmatrix} \tag{A.11}$$

By combining the expressions (A.10) and (A.11) the final form of the moment equation, in terms of the components along the fixed axes X, Y, Z will read as follows:

$$
\begin{bmatrix} M_X \\ M_Y \\ M_Z \end{bmatrix}_i = [D]_i \begin{bmatrix} \ddot{\theta} \\ \ddot{\psi} \\ \ddot{\phi} \end{bmatrix}_i + [E]_i \begin{bmatrix} \dot{\theta}^2 \\ \dot{\psi}^2 \\ \dot{\phi}^2 \end{bmatrix}_i + [F]_i \begin{bmatrix} \dot{\psi}\dot{\phi} \\ \dot{\phi}\dot{\theta} \\ \dot{\theta}\dot{\psi} \end{bmatrix}_i \tag{A.12}
$$

Here the subscript i denotes the moment on the i-th link due to its central rotation. The transformation matrices contain Euler's angles as well as the inertia properties of the link. They are defined for each link by:

$$
[D] = [\Theta]^T[\Psi]^T[\Phi]^T[J][\P]
$$

$$
[E] = [\Theta]^T[\Psi]^T[\Phi]^T([J_1][\P_2] + [J_2][\P_4]) \tag{A.13}
$$

$$
[F] = [\Theta]^T[\Psi]^T[\Phi]^T([J][\P_1] + [J_1][\P_3] + [J_2][\P_5])
$$

If the principal axes of the link are used, the inertia matrix $[J_1]$ will be equal to zero, and the other two, $J$ and $J_2$, will reduce to the diagonal form, i.e., $[\ \check{J}\ ]$ and $[\ \check{J}_2\ ]$.


*Linear motion of links*


The linear motion of a link is described by considering the position of its mass centre as given by different lengths $\vec{l}$ and a distance $\vec{d}$. The length $\vec{l}$ on a link is defined as the vector $\vec{l}$ extended from the first link's joint to some other joint on the same link. Each link has none, one or more lengths $\vec{l}$; their subscripts are independent of their link's subscript and are denoted by $\vec{l}_j (j=1,2,\ldots,m)$. The distance d is defined as the vector $\vec{d}$ from the first link's joint to its mass centre. Each link i has one distance d denoted by $\vec{d}_i (i=1,2,\ldots,n)$. The components of vectors $\vec{l}$ and $\vec{d}$ expressed in the coordinate frame x, y, z fixed to the link will be constant with respect to time. However, their components in the coordinate frame X, Y, Z fixed in space will depend on instantaneous magnitudes of Euler's angles and will therefore change with time. The transformation of their components from moving to fixed coordinate frame will be given by:

38

$$\begin{bmatrix} \ell_X \\ \ell_Y \\ \ell_Z \end{bmatrix} = [\Theta]^T [\Psi]^T [\Phi]^T \begin{bmatrix} \ell_x \\ \ell_y \\ \ell_z \end{bmatrix} \qquad (A.14)$$

$$\begin{bmatrix} d_X \\ d_Y \\ d_Z \end{bmatrix} = [\Theta]^T [\Psi]^T [\Phi]^T \begin{bmatrix} d_x \\ d_y \\ d_z \end{bmatrix} \qquad (A.15)$$

The second time rate of change of the components of vector $\vec{\ell}$ in the directions of fixed axes X, Y, Z can be found by expanding the following expression:

$$\begin{bmatrix} \ddot{\ell}_X \\ \ddot{\ell}_Y \\ \ddot{\ell}_Z \end{bmatrix} = \frac{d^2}{dt^2} \left( [\Theta]^T [\Psi]^T [\Phi]^T \right) \cdot \begin{bmatrix} \ell_x \\ \ell_y \\ \ell_z \end{bmatrix}$$

that gives

$$\begin{bmatrix} \ddot{\ell}_X \\ \ddot{\ell}_Y \\ \ddot{\ell}_Z \end{bmatrix} = \left( [\Theta_2]^T [\Psi]^T [\Phi]^T \dot{\theta}^2 + [\Theta]^T [\Psi_2]^T [\Phi]^T \dot{\psi}^2 + [\Theta]^T [\Psi]^T [\Phi_2]^T \dot{\phi}^2 + \right.$$

$$+ 2[\Theta]^T [\Psi_1]^T [\Phi_1]^T \dot{\psi}\dot{\phi} + 2[\Theta_1]^T [\Psi]^T [\Phi_1]^T \dot{\phi}\dot{\theta} + 2[\Theta_1]^T [\Psi]^T [\Phi]^T \dot{\theta}\dot{\psi} +$$

$$\left. + [\Theta_1]^T [\Psi]^T [\Phi]^T \ddot{\theta} + [\Theta]^T [\Psi_1]^T [\Phi]^T \ddot{\psi} + [\Theta]^T [\Psi]^T [\Phi_1]^T \ddot{\phi} \right) \begin{bmatrix} \ell_x \\ \ell_y \\ \ell_z \end{bmatrix}$$

$$(A.16)$$

The matrices $[\Theta_1]$, $[\Theta_2]$, $[\Psi_1]$, $[\Psi_2]$, $[\Phi_1]$, and $[\Phi_2]$ denote the derivatives of matrices $[\Theta]$, $[\Psi]$, and $[\Phi]$, respectively. They are given as follows:

$$[\Theta_1] = \frac{d}{d} [\Theta] = \begin{bmatrix} 0 & 0 & 0 \\ 0 & -\sin\theta & \cos\theta \\ 0 & -\cos\theta & -\sin\theta \end{bmatrix} ; [\Theta_2] = \frac{d^2}{d\theta^2} [\Theta] = \begin{bmatrix} 0 & 0 & 0 \\ 0 & -\cos\theta & -\sin\theta \\ 0 & \sin\theta & -\cos\theta \end{bmatrix}$$

$$[\Psi_1] = \frac{d}{d\psi}[\Psi] = \begin{bmatrix} -\sin\psi & 0 & -\cos\psi \\ 0 & 0 & 0 \\ \cos\psi & 0 & -\sin\psi \end{bmatrix} ; \quad [\Psi_2] = \frac{d^2}{d\psi^2}[\Psi] = \begin{bmatrix} -\cos\psi & 0 & \sin\psi \\ 0 & & 0 \\ -\sin\psi & 0 & -\cos\psi \end{bmatrix}$$

$$[\Phi_1] = \frac{d}{d\phi}[\Phi] = \begin{bmatrix} 0 & 0 & 0 \\ 0 & -\sin\phi & \cos\phi \\ 0 & -\cos\phi & -\sin\phi \end{bmatrix} ; \quad [\Phi_2] = \frac{d^2}{d\phi^2}[\Phi] = \begin{bmatrix} 0 & 0 & 0 \\ 0 & -\cos\phi & -\sin\phi \\ 0 & \sin\phi & -\cos\phi \end{bmatrix}$$

$$(A.17)$$

The expression (A.16) can be rewritten in a more convenient and compact form, i.e.

$$\begin{bmatrix} \ddot{\ell}_X \\ \ddot{\ell}_Y \\ \ddot{\ell}_Z \end{bmatrix} = [A][\Lambda] \begin{bmatrix} \ddot{\theta} \\ \ddot{\psi} \\ \ddot{\phi} \end{bmatrix} + [B][\Lambda] \begin{bmatrix} \dot{\theta}^2 \\ \dot{\psi}^2 \\ \dot{\phi}^2 \end{bmatrix} + [C][\Lambda] \begin{bmatrix} \dot{\psi}\dot{\phi} \\ \dot{\phi}\dot{\theta} \\ \dot{\theta}\dot{\psi} \end{bmatrix} \qquad (A.18)$$

The matrices [A], [B], [C], and [$\Lambda$] stand for

$$[A] = \begin{bmatrix} [\Theta_1]^T[\Psi]^T[\Phi]^T & [\Theta]^T[\Psi_1]^T[\Phi]^T & [\Theta]^T[\Psi]^T[\Phi_1]^T \end{bmatrix}$$

$$[B] = \begin{bmatrix} [\Theta_2]^T[\Psi]^T[\Phi]^T & [\Theta]^T[\Psi_2]^T[\Phi]^T & [\Theta]^T[\Psi]^T[\Phi_2]^T \end{bmatrix} \qquad (A.19)$$

$$[C] = 2\begin{bmatrix} [\Theta][\Psi_1]^T[\Phi_1]^T & [\Theta_1]^T[\Psi]^T[\Phi_1]^T & [\Theta_1]^T[\Psi_1]^T[\Phi]^T \end{bmatrix}$$

and

$$[\Lambda] = \begin{bmatrix} \begin{matrix} \ell_x \\ \ell_y \\ \ell_z \end{matrix} & 0 & 0 \\ 0 & \begin{matrix} \ell_x \\ \ell_y \\ \ell_z \end{matrix} & 0 \\ 0 & 0 & \begin{matrix} \ell_x \\ \ell_y \\ \ell_z \end{matrix} \end{bmatrix} \qquad (A.20)$$

The first three matrices are functions of Euler's angles $\theta$, $\psi$, $\phi$ that belong to the member i on which the length $\ell_j$ is located. Hence, the subscript j has to be used throughout the expression (A.18).

The second time rate of change of the components of vector $\vec{d}$ in the directions of fixed axes X, Y, Z will be

$$\begin{bmatrix} \ddot{d}_x \\ \ddot{d}_y \\ \ddot{d}_z \end{bmatrix} = \frac{d^2}{dt^2} \left([\theta]^T [\psi]^T [\phi]^T\right) \cdot \begin{bmatrix} d_x \\ d_y \\ d_z \end{bmatrix}$$

as the components $d_x$, $d_y$, $d_z$ do not change with time. A similar derivation as for length $\ell$ gives the above expression in the following form:

$$\begin{bmatrix} \ddot{d}_X \\ \ddot{d}_Y \\ \ddot{d}_Z \end{bmatrix} = [A][\Delta] \begin{bmatrix} \ddot{\theta} \\ \ddot{\psi} \\ \ddot{\phi} \end{bmatrix} + [B][\Delta] \begin{bmatrix} \dot{\theta}^2 \\ \dot{\psi}^2 \\ \dot{\phi}^2 \end{bmatrix} + [C][\Delta] \begin{bmatrix} \dot{\psi}\dot{\phi} \\ \dot{\phi}\dot{\theta} \\ \dot{\theta}\dot{\psi} \end{bmatrix} \qquad (A.21)$$

The matrices [A], [B], and [C] are as given by expressions (A.19) and the matrix [Δ] stands for

$$[\Delta] = \begin{bmatrix} \begin{matrix} d_x \\ d_y \\ d_z \end{matrix} & 0 & 0 \\ 0 & \begin{matrix} d_x \\ d_y \\ d_z \end{matrix} & 0 \\ 0 & 0 & \begin{matrix} d_x \\ d_y \\ d_z \end{matrix} \end{bmatrix} \qquad (A.22)$$

The Euler's angles in matrices [A], [B], and [C] correspond to the member i. Hence the subscript i should be used throughout the expression (A.21).

The mass centre of a link i will be at a distance $(X, Y, Z)_{ip}$ from the joint p whose torque will be considered. This distance can be found by summing all lengths $\ell$ from the link's first joint to the considered joint p, and adding the distance d that belongs to that particular link. By using the structural matrices $[\epsilon]$ and $[\delta]$ it can be written in the following form:

$$\begin{bmatrix} X \\ Y \\ Z \end{bmatrix}_{ip} = \epsilon_{ip} \begin{bmatrix} d_X \\ d_Y \\ d_Z \end{bmatrix}_i + \sum_{j=1}^{m} \delta_{ijp} \begin{bmatrix} \ell_X \\ \ell_Y \\ \ell_Z \end{bmatrix}_j \qquad (A.23)$$

The terms of the structural matrices are defined as follows: $\epsilon_{ip} = 1$ if the member i contributes to the moment about the joint p; $\epsilon_{ip} = 0$ otherwise; $\delta_{ijp} = 1$ if the length $\ell_j$ lies on the positive path extended from the considered joint p to the first joint of the link i, $\delta_{ijp} = 0$ otherwise.

The main support point, that is fixed in space, is always denoted as the zero joint, i.e., the joint with p = 0. The distance from the fixed main support point is according to expression (A.23) equal to

$$\begin{bmatrix} X \\ Y \\ Z \end{bmatrix}_{io} = \begin{bmatrix} d_X \\ d_Y \\ d_Z \end{bmatrix}_i + \sum_{j=1}^{m} \delta_{ijo} \begin{bmatrix} \ell_X \\ \ell_Y \\ \ell_Z \end{bmatrix}_j \qquad (A.24)$$

as the $\epsilon_{io}$ is always equal to unity. The total acceleration of the mass centre of a link i can be obtained by differentiating the distance (A.24), i.e.

$$\begin{bmatrix} \ddot{X} \\ \ddot{Y} \\ \ddot{Z} \end{bmatrix}_i = \begin{bmatrix} \ddot{d}_X \\ \ddot{d}_Y \\ \ddot{d}_Z \end{bmatrix}_i + \sum_{j=1}^{m} \delta_{ijo} \begin{bmatrix} \ddot{\ell}_X \\ \ddot{\ell}_Y \\ \ddot{\ell}_Z \end{bmatrix}_j \qquad (A.25)$$

By substituting expressions (A.21) and (A.18) for $\ddot{d}$ and $\ddot{\ell}$ in equation (A.25) we obtain

$$\begin{bmatrix} \ddot{X} \\ \ddot{Y} \\ \ddot{Z} \end{bmatrix}_i = [A]_i [\Delta]_i \begin{bmatrix} \ddot{\theta} \\ \ddot{\psi} \\ \ddot{\phi} \end{bmatrix}_i + [B]_i [\Delta]_i \begin{bmatrix} \dot{\theta}^2 \\ \dot{\psi}^2 \\ \dot{\phi}^2 \end{bmatrix}_i + [C]_i [\Delta]_i \begin{bmatrix} \dot{\psi}\dot{\phi} \\ \dot{\phi}\dot{\theta} \\ \dot{\theta}\dot{\psi} \end{bmatrix}_i +$$

$$+ \sum_{j=1}^{m} \delta_{ijo}[A]_j[\Lambda]_j \begin{bmatrix} \ddot{\theta} \\ \ddot{\psi} \\ \ddot{\phi} \end{bmatrix}_j + \sum_{j=1}^{m} \delta_{ijo}[B]_j[\Lambda]_j \begin{bmatrix} \dot{\theta}^2 \\ \dot{\psi}^2 \\ \dot{\phi}^2 \end{bmatrix}_j +$$

$$+ \sum_{j=1}^{m} \delta_{ijo}[C]_j[\Lambda]_j \begin{bmatrix} \dot{\psi}\dot{\phi} \\ \dot{\phi}\dot{\theta} \\ \dot{\theta}\dot{\psi} \end{bmatrix}_j \qquad\qquad (A.26)$$

The sums over the subscript j can be written in a more convenient form if we introduce a structural matrix $\gamma$ defined as follows: $\gamma_{ij} = 1$ if the length $\ell_j$ lies on the link i; $\gamma_{ij} = 0$ otherwise. The following transformations are then obvious:

$$[A]_j[\Lambda]_j \begin{bmatrix} \ddot{\theta} \\ \ddot{\psi} \\ \ddot{\phi} \end{bmatrix}_j = \sum_{k=1}^{n} \gamma_{kj}[A]_k[\Lambda]_j \begin{bmatrix} \ddot{\theta} \\ \ddot{\psi} \\ \ddot{\phi} \end{bmatrix}_k$$

$$[B]_j[\Lambda]_j \begin{bmatrix} \dot{\theta}^2 \\ \dot{\psi}^2 \\ \dot{\phi}^2 \end{bmatrix}_j = \sum_{k=1}^{n} \gamma_{kj}[B]_k[\Lambda]_j \begin{bmatrix} \dot{\theta}^2 \\ \dot{\psi}^2 \\ \dot{\phi}^2 \end{bmatrix}_k$$

$$[C]_j[\Lambda]_j \begin{bmatrix} \dot{\psi}\dot{\phi} \\ \dot{\phi}\dot{\theta} \\ \dot{\theta}\dot{\psi} \end{bmatrix}_j = \sum_{k=1}^{n} \gamma_{kj}[C]_k[\Lambda]_j \begin{bmatrix} \dot{\psi}\dot{\phi} \\ \dot{\phi}\dot{\theta} \\ \dot{\theta}\dot{\psi} \end{bmatrix}$$

Expression (A.26) for the total acceleration of the mass centre of a link i can be now written in the form convenient for later use:

$$\begin{bmatrix} \ddot{X} \\ \ddot{Y} \\ \ddot{Z} \end{bmatrix}_i = [A]_i[\Delta]_i \begin{bmatrix} \ddot{\theta} \\ \ddot{\psi} \\ \ddot{\phi} \end{bmatrix}_i + [B]_i[\Delta]_i \begin{bmatrix} \dot{\theta}^2 \\ \dot{\psi}^2 \\ \dot{\phi}^2 \end{bmatrix}_i + [C]_i[\Delta]_i \begin{bmatrix} \dot{\psi}\dot{\phi} \\ \dot{\phi}\dot{\theta} \\ \dot{\theta}\dot{\psi} \end{bmatrix}_i +$$

$$+ \sum_{j=1}^{n} \delta_{ijo} \sum_{k=1}^{n} \gamma_{kj}[A]_k[\Lambda]_j \begin{bmatrix} \ddot{\theta} \\ \ddot{\psi} \\ \ddot{\phi} \end{bmatrix}_k + \sum_{j=1}^{m} \delta_{ijo} \sum_{k=1}^{n} \gamma_{kj}[B]_k[\Lambda]_j \begin{bmatrix} \dot{\theta}^2 \\ \dot{\psi}^2 \\ \dot{\phi}^2 \end{bmatrix}_k +$$

$$+ \sum_{j=1}^{m} \delta_{ijo} \sum_{k=1}^{n} \gamma_{kj} [C]_k [\Lambda]_j \begin{bmatrix} \dot{\psi}\dot{\phi} \\ \dot{\phi}\dot{\theta} \\ \dot{\theta}\dot{\psi} \end{bmatrix}_k \tag{A.27}$$

The moment about a joint p necessary to produce the linear acceleration of a link i will be

$$\begin{bmatrix} M_X \\ M_Y \\ M_Z \end{bmatrix}_{ip} = m_i \begin{bmatrix} 0 & -Z & Y \\ Z & 0 & -X \\ -Y & X & 0 \end{bmatrix}_{ip} \begin{bmatrix} \ddot{X} \\ \ddot{Y} \\ \ddot{Z} \end{bmatrix} \tag{A.28}$$

where the X, Y and Z are given by equation (A.23) and the accelera-
tions $\ddot{X}$, $\ddot{Y}$ and $\ddot{Z}$ by equation (A.27). By introducing the short notation
for the lever-arm matrix

$$[X]_{ip} = \begin{bmatrix} 0 & -Z & Y \\ Z & 0 & -X \\ -Y & X & 0 \end{bmatrix}_{ip} = \varepsilon_{ip} \begin{bmatrix} 0 & -d_Z & d_Y \\ d_Z & 0 & -d_X \\ -d_Y & d_X & 0 \end{bmatrix}_i +$$

$$+ \sum_{j=1}^{m} \delta_{ijp} \begin{bmatrix} 0 & -\ell_Z & \ell_Y \\ \ell_Z & 0 & -\ell_X \\ -\ell_Y & \ell_X & 0 \end{bmatrix}_j \tag{A.29}$$

the moment equation (A.28) gets the following form:

$$\begin{bmatrix} M_X \\ M_Y \\ M_Z \end{bmatrix}_{ip} = m_i [X]_{ip} [A]_i [\Delta]_i \begin{bmatrix} \ddot{\theta} \\ \ddot{\psi} \\ \ddot{\phi} \end{bmatrix}_i + m_i [X]_{ip} [B]_i [\Delta]_i \begin{bmatrix} \dot{\theta}^2 \\ \dot{\psi}^2 \\ \dot{\phi}^2 \end{bmatrix}_i +$$

$$+ m_i [X]_{ip} [C]_i [\Delta]_i \begin{bmatrix} \dot{\psi}\dot{\phi} \\ \dot{\phi}\dot{\theta} \\ \dot{\theta}\dot{\psi} \end{bmatrix}_i + m_i [X]_{ip} \sum_{j=1}^{m} \delta_{ijo} \sum_{k=1}^{n} \gamma_{kj} [A]_k [\Lambda]_j \begin{bmatrix} \ddot{\theta} \\ \ddot{\psi} \\ \ddot{\phi} \end{bmatrix}_k +$$

$$+ \, m_i [X]_{ip} \sum_{j=1}^{m} \delta_{ijo} \sum_{k=1}^{n} \gamma_{kj} [B]_k [\Delta]_j \begin{bmatrix} \dot{\theta}^2 \\ \dot{\psi}^2 \\ \dot{\phi}^2 \end{bmatrix}_k +$$

$$+ \, m_i [X]_{ip} \sum_{j=1}^{m} \delta_{ijo} \sum_{k=1}^{n} \gamma_{kj} [C]_k [\Delta]_j \begin{bmatrix} \dot{\psi}\dot{\phi} \\ \dot{\phi}\dot{\theta} \\ \dot{\theta}\dot{\psi} \end{bmatrix}_k \qquad (A.30)$$

In this equation the lever-arm matrix $[X]_{ip}$ given by expression (A.29) is equal zero for a link not contributing to the moment about the joint p. With this additional condition the moment expression (A.30) is valid for any link i. This is the moment about the joint p, necessary for the linear motion of the link i.

*Joint moments*

The total moment about a joint p will be equal to the sum of all the moments necessary for angular and linear acceleration of the links influencing the moment about the considered joint, plus the moments necessary to sustain gravity forces. The moments due to gravity force of the link i about the joint p will be equal to

$$-m_i [X]_{ip} \begin{bmatrix} 0 \\ 0 \\ g \end{bmatrix}$$

Hence, the total moment about a given p including the gravity compensation will be

$$\begin{bmatrix} M_X \\ M_Y \\ M_Z \end{bmatrix}_p = \sum_{i=1}^{n} \begin{bmatrix} M_X \\ M_Y \\ M_Z \end{bmatrix}_{ip} + \sum_{i=1}^{n} \varepsilon_{ip} \begin{bmatrix} M_X \\ M_Y \\ M_Z \end{bmatrix}_i + \sum_{i=1}^{n} m_i [X]_{ip} \begin{bmatrix} 0 \\ 0 \\ g \end{bmatrix} \qquad (A.31)$$

The moment expressions for the link i in equation (A.31) are to be taken according to expressions (A.12) and (A.30). After substitution we get the following:

$$
\begin{bmatrix} M_X \\ M_Y \\ M_Z \end{bmatrix}_p = \sum_{i=1}^{n} m_i [X]_{ip} \sum_{j=1}^{m} \delta_{ijo} \sum_{k=1}^{n} \gamma_{kj} [A]_k [\Lambda]_j \begin{bmatrix} \ddot{\theta} \\ \ddot{\psi} \\ \ddot{\phi} \end{bmatrix}_k +
$$

$$
+ \sum_{i=1}^{n} m_i [X]_{ip} [A]_i [\Delta]_i \begin{bmatrix} \ddot{\theta} \\ \ddot{\psi} \\ \ddot{\phi} \end{bmatrix}_i +
$$

$$
+ \sum_{i=1}^{n} m_i [X]_{ip} \sum_{j=1}^{m} \delta_{ijo} \sum_{k=1}^{n} \gamma_{kj} [B]_k [\Lambda]_j \begin{bmatrix} \dot{\theta}^2 \\ \dot{\psi}^2 \\ \dot{\phi}^2 \end{bmatrix}_k +
$$

$$
+ \sum_{i=1}^{n} m_i [X]_{ip} [B]_i [\Delta]_i \begin{bmatrix} \dot{\theta}^2 \\ \dot{\psi}^2 \\ \dot{\phi}^2 \end{bmatrix}_i +
$$

$$
+ \sum_{i=1}^{n} m_i [X]_{ip} \sum_{j=1}^{m} \delta_{ijo} \sum_{k=1}^{n} \gamma_{kj} [C]_k [\Lambda]_j \begin{bmatrix} \dot{\psi}\dot{\phi} \\ \dot{\phi}\dot{\theta} \\ \dot{\theta}\dot{\psi} \end{bmatrix}_k +
$$

$$
+ \sum_{i=1}^{n} m_i [X]_{ip} [C]_i [\Delta]_i \begin{bmatrix} \dot{\psi}\dot{\phi} \\ \dot{\phi}\dot{\theta} \\ \dot{\theta}\dot{\psi} \end{bmatrix}_i + \sum_{i=1}^{n} \varepsilon_{ip} [D]_i \begin{bmatrix} \ddot{\theta} \\ \ddot{\psi} \\ \ddot{\phi} \end{bmatrix}_i +
$$

$$
+ \sum_{i=1}^{n} \varepsilon_{ip} [E]_i \begin{bmatrix} \dot{\theta}^2 \\ \dot{\psi}^2 \\ \dot{\phi}^2 \end{bmatrix}_i + \sum_{i=1}^{n} \varepsilon_{ip} [F]_i \begin{bmatrix} \dot{\psi}\dot{\phi} \\ \dot{\phi}\dot{\theta} \\ \dot{\theta}\dot{\psi} \end{bmatrix}_i + \sum_{i=1}^{n} m_i [X]_{ip} \begin{bmatrix} 0 \\ 0 \\ g \end{bmatrix}
$$

The terms with triple sums can be transformed by rearranging the order
of summation and introducing the short notation

$$
[S]_{jp} = \sum_{i=1}^{n} \delta_{ijo} m_i [X]_{ip} \tag{A.32}
$$

The other six terms can be transformed by changing the dummy subscript
i to k. The resultant form for the total moment about a joint p will
be then as follows:

46

$$\begin{bmatrix} M_X \\ M_Y \\ M_Z \end{bmatrix}_p = \sum_{k=1}^{n} (m_k [X]_{kp} [A]_k [\Delta]_k + \sum_{j=1}^{m} \gamma_{kj} [S]_{jp} [A]_k [\Lambda]_j +$$

$$+ \varepsilon_{kp} [D]_k) \begin{bmatrix} \ddot{\theta} \\ \ddot{\psi} \\ \ddot{\phi} \end{bmatrix}_k + \sum_{k=1}^{n} (m_k [X]_{kp} [B]_k [\Delta]_k + \sum_{j=1}^{m} \gamma_{kj} [S]_{jp} [B]_k [\Lambda]_j +$$

$$+ \varepsilon_{kp} [E]_k) \begin{bmatrix} \dot{\theta}^2 \\ \dot{\psi}^2 \\ \dot{\phi}^2 \end{bmatrix}_k + \sum_{k=1}^{n} (m_k [X]_{kp} [C]_k [\Delta]_k + \sum_{j=1}^{m} \gamma_{kj} [S]_{jp} [C]_k [\Lambda]_j +$$

$$+ \varepsilon_{kp} [F]_k) \begin{bmatrix} \dot{\psi}\dot{\phi} \\ \dot{\phi}\dot{\theta} \\ \dot{\theta}\dot{\psi} \end{bmatrix}_k + \sum_{i=1}^{n} m_i [X]_{ip} \begin{bmatrix} 0 \\ 0 \\ g \end{bmatrix}$$

The joint moments can be referred to some coordinate frame that is different from the fixed X, Y, Z coordinate frame. The two axes of that new coordinate frame, the "joint coordinate frame", are fixed to two neighbouring links, one to each link. The third axis is taken perpendicular to the first two. This frame is not necessarily rectangular. The introduction of the joint coordinate frame makes it possible to refer the moments about the joint hardware axes (Fig. A.3). If the direction cosines of the first two axes of the joint p, relative to the moving axes of the corresponding link, are described with $(\ell\ m\ n)_{p,1}$ and $(\ell\ m\ n)_{p,2}$, respectively, the direction cosines of these axes relative to the absolute coordinate frame X, Y, Z will be

$$[\ell\ m\ n]_{p,1} ([\Phi]^T [\Psi]^T [\Theta]^T)_{\beta(p,1)} = [\alpha_{11}\ \alpha_{12}\ \alpha_{13}]_p$$

$$[\ell\ m\ n]_{p,2} ([\Phi]^T [\Psi]^T [\Theta]^T)_{\beta(p,2)} = [\alpha_{21}\ \alpha_{22}\ \alpha_{23}]_p \tag{A.33}$$

The subscripts $\beta(p,1)$ and $\beta(p,2)$ are defined as the corresponding subscripts of the first and the second link connected by the joint p. The direction cosines for the third axis can be determined by the orthogonality conditions that give

$$[\alpha_{31}\ \alpha_{32}\ \alpha_{33}] = a[(\alpha_{12}\alpha_{12} - \alpha_{22}\alpha_{13}) (\alpha_{13}\alpha_{21} - \alpha_{23}\alpha_{11}) (\alpha_{11}\alpha_{22} - \alpha_{21}\alpha_{12})]$$

where

$$a^{-2} = (\alpha_{12}\alpha_{23} - \alpha_{22}\alpha_{13})^2 + (\alpha_{13}\alpha_{21} - \alpha_{23}\alpha_{11})^2 + (\alpha_{11}\alpha_{22} - \alpha_{21}\alpha_{12})^2$$

All direction cosines $\alpha_{ij}$, assembled into a matrix $[\alpha]_p$, are used to transform the moments at a joint about the fixed coordinate frame into the moments about the given joint coordinate axes. The resultant moments denoted by $M_1$, $M_2$, $M_3$, will be

$$\begin{bmatrix} M_1 \\ M_2 \\ M_3 \end{bmatrix} = [\alpha]_p^{-1} \begin{bmatrix} M_X \\ M_Y \\ M_Z \end{bmatrix}$$

Should the referent axes remain fixed in space, the corresponding subscripts $\beta(p,1)$ and $\beta(p,2)$ have to be taken equal to zero, and the following matrix defined:

$$([\Phi]^T [\Psi]^T [\Theta]^T)_o = [\,^{\backprime}I_{\backprime}]$$

This matrix will be used also for the joint $p=0$ where $\beta(0,1)=\beta(0,2)=0$.

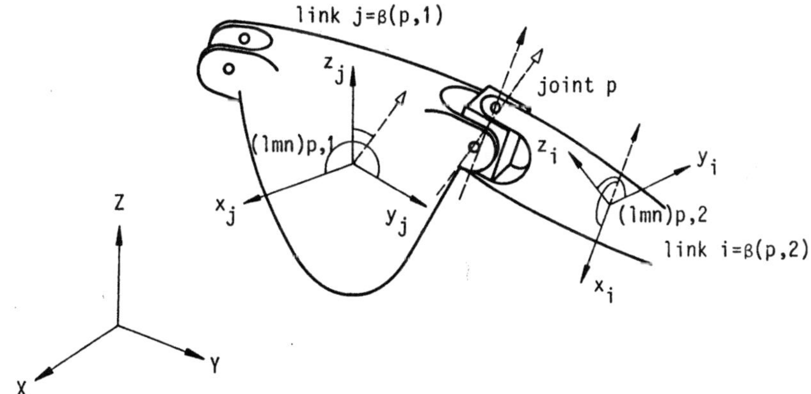


Fig. A.3. Joint coordinate frame fixed to two neighbouring links

The moments can now be written for the whole set of joints p and about the arbitrary axes either fixed in space or attached to the neightbouring link. To keep the expressions short, the following notation will be introduced:

$$[\alpha]_p^{-1} (m_k [X]_{kp} [A]_k [\Delta]_k + \sum_{j=1}^{m} \gamma_{kj} [S]_{jp} [A]_k [\Lambda]_j + \epsilon_{kp} [D]_k) = [AA]_{pk}$$

$$[\alpha]_p^{-1}(m_k \; X_{\;kp}[B]_k[\Delta]_k \; + \; \sum_{j=1}^{m}\gamma_{kj}[S]_{jp}[B]_k[\Lambda]_j + \varepsilon_{kp}[E]_k) \; = \; [BB]_{pk}$$

$$[\alpha]_p^{-1}(m_k[X]_{kp}[C]_k[\Delta]_k \; + \; \sum_{j=1}^{m}\gamma_{kj}[S]_{jp}[C]_k[\Lambda]_j + \varepsilon_{kp}[F]_k) \; = \; [CC]_{pk}$$

$$\sum_{i=1}^{n} m_i[X]_{ip}\begin{bmatrix} 0 \\ 0 \\ g \end{bmatrix} = \{G\}_p \qquad \begin{bmatrix} M_1 \\ M_2 \\ M_3 \end{bmatrix}_p = \{M\}_p \qquad\qquad (A.34)$$

The moment equations for all joints p take now the following short form:

$$\begin{bmatrix} \overline{\phantom{-----}} \\ \{M\}_p \\ \overline{\phantom{-----}} \end{bmatrix} = \sum_{k=1}^{n}\begin{bmatrix} \overline{\phantom{--------}} \\ [AA]_{pk} \\ \overline{\phantom{--------}} \end{bmatrix}\begin{bmatrix} \ddot{\theta} \\ \ddot{\psi} \\ \ddot{\phi} \end{bmatrix}_k + \sum_{k=1}^{n}\begin{bmatrix} \overline{\phantom{--------}} \\ [BB]_{pk} \\ \overline{\phantom{--------}} \end{bmatrix}\begin{bmatrix} \dot{\theta}^2 \\ \dot{\psi}^2 \\ \dot{\phi}^2 \end{bmatrix}_k +$$

$$+ \sum_{k=1}^{n}\begin{bmatrix} \overline{\phantom{--------}} \\ [CC]_{pk} \\ \overline{\phantom{--------}} \end{bmatrix}\begin{bmatrix} \dot{\psi}\dot{\phi} \\ \dot{\phi}\dot{\theta} \\ \dot{\theta}\dot{\psi} \end{bmatrix}_k + \begin{bmatrix} \overline{\phantom{-----}} \\ \{G\}_p \\ \overline{\phantom{-----}} \end{bmatrix}$$

The sums in the above equations can be avoided by expanding the matrices to the system size, i.e.

$$\begin{bmatrix} \overline{\phantom{-----}} \\ \{M\}_p \\ \overline{\phantom{-----}} \end{bmatrix} = \begin{bmatrix} [AA]_{pk} \end{bmatrix}\begin{bmatrix} \begin{bmatrix} \ddot{\theta} \\ \ddot{\psi} \\ \ddot{\phi} \end{bmatrix}_k \end{bmatrix} + \begin{bmatrix} [BB]_{pk} \end{bmatrix}\begin{bmatrix} \begin{bmatrix} \dot{\theta}^2 \\ \dot{\psi}^2 \\ \dot{\phi}^2 \end{bmatrix}_k \end{bmatrix} +$$

$$+ \begin{bmatrix} [CC]_{pk} \end{bmatrix}\begin{bmatrix} \begin{bmatrix} \dot{\psi}\dot{\phi} \\ \dot{\phi}\dot{\theta} \\ \dot{\theta}\dot{\psi} \end{bmatrix}_k \end{bmatrix} + \begin{bmatrix} \overline{\phantom{-----}} \\ \{G\}_p \\ \overline{\phantom{-----}} \end{bmatrix} \qquad\qquad (A.35)$$

When all the joints, i.e. p=0 to n-1, are used in the analysis, the full size matrices [AA], [BB], and [CC] are square matrices. The resultant equation obtained describes the moments about all the joint axes in terms of Euler's angles and their derivatives.

If the motion of all the links is prescribed, the moment components
about the joint axes, the joint torques, can be easily found. Being
the first problem of mechanics, it results in a set of algebraic equations. In the case when the moments at all the joints are given, the
motion of all the links can be found by integration of a set of resulting simultaneous differential equations. In practical application,
however, the task is of a mixed type; the motion of some links is
usually prescribed and a complement of moments is known. The unknown
motion and unknown moments can be determined by solving a resultant
system that it composed of algebraic and differential equations.


*Equations of motion*


To discuss some different cases in practical application, the moment
equations (A.35) are first rewritten in a shorter form by introducing
the following notation:

$$\left[ [AA]_{pk} \right] = [A] \qquad \left[ [BB]_{pk} \right] = [B] \qquad \left[ [CC]_{pk} \right] = [C]$$

$$\left[ \begin{bmatrix} \ddot{\theta} \\ \ddot{\psi} \\ \ddot{\phi} \end{bmatrix}_k \right] = \{\ddot{\Xi}\} \qquad \left[ \begin{bmatrix} \dot{\theta}^2 \\ \dot{\psi}^2 \\ \dot{\phi}^2 \end{bmatrix}_k \right] = \{\dot{\Xi}^2\} \qquad \left[ \begin{bmatrix} \dot{\psi}\dot{\phi} \\ \dot{\phi}\dot{\theta} \\ \dot{\theta}\dot{\psi} \end{bmatrix}_k \right] = \{\dot{\Xi}\dot{\Xi}\}$$

$$\left[ \{M\}_p \right] = \{M\} \qquad \left[ \{G\}_p \right] = \{G\} \tag{A.36}$$

This notation gives the final moment equations the following short
form:

$$\{M\} = [A]\{\ddot{\Xi}\} + [B]\{\dot{\Xi}^2\} + [C]\{\dot{\Xi}\dot{\Xi}\} + \{G\} \tag{A.37}$$

The coefficient matrices $[A]$, $[B]$, and $[C]$ as well as the column mat-
rix $\{G\}$ depend on Euler's angles $\theta_i$, $\psi_i$, and $\phi_i$, $(i=1,2,\ldots,n)$, that
will be denoted by the vector $\{\Xi\}$.

50

In the preceding chapter, the so-called mixed type of the task was mentioned, when the link's motion and the driving torques are both partly known and their complements are sought (eqs. (1.3.7) and (1.3.8). When applied to the problem of investigating the dynamics of biped structure, the links' motion is partly known, performing a given type of gait, while the known moments are equal to zero. The vanishing of the given moments results from the equilibrium conditions about the supporting point and about the joints of passive links. In that case the moments $\{M_o\}$ are equal to zero and the set of differential equations in the coordinates (1.3.7), in our case, the unknown Euler's angles, gets the following form:

$$\{\ddot{\Xi}_x\} = -[A_{ox}]^{-1}([A_{oo}]\{\ddot{\Xi}\} + [T_o]([B]\{\dot{\Xi}^2\} + [C]\{\dot{\Xi}\dot{\Xi}\} + \{G\})) \quad (A.38)$$

The corresponding moments $\{M\}$ can be found in full by using equation (1.3.1) where Euler's angles are arranged in the sequence according to equation (1.3.3), i.e.

$$\{\Xi\} = \left[ R_o \mid R_x \right] \left\{ \frac{\Xi_o}{\Xi_x} \right\}$$

By using the equation (1.2.1) the obtained motion will also be checked against the vanishing moments.

The unknown Euler's angles $\{\Xi_x\}$ describe in this case the motion of the compensating links necessary to achieve the equilibrium about the supporting point or to restore the repeatability of the stride, as well as the free motion of the passive links.


*Boundary conditions*


When a steady locomotion is being considered, the initial and the final conditions of the unknown motion, i.e. the unknown Euler's angles at the beginning and at the end of a step, are not known. The initial and the final states of the unknown angles have to follow from the repeatability conditions. These conditions can be stated as follows: All the angles $\{\Xi\}$ and their derivatives $\{\dot{\Xi}\}$ at the end of a step period have to be equal to those at the beginning of the step period, i.e.

$$\{\Xi\}_T = \{\Xi\}_o \quad \text{and} \quad \{\dot{\Xi}\}_T = \{\dot{\Xi}\}_o$$

The known part of the angles $\{\Xi\}$, i.e. $\{\Xi_o\}$ will have these conditions
satisfied when the gait data used correspond to a steady walk i.e.

$$
\begin{bmatrix} \Xi_x \\ \hline \dot{\Xi}_x \end{bmatrix}_T = \begin{bmatrix} \Xi_x \\ \hline \dot{\Xi}_x \end{bmatrix}_o
\tag{A.39}
$$

represent the boundary conditions for the problem at hand.

The solution to such a boundary-condition problem can be found by
linearization in the vicinity of an approximate solution followed by
successive corrections obtained by using sensitivity matrices. If the
corrected initial conditions have to meet the corrected final conditions, the following must hold:

$$
\begin{bmatrix} \Xi_x \\ \hline \dot{\Xi}_x \end{bmatrix}_T + \begin{bmatrix} \Delta\Xi_x \\ \hline \Delta\dot{\Xi}_x \end{bmatrix}_T = \begin{bmatrix} \Xi_x \\ \hline \dot{\Xi}_x \end{bmatrix}_o + \begin{bmatrix} \Delta\Xi_x \\ \hline \Delta\dot{\Xi}_x \end{bmatrix}_o
\tag{A.40}
$$

The change in final conditions can be found through the sensitivity
matrix from the change in initial conditions, i.e.

$$
\begin{bmatrix} \Delta\Xi_x \\ \hline \Delta\dot{\Xi}_x \end{bmatrix}_T = [U] \begin{bmatrix} \Delta\Xi_x \\ \hline \Delta\dot{\Xi}_x \end{bmatrix}_o
\tag{A.41}
$$

Here the matrix [U] represent the sensitivity matrix

$$
[U] = \begin{bmatrix} \dfrac{\partial(\Xi_i)}{\partial(\Xi_j)} & \dfrac{\partial(\Xi_i)}{\partial(\dot{\Xi}_j)} \\ \hline \dfrac{\partial(\dot{\Xi}_i)}{\partial(\Xi_j)} & \dfrac{\partial(\dot{\Xi}_i)}{\partial(\dot{\Xi}_j)} \end{bmatrix}
\tag{A.42}
$$

with its submatrices expressed in a Jacobian form. The equations (A.40)
and (A.41) give the necessary correction of the initial conditions,
i.e.

$$
\begin{bmatrix} \Delta\Xi_x \\ \hline \Delta\dot{\Xi}_x \end{bmatrix}_o = ([U]-[I])^{-1} \left( \begin{bmatrix} \Xi_x \\ \hline \dot{\Xi}_x \end{bmatrix}_o - \begin{bmatrix} \Xi_x \\ \hline \dot{\Xi}_x \end{bmatrix}_T \right)
\tag{A.43}
$$

52

When a human biped and its walk are being simulated, the extent of a symmetry inherent in the system and its gait reduces the necessary study of a full step to the study of a half step only. In that case the repeatability conditions can be established for a half of the step period by considering the symmetry of the system. The symmetry here is of an odd type and is relative to the x-z plane. Hence, the following requirement will hold: (a) All the motion in the x-z plane will be repeated after each half-step by a central link and by the opposite analogue of a limb link; (b) All the motion out of the x-z plane will be repeated in the opposite direction after a half-step by the trunk and by the opposite analogue of a limb link. By using this symmetry the repeatability conditions (A.39) get the form

$$
\begin{bmatrix} \Xi_x \\ \hline \dot{\Xi}_x \end{bmatrix}_{T/2} = [W] \begin{bmatrix} \Xi_x \\ \hline \dot{\Xi}_x \end{bmatrix}_0 \tag{A.44}
$$

where the transformation matrix [W] implements the symmetry requirements stated above. The correction in the initial conditions in this case will be, analogous to Eq. (A.43),

$$
\begin{bmatrix} \Delta\Xi_x \\ \hline \Delta\dot{\Xi}_x \end{bmatrix}_0 = ([U]-[W])^{-1} ([W] \begin{bmatrix} \Xi_x \\ \hline \dot{\Xi}_x \end{bmatrix}_0 - \begin{bmatrix} \Xi_x \\ \hline \dot{\Xi}_x \end{bmatrix}_{T/2}) \tag{A.45}
$$

The reduction of the studied part of a step makes it further possible, by a wise choice of the beginning point of the half-step considered, to have the same biped structure for the complete study. Namely, if the half-step considered is cut out of a full step in such a way that the same foot is in contact with the ground all the time, the structural matrices [δ] will not need to be changed during the half of the step considered.

# Chapter 2:
# Synthesis of Nominal Dynamics

## 2.1. Introduction

The active spatial mechanism for realization of the artificial anthropomorphic gait belongs to the class of complex kinematic chains. During the single-support gait phase, the mechanism is a complex open kinematic chain, while in the double-support phase one of the chains becomes closed. This chapter is devoted to the description of a computer procedure for forming the dynamic equations of motion of such mechanism, comprising both the open and closed complex kinematic chains. The procedure is based on the methods by which the computer forms the mathematical models for simple open kinematic chains. These methods were described in detail in the previous books of this series [1-3].

In addition, it is described the algorithmic structure of a programme package for the synthesis of nominal dynamics (functional movements) by the semi-inverse method [4-6]. The programme package is given in the Appendix. The input data block of this programme package contains: the information on kinematic scheme of the mechanism for realization of the artificial anthropomorphic gait, the mechanism parameters, the prescribed part of the system dynamics, the law of the zero-moment point (ZMP) displacement, the repeatibility conditions, and the step of integration of the dynamic equations. On the basis of these data, the programme is automatically adjusted for the synthesis of compensating part of the dynamics. This part of dynamics is obtained as the solution of the differential equations for the anthropomorphic mechanism describing compensating motion under the prescribed boundary conditions.

## 2.2. Modelling the Robotic System Dynamics

For the sake of better understanding of the modelling procedure, let us introduce the following definitions [3]:

*Structure* of the mechanism is a set of rigid bodies interconnected by the revolute and/or prismatic joints. Since the structure of an anthropomorphic mechanism is only composed of revolute joints, our consideration will be confined to the mechanisms comprising solely this type of joints.

*Link* of the mechanism is defined by an arranged set of parameters $C_i(K_i, \mathcal{D}_i)$, where $K_i$ represents a set of kinematic, and $\mathcal{D}_i$ a set of dynamic parameters, whereas the subscript $i$ denotes the i-th link of the mechanism.

The sets $K_i$ and $\mathcal{D}_i$ have the following form

$$K_i = (Q_i, R_i, E_i), \qquad \mathcal{D}_i = (m_i \; \underline{\underline{J}}_i)$$

with $Q_i = (\vec{q}_{i1}, \vec{q}_{i2}, \vec{q}_{i3})$ — the internal (local) orthonormal coordinate frame attached to the link i;

$\tilde{R}_i = \{\vec{\tilde{r}}_{ik}{}^{*)}\}$       — the set of the distance vectors from points $Z_{ik}$ to the origin of coordinate frame $Q_i$, where the points $Z_{ik}$ represent the centres of joints between the i-th and k-th mechanisms links, $k \in \{1, \ldots, k_i\}$;

$\tilde{E}_i = \{\vec{\tilde{e}}_{ik}\}$       — the set of the unit vectors of joint axes by which the i-th link $C_i$ is connected to the remaining links $C_k$ in points $Z_{ik}$;

$m_i$       — the mass of the i-th link;

$\underline{\underline{J}}_i = (J_{i1}, J_{i2}, J_{i3})$ — inertia tensor of the i-th link defined with respect to the local frame $Q_i$.

A kinematic pair $P_{ik}$ represents a set of two adjacent links $\{C_i, C_k\}$ interconnected by a joint at point $Z_{ik}$.

The notion of class and subclass of a kinematic pair is introduced depending on the type of joint connection. A j-th class kinematic pair (j=1,...,5) is defined as a set of two adjacent links interconnected by a joint with n=6-j degrees of freedom. Further, the $\ell$-th subclass ($\ell$=1,2,3) is defined in such a way that the maximal number of allowed rotations in it is 4-$\ell$. Thus the classes 1, 2, and 3 contain three possible rotations, class 4, two, and class 5 only one rotation.

---

*)  ~ denotes the vector projections with respect to the link coordinate frame $Q_i$.

A kinematic chain $\Lambda_n$ is a set of n interconnected kinematic pairs, $\Lambda_n = \{P_{ik}\}$, i∈N, k∈N, where $N = \{1,2,...,n\}$.

According to the structure of connections, chains are classified into simple, complex, open and closed.

*Simple* kinematic chain is such a chain in which none of the links $C_i$, ∀i∈N is constituent of more than two kinematic pairs. On the other hand, a *complex* kinematic chain contains at least one link $C_i$, ∃i∈N which enters into more than two kinematic pairs.

A simple kinematic chain is *open* if it possesses at least one link $C_i$ i∈N which belongs only to one kinematic pair, whereas a *closed* chain is composed of the links $C_i$, ∀i∈N, each of which enters into at least two kinematic pairs.

*Joint coordinates* are scalar quantities $q_{ik}^{\ell}$ which define, in a unique manner, the relative position of the links of kinematic pair $P_{ik} = \{C_i, C_k\}$. The superscript $\ell$∈$\{1,...,s\}$, where s=6-j is the number of degrees of freedom, (d.o.f.) and j is the class of pair $P_{ik}$.

The definition of joint coordinates may be simplified if we consider fifth-class pairs only, i.e., pairs with a single d.o.f. of relative motion. This assumption does not reduce the generality of consideration, since a pair of any class may be represented by superposition of virtual fifth-class pairs. It is therefore particularly significant to introduce, for such pairs  a unique manner of defining the joint coordinates. Fig. 2.1. shows a revolute kinematic pair whose joint angle (coordinate) is denoted by $q^i$.

Scalar quantities $x_{ek}$, k∈$\{1,...,m\}$, which determine the position and (partially or completely) orientation of the n-th link of chain $\Lambda_n$ with respect to a reference coordinate frame are said to be the *external (Cartesian) coordinates* of the mechanism.

An *active mechanism* represents a system comprising: a) a mechanical part which may be modelled by an appropriate kinematic chain, and b) a set of actuators by means of which driving torques (forces) are realized.

To form the mathematical model of an active mechanism, it is necessary to:

1) identify the parameters of kinematic chain and actuators,

2) form the dynamic model of mechanism, and

3) form the models of actuators.

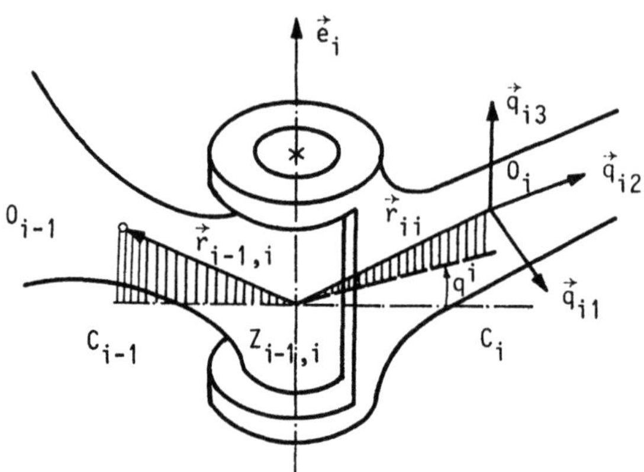

Fig. 2.1. Joint coordinate of a revolute kinematic pair $\{C_{i-1}, C_i\}$

The dynamic model of an open active mechanism represents a set of n non-linear differential equations which describe system motion in the space of joint coordinates assuming that all joints are of the 5-th class.

## 2.3. The Method Based on Fundamental Theorems of Mechanics as a Tool for Forming Dynamic Equations of Motion for Open Kinematic Chain

This method was set in the early period of computer-oriented mathematical modelling development of open-chain manipulation structures [6, 7, 8]. We shall describe this method which is, in nature, a recursive numerical method. This method consists of the following stages: determination of the "home" position, determination of the actual position of links with respect to the reference frame, a kinematic stage, and a dynamic stage. We will use the following notations in describing these stages. $N$ and $I$ denote the sets of indices

$$N = \{i:\ i \in (1,\ldots,n)\}, \qquad I = \{j:\ j \in (1,\ldots,i)\}$$

where n is the number of the kinematic pairs.

*Stage 1: Determination of mechanism initial position*

It is assumed that the active mechanism under consideration may be described by a simple open kinematic chain $\Lambda_n$ consisting of a set of kinematic pairs $\{C_{i-1}, C_i\}$, $i \in N$. The kinematic pair $\{C_o, C_1\}$ represents in fact the first link of the mechanism which is connected to the support $C_o$ by joint $Z_{1o}$. It is necessary to determine the position of all links with respect to the reference frame, under the condition that all joint coordinates are equal zero, $q^i = 0$, $i \in N$.

Let us assume all kinematic parameters $K_j^o = (Q_j^o, R_j^o, E_j^o)$ for $j \in (1, \ldots \ldots, i-1)$ to be known, where the superscript o means that $q^j = 0$, $j \in N$. The task of determining the "home" position reduces now to determining $K_i^o = (Q_i^o, R_i^o, E_i^o)$. The sets $R_i$ and $E_i$ for the case of a simple kinematic chain have the following form

$$R_i = \{\vec{r}_{ii}, \vec{r}_{i,i+1}\} \quad \text{and} \quad E_i = \{\vec{e}_i, \vec{e}_{i+1}\},$$

where $\vec{e}_i$ and $\vec{e}_{i+1}$ stand for vectors $\vec{e}_{ii}$ and $\vec{e}_{i,i+1}$.

On the basis of the assumption that $K_{i-1}^o$ is known, it follows that $\vec{e}_i^o$ and $\vec{r}_{i-1,i}^o$ are also known. Further, let us note the set of 3 orthogonal unit vectors at point $Z_{i-1,i}$: $\vec{e}_i^o$, $\vec{a}_i^o$ and $\vec{e}_i^o \times \vec{a}_i^o$, where $\vec{a}_i^o = \text{ort}^{*)} (\vec{e}_i^o \times (\vec{r}_{i-1,i}^o \times \vec{e}_i^o))$, whose components are known with respect to the reference coordinate frame (Fig. 2.2). On the other hand, let us note the set of vectors $\vec{e}_i$, $\vec{a}_i$ and $\vec{e}_i \times \vec{a}_i$, where $\vec{a}_i = -\text{ort}(\vec{e}_i \times (\vec{r}_{ii} \times \vec{e}_i))$, whose components are known with respect to the local coordinate frame $Q_i$. Under the condition $q^i = 0$, these two sets of vectors coincide. Since $Q_i^o = [\vec{q}_{i1}^o \ \vec{q}_{i2}^o \ \vec{q}_{i3}^o]$ represents transformation matrix of the i-th link coordinate frame into the reference frame, it follows that

$$\vec{e}_i^o = Q_i^o \vec{e}_i$$

$$\vec{a}_i^o = Q_i^o \vec{a}_i \qquad\qquad (2.3.1)$$

$$\vec{e}_i^o \times \vec{a}_i^o = Q_i^o (\vec{e}_i \times \vec{a}_i)$$

and this completely determines the matrix $Q_i^o$:

---

*) Ort $(\cdot)$ denotes the unit vector of $(\cdot)$.

$$Q_i^O = [\vec{e}_i^O \ \vec{a}_i^O \ \vec{e}_i^O \times \vec{a}_i^O][\vec{\tilde{e}}_i \ \vec{\tilde{a}}_i \ \vec{\tilde{e}}_i \times \vec{\tilde{a}}_i]^T \qquad (2.3.2)$$

Let us note that matrix transposition has been used instead of its inverse, due to the orthogonality of vectors $\vec{\tilde{e}}_i$, $\vec{\tilde{a}}_i$ and $\vec{\tilde{e}}_i \times \vec{\tilde{a}}_i$.

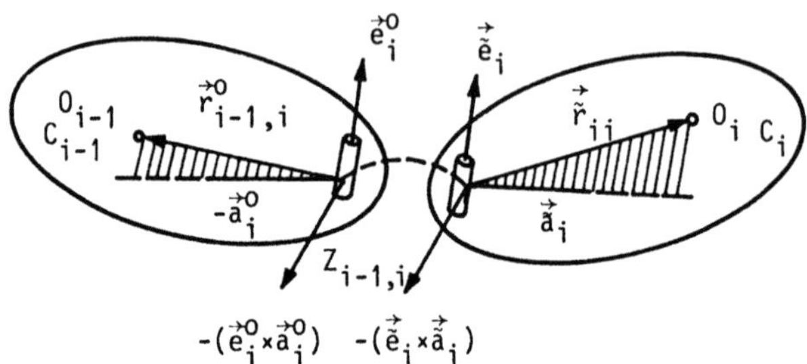


Fig. 2.2. Relative position of link $C_i$ with respect to $C_{i-1}$ with $q^i = 0$

Now, it is necessary to determine the remaining elements of set $K_i^O$, i.e., sets $R_i^O$ and $E_i^O$. Since the transformation matrix $Q_i^O$ has been determined, the following relations evidently hold

$$\vec{r}_{ii}^O = Q_i^O \vec{\tilde{r}}_{ii}$$

$$\vec{r}_{i,i+1}^O = Q_i^O \vec{\tilde{r}}_{i,i+1} \qquad (2.3.3)$$

$$\vec{e}_{i+1}^O = Q_i^O \vec{\tilde{e}}_{i+1}$$

The recursiveness required to calculate $Q_i^O$ for $\forall i \in N$, and all the vectors contained in $R_i^O$ and $E_i^O$, has thus been established.

*Stage 2: Mechanism position*

This stage consists of determining the elements of the set of kinematic variables $K_i = (Q_i, R_i, E_i)$, where the mechanism position is determined by joint coordinates $q^i$, $i \in N$.

Applying the theorem of finite rotations (Rodrigue's formula) for the revolute joints one obtains

$$\vec{q}_{ij} = \begin{bmatrix} \vec{e}_i \times (\vec{q}^{\,O}_{ij} \times \vec{e}_i) \\[4pt] (\vec{e}_i \times \vec{q}^{\,O}_{ij}) \\[4pt] (\vec{e}_i \cdot \vec{q}^{\,O}_{ij})\,\vec{e}_i \end{bmatrix}^T \begin{bmatrix} \cos q^i \\[4pt] \sin q^i \\[4pt] 1 \end{bmatrix} , \qquad j=1,2,3. \qquad (2.3.4)$$

We have thus determined the matrix $Q_i = [\vec{q}_{i1}, \vec{q}_{i2}, \vec{q}_{i3}]$ which represents transformation matrix from the i-th link coordinate system $Q_i$ into the reference system. Now, we directly obtain

$$R_i = \{\vec{r}_{ii}, \vec{r}_{i,i+1}\} = \{Q_i \vec{r}_{ii}, Q_i \vec{r}_{i,i+1}\}$$

$$E_i = \{\vec{e}_i, \vec{e}_{i+1}\} = \{\vec{e}_i, Q_i \vec{e}_{i+1}\} \qquad (2.3.5)$$

where $\vec{r}_{ij} = \vec{r}_{i-1,j} - \vec{r}_{i-1,i} + \vec{r}_{ii}$, $j \in I$, $i \in N$, evidently holds. Determination of $\vec{e}_{i+1}$ provides for the recursiveness of calculating the transformation matrix $Q_i$ according to (2.3.4), and thus the elements of sets (2.3.5).


*Stage 3: Mechanism kinematics*

This stage consists of forming the set of kinematic quantities $\mathcal{L}_i = \{\Omega_i, W_i\}$, where

$$\Omega_i = \{\vec{\omega}_i, \vec{v}_i\}$$

$$W_i = \{\vec{\varepsilon}_i, \vec{w}_i\}$$

with

$\vec{\omega}_i$ - angular velocity of the i-th link

$\vec{v}_i$ - linear velocity of the i-th link

$\vec{\varepsilon}_i$ - angular acceleration of the i-th link, and

$\vec{w}_i$ - linear acceleration of the i-th link.

The kinematic quantities are functions of the velocities and accelerations of joint coordinates $\dot{q}^i$ and $\ddot{q}^i$ and variables formed in the preceding stage: $K_i = (Q_i, R_i, E_i)$. All relations considered in this stage are recursive. Applying the basic theorems of rigid body kinematics, one obtains

$$\vec{\omega}_i = \vec{\omega}_{i-1} + \dot{q}^i \vec{e}_i$$

$$\vec{v}_i = \vec{v}_{i-1} - \vec{\omega}_{i-1} \times \vec{r}_{i-1,i} + \vec{\omega}_i \times \vec{r}_{ii}$$

$$\vec{\varepsilon}_i = \vec{\varepsilon}_{i-1} + \dot{q}^i \vec{\omega}_{i-1} \times \vec{e}_i + \ddot{q}^i \vec{e}_i$$

$$\vec{w}_i = \vec{w}_{i-1} - \vec{\varepsilon}_{i-1} \times \vec{r}_{i-1,i} - \vec{\omega}_{i-1} \times (\vec{\omega}_{i-1} \times \vec{r}_{i-1,i}) +$$
$$+ \vec{\varepsilon}_i \times \vec{r}_{ii} + \vec{\omega}_i \times (\vec{\omega}_i \times \vec{r}_{ii})$$

$$(2.3.6)$$

It follows that angular and linear accelerations are functions of the second derivatives of joint coordinates $\ddot{q}^j$, $j \in I$. However, to form the dynamic model matrices, expressions $\vec{\varepsilon}_i$ and $\vec{w}_i$ should be rearranged so that accelerations $\ddot{q}^j$, $j \in N$, appear in an explicit form

$$\vec{\varepsilon}_i = [\vec{\alpha}_{i1} \cdots \vec{\alpha}_{ii} \ 0 \cdots 0] \ddot{q} + \vec{\alpha}_i^O$$

$$\vec{w}_i = [\vec{\beta}_{i1} \cdots \vec{\beta}_{ii} \ 0 \cdots 0] \ddot{q} + \vec{\beta}_i^O$$

$$(2.3.7)$$

where $\ddot{q} = [\ddot{q}^1 \cdots \ddot{q}^n]$. The set of kinematic quantities $W_i$ thus reduces to

$$W_i = \{\vec{\alpha}_{ij}, \ \vec{\alpha}_i^O, \ \vec{\beta}_{ij}, \ \vec{\beta}_i^O; \ j \in I\} \ .$$

From (2.3.6) and (2.3.7) we obtain the recursive relations determining the elements of set $W_i$:

$$\vec{\alpha}_{ij} = \vec{e}_j, \ j \in I, \ i \in N; \quad \vec{\alpha}_{ij} = \vec{\alpha}_{i-1,j}, \ j \neq i; \quad \vec{\alpha}_i^O = \vec{\alpha}_{i-1}^O + \dot{q}^i (\vec{\omega}_{i-1} \times \vec{e}_i)$$

$$\vec{\beta}_{ij} = \vec{\beta}_{i-1,j} + \vec{e}_j \times \vec{R}_{i,i-1}, \qquad j \in \{1, \ldots, i-1\}$$

$$\vec{\beta}_{ii} = (\vec{e}_i \times \vec{r}_{ii})$$

$$(2.3.8)$$

$$\vec{\beta}_i^O = \vec{\beta}_{i-1}^O + \vec{\alpha}_{i-1}^O \times \vec{R}_{i,i-1} + \dot{q}^i (\vec{\omega}_{i-1} \times \vec{e}_i) \times \vec{r}_{ii} -$$
$$- \vec{\gamma}_{i-1,i} + \vec{\gamma}_{ii}$$

where $\vec{\gamma}_{ij} = \vec{\omega}_i \times (\vec{\omega}_i \times \vec{r}_{ij})$, $j \in I$, $i \in N$. The vector $\vec{R}_{i,i-1} = \vec{r}_{ii} - \vec{r}_{i-1,i}$ represents the distance vector between the (i-1)-th and i-th link centres of masses. All kinematic quantities required for forming the dynamic model have thus been determined.

*Stage 4: Mechanism dynamics*

This stage includes evaluation of dynamic quantities $\Delta_i = \{F_i, M_i\}$, $i \in N$, where

$$F_i = \{\vec{F}_i\}, \qquad M_i = \{\vec{M}_i\}$$

with

$\vec{F}_i$ - inertial force in the centre of mass of the i-th link, and

$\vec{M}_i$ - moment of the inertial force of the i-th link.

The inertial force may be calculated using Newton's law

$$\vec{F}_i = -m_i \vec{w}_i = [\vec{a}_{i1} \cdots \vec{a}_{ii} \; 0 \cdots 0]\ddot{q} + \vec{a}_i^O \tag{2.3.9}$$

where $m_i$ - is the mass of the i-th link. Comparing this with (2.3.7), we obtain

$$\vec{a}_{ij} = -m_i \vec{\beta}_{ij}, \qquad j \in I$$
$$\vec{a}_i^O = -m_i \vec{\beta}_i^O. \tag{2.3.10}$$

The moments of inertial forces are determined from Euler's dynamic equations and may be represented in the form

$$\vec{M}_i = [\vec{b}_{i1} \cdots \vec{b}_{ii} \; 0 \cdots 0]\ddot{q} + \vec{b}_i^O \tag{2.3.11}$$

where

$$\vec{b}_{ij} = -T_i \vec{\alpha}_{ij}, \qquad j \in I$$
$$\vec{b}_i^O = -T_i \vec{\alpha}_i^O + \vec{\lambda}_i \tag{2.3.12}$$

with

$$T_i = \sum_{\ell=1}^{3} Q_{i\ell} J_{i\ell}$$

$$Q_{i\ell} = [q_{i\ell}^1 \vec{q}_{i\ell} \; q_{i\ell}^2 \vec{q}_{i\ell} \; q_{i\ell}^3 \vec{q}_{i\ell}] \tag{2.3.13}$$

$$\vec{\lambda}_i = Q_i \begin{bmatrix} (\vec{\omega}_i \cdot \vec{q}_{i2})(\vec{\omega}_i \cdot \vec{q}_{i3})(J_{i2}-J_{i3}) \\ (\vec{\omega}_i \cdot \vec{q}_{i3})(\vec{\omega}_i \cdot \vec{q}_{i1})(J_{i3}-J_{i1}) \\ (\vec{\omega}_i \cdot \vec{q}_{i1})(\vec{\omega}_i \cdot \vec{q}_{i2})(J_{i1}-J_{i2}) \end{bmatrix}$$

$q_{i\ell}^j$ (j=1,2,3) denoting the j-th component of vector $\vec{q}_{i\ell}$.

Apart from the inertial forces and moments, external forces and moments $\vec{G}_i$ and $M_i^G$ also act upon the links so that the total forces and moments can be expressed in the form

$$\vec{F}_i^u = \vec{F}_i + \vec{G}_i, \qquad \vec{M}_i^u = \vec{M}_i + \vec{M}_i^G \tag{2.3.14}$$

Substituting expressions (2.3.9) and (2.3.11) into (2.3.14) one obtains,

$$\vec{F}_j^u = \sum_{k=1}^{j} \vec{a}_{jk}\ddot{q}^k + \vec{a}_j^o + \vec{G}_j, \qquad \vec{M}_j^u = \sum_{k=1}^{j} \vec{b}_{jk}\ddot{q}^k + \vec{b}_j^o + \vec{M}_j^G \tag{2.3.15}$$

Let the kinematic chain be fictively ruptured at the i-th joint and consider the equilibrium of the mechanism free end (Fig. 2.3). The action of the rejected mechanism part is substituted by a force $\vec{R}_i$ and moment $\vec{M}_i^*$. This force and moment will be termed as the total reaction force at the i-th joint. In determining the overal reaction the external forces $\vec{F}_j^u$ and moments $\vec{M}_j^u$ (j=i,i+1,...,n) are reduced to the centre of the i-th joint. Reactions of drives of all subsequent joints need not be taken into account since each drive acts upon two adjacent links with forces and moments of equal magnitude but of opposite sense which are thus vanishing in the summation process.

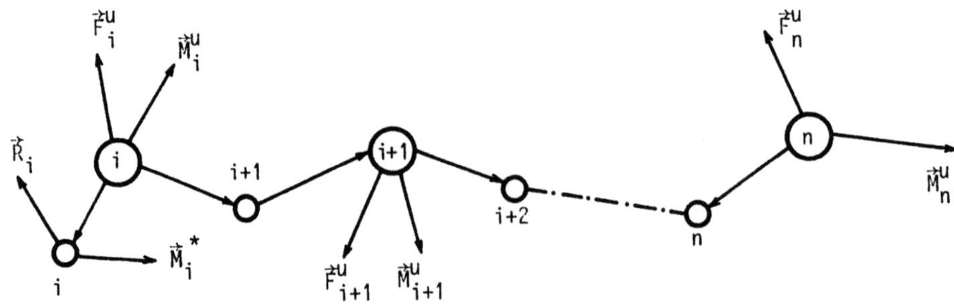


Fig. 2.3. Equilibrium of the free chain

The overall reactions can therefore be written as [6-8]:

$$\vec{R}_i = - \sum_{j=i}^{n} \vec{F}_j^u = - \sum_{j=i}^{n} ( \sum_{k=1}^{j} \vec{a}_{jk}\ddot{q}^k + \vec{a}_j^o + \vec{G}_j) \qquad (2.3.16)$$

$$\vec{M}_i^* = - \sum_{j=i}^{n} (\vec{M}_j^u + \vec{r}_{ji} \times \vec{F}_j^u) = - \sum_{j=i}^{n} [ \sum_{k=1}^{j} (\vec{b}_{jk} + \vec{r}_{ji} \times \vec{a}_{jk})\ddot{q}^k +$$

$$+ \vec{r}_{ji} \times \vec{a}_j^o + \vec{b}_j^o + \vec{M}_j^G + \vec{r}_{ji} \times \vec{G}_j ]$$

where n is the number of links in the chain.

To obtain the reactions, it is appropriate to establish the recursive relations between the reactions of two adjacent joints. From Fig. 2.4(a) which illustrates a single mechanism link, it is evident that

$$\vec{R}_i = \vec{R}_{i+1} - \vec{F}_i^u$$

$$\vec{M}_i^* = \vec{M}_{i+1}^* + (\vec{r}_{ii} - \vec{r}_{i+1,i}) \times \vec{R}_{i+1} + \vec{r}_{ii} \times \vec{F}_i^u - \vec{M}_i^u \qquad (2.3.17)$$

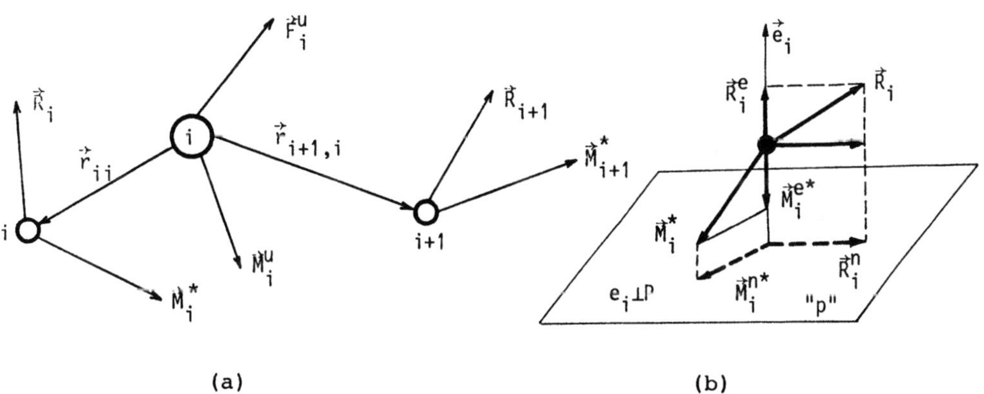


(a)         (b)

Fig. 2.4. Reaction at the i-th link

Following the chain from the last to the first link according to (2.3.17) all the reactions can be determined.

Let us decompose the overall reaction force $\vec{R}_i$ and moment $\vec{M}_i^*$ at the i-th joint, to their components parallel and perpendicular to vector $\vec{e}_i$ (Fig. 2.4 b). Because the perpendicular components cannot cause any movement

64

in the mechanism, they are acting only as load on the mechanism joint,
and are necessary when considering the forces of friction. One of the
parallel components (forces for revolute kinematic pairs and moments
for prismatic pairs) also contributes to the force of friction present
at the i-th joint. In order to maintain the mechanism equilibrium, the
other parallel component of the reaction must be balanced by a reaction at the i-th joint. In our case (revolute kinematic pair)

$$P_i^M = \vec{M}_i^* \cdot \vec{e}_i \tag{2.3.18}$$

Conditions (2.3.17) and (2.3.18) enable the determination of the required moments at each joint of the robot mechanism:

$$P_i^M = -\vec{e}_i \cdot \sum_{k=1}^{n} \sum_{j=\max(i,k)}^{n} (\vec{b}_{jk} + \vec{r}_{ji} \times \vec{a}_{jk}) \ddot{q}^k -$$

$$- \vec{e}_i \cdot \sum_{j=i}^{n} (\vec{r}_{ji} \times (\vec{a}_j + \vec{G}_j) + \vec{b}_j^O) \tag{2.3.19}$$

This expression can be written in matrix form

$$P = H(q, \theta)\ddot{q} + h(q, \dot{q}, \theta) \tag{2.3.20}$$

where

$$P = [P_1 \cdots P_n]^T - \text{vector of driving moments } P^M,$$

$$q = [q^1 \cdots q^n]^T - \text{joint coordinates,}$$

$$\theta = [\theta_1 \cdots \theta_n]^T - \text{geometric and dynamic parameters vector.}$$

According to the expression  given above, the elements of matrices in
(2.3.20) take the form:

$$[H_{ik}] = -\vec{e}_i \cdot \sum_{j=\max(i,k)}^{n} (\vec{b}_{jk} + \vec{r}_{ji} \times \vec{a}_{jk}) \tag{2.3.21}$$

$$[h_i] = -\vec{e}_i \cdot \sum_{j=i}^{n} (\vec{r}_{ji} \times (\vec{a}_j^O + \vec{G}_j) + \vec{b}_j^O)$$

## 2.4. Forming the Dynamic Equations of Motion for Complex Kinematic Chains

A link of a complex kinematic chain belonging to more than two kinematic pairs will be termed the *branching* link.

As already mentioned, the structure of an anthropomorphic active mechanism is a complex kinematic chain. In order to make use of the procedure described in the preceding section, such a structure will be partitioned into three simple kinematic chains (see Fig. 2.5a). The first chain represents "legs" (links 1-7) second, the uper part of "body" (links 8-10) and the third, the right "hand" (links 11 and 12).

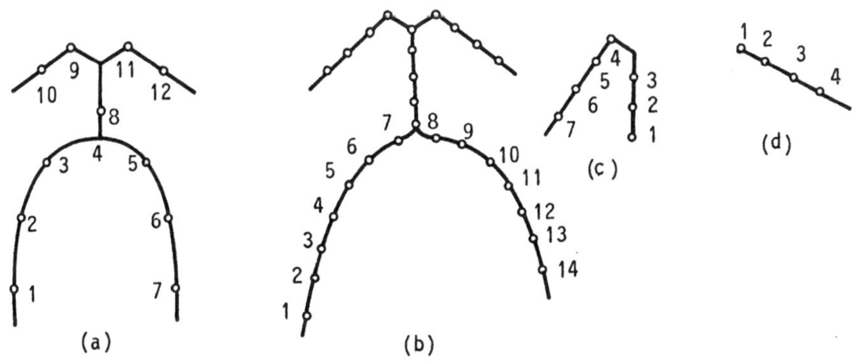

Fig. 2.5. Structure of a complex kinematic chain

After replacing the spherical joints by corresponding number of revolute joints of the fifth class, the following structure is obtained: The first chain contains 14 kinematic pairs of the fifth class (Fig. 2.5b), the second 7 (Fig. 2.5c) and the third one 4 (Fig. 2.5d). Such complex kinematic chain possesses 25 d.o.f. The links 4 and 8 are branching links.

During the single-support phase of gait, the mechanism is supported only on one leg. The corresponding kinematic scheme of the mechanism can be considered as a mechanism composed of more (three) open kinematic chains. The kinetostatic procedure for forming the differential equations for motion of complex kinematic chains is analogous to that for simple kinematic chains (see Section 2.3). A fictitious "rupture" of the complex kinematic chain at the i-th joint is illustrated in Fig.

2.6. The rejected part of the mechanism may be either a simple or a complex kinematic chain. After some rearrangement of the corresponding expressions, the differential equations of motion are also obtained in the form of (2.3.20).

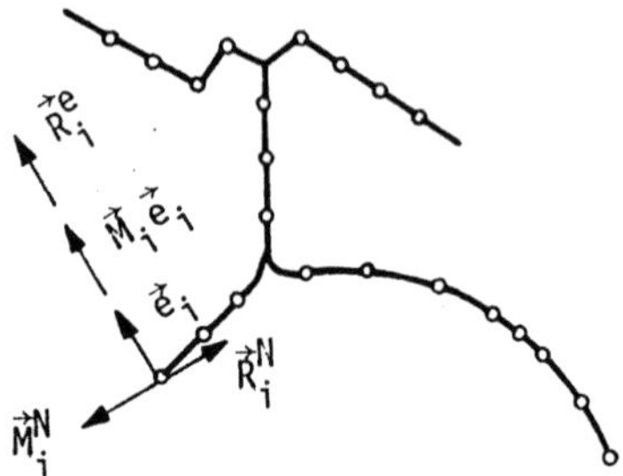


Fig. 2.6. Equilibrium of the "ruptured" complex kinematic chain

Calculation of the elements of matrix H and vector h of complex kinematic chains can be carried out by introducing the corresponding number of series of "+" joints. A series of "+" joints is formed in such a way that, when the chain is ruptured at a certain "+" joint, the j-th link should remain in the external part (not connected to the support) of the mechanism. Then, in the procedure of forming differential equations, the inertial force and moment of the j-th mechanism link is reduced only to its "own" "+" joint. Consequently, the following procedure is possible: the quantities $\vec{F}_j^u$ and $\vec{M}_j^u$ corresponding to the j-th link are successively reduced to all "+" joints going from the j-th link towards the support, as shown in Fig. 2.7. Then, after projecting the values $\vec{F}_j^u$ and $\vec{M}_j^u$ onto the axis of the i-th joint, the resulting quantities, denoted by $\Delta H_{ik}^j$ and $\Delta h_i^j$, can be calculated in the following way [6-8]:

$$\Delta H_{ik}^j = -\vec{e}_i(\vec{b}_{jk}+\vec{r}_{ji}\times\vec{a}_{jk})$$

$$\Delta h_i^j = -\vec{e}_i(\vec{r}_{ji}\times(\vec{a}_j^O+\vec{G}_j)+\vec{b}_j^O)$$

$$(2.4.1)$$

Now, the components of matrix H and vector h are obtained by summing up the corresponding values from (2.4.1) with respect to all series of "+" joints

$$H_{ik} = \sum_{(j)} \Delta H_{ik}^j; \qquad h_i = \sum_{(j)} \Delta h_i^j \qquad (2.4.2)$$

Fig. 2.8. illustrates a branching link having three kinematic pairs and which is the consituent of two series of "+" joints. In such case the

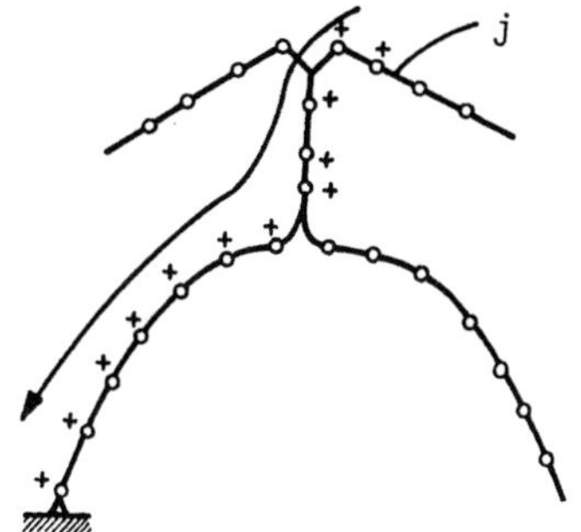

Fig. 2.7. Series of "+" joints of the complex kinematic chain

topological structure of the complex kinematic chain can be represented
by the matrix MS. Each row of this matrix contains ordinal numbers of
the corresponding series of "+" joints. The element MS (i,j) is the
j-th joint in the i-th series of "+" joints. In addition, for each series of "+" joints, the ordinal number of the initial joint is also defined. The *initial* joint of the i-th series of "+" joints is the first
joint of the i-th series differing from joints of the (i-1)-th series
of "+" joints. The initial joint of the first series of "+" joints is
MS(1,1).

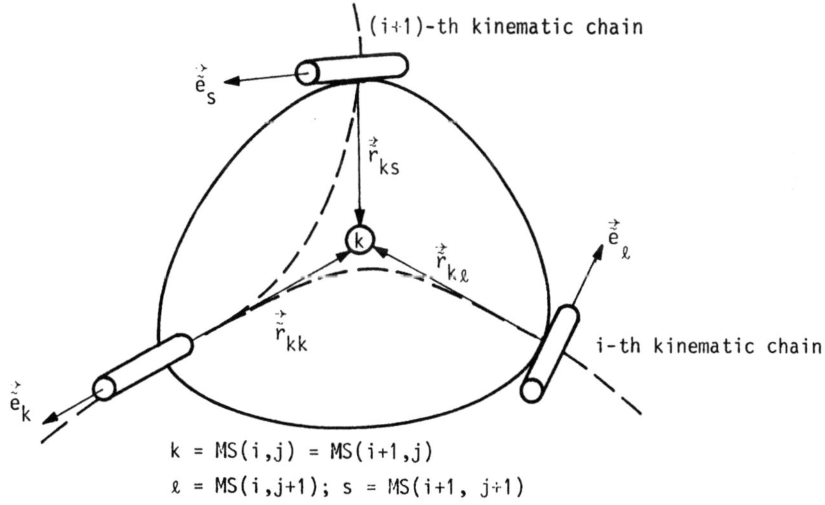



Fig. 2.8. Branching link of a complex kinematic chain

For the first link appearing in the first series of "+" joints, the
matrix $Q_o^o$ should be known. If a fixed support serves as the basis, the
matrix $Q_o^o$ is a unit matrix. If we proceed to another chain, then the
matrices $Q_i^o$ should be either stored or formed on the basis of (2.3.2).

The transformation matrix of the branching link serves to calculate the vectors $\vec{r}$ and $\vec{e}$. Besides, for branching links the information is needed on their velocities $(\vec{v})$ and accelerations $(\vec{w})$ and the vector coefficients $\vec{\alpha}$, $\vec{\beta}$, $\vec{\alpha}^O$, $\vec{\beta}^O$. It should be noted that all support vectors are equal to zero. These quantities for the mobile branching links should be stored when the preceding chain is analysed.

Fig. 2.9. illustrates a global flow-chart for the computer synthesis of the mathematical model of a complex kinematic chain. The algorithm consists of the following stages: definition of the mechanism parameters (block 1); "assembling" the mechanism, i.e., determining the transformation matrices between the coordinate frames connected to particular links and the absolute coordinate frame for zero values of the local coordinates (Matrices $Q_i$, block 2); calculation of the mechanism position in the fixed coordinate frame when the local coordinates are changed for the corresponding values (block 3); calculation of the angular velocities and angular accelerations of mass centres of the links (block 4); determination of the linear velocities and accelerations and the corresponding vector coefficients (block 5). In the subsequent stage, it is calculated the inertial forces and moments of inertial forces, i.e., the corresponding acceleration coefficients, as well as, the remaining non-inertial effects: centrifugal and Coriolis' effects (blocks 6, 7 and 8). The case when a link may be considered as a "cane" is treated separately, using another procedure for determination of the moments of inertial forces. In block 9 are calculated the elements of matrices H and h for the $\ell$-th joint in the j-th series of "+" joints. In block 10, by means of the transformation matrix of branching link, are calculated the vectors $\vec{r}$ and $\vec{e}$ which determine the position of the initial joint in the subsequent series of "+" joints.

*Forming the equations of dynamic connections*

To perform relative displacement, the robot's joints are equiped with suitable actuators. Conditionally, we shall distinguish two types of actuators: "kinematic" and "dynamic". The task of the former type of actuator is to ensure realization of the prescribed laws of the relative motion, while a dynamic actuator is responsible for the motion laws of kinematic pairs related to the compensating dynamics of the mechanism. An actuator acts upon the adjacent members of the kinematic pair by equal, but oppositely directed, reactions. Let $\{\bar{q}\}$ be a set of

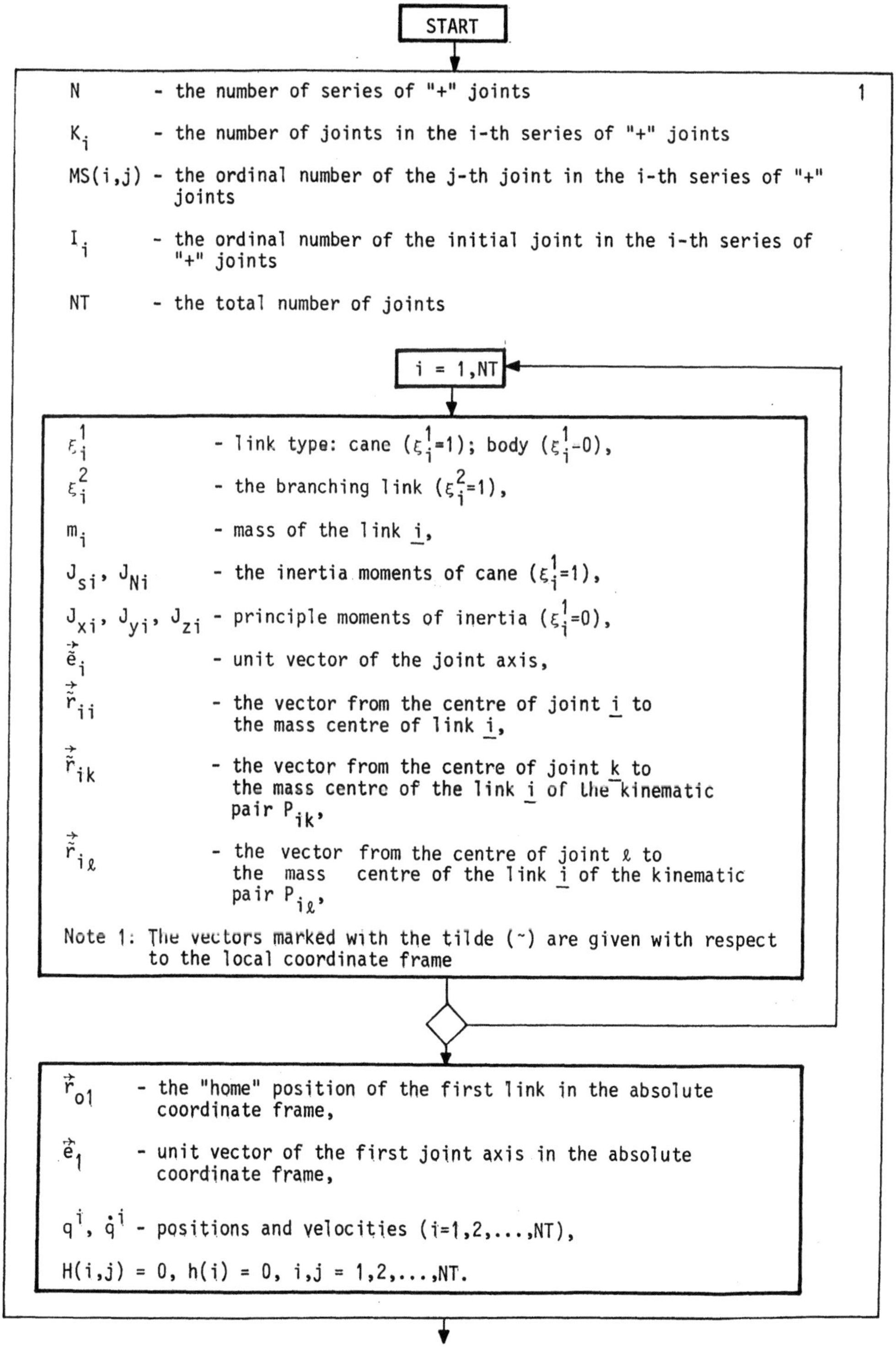

START

| N | - the number of series of "+" joints | 1 |

$K_i$ - the number of joints in the i-th series of "+" joints

$MS(i,j)$ - the ordinal number of the j-th joint in the i-th series of "+" joints

$I_i$ - the ordinal number of the initial joint in the i-th series of "+" joints

$NT$ - the total number of joints

$i = 1,NT$

$\xi_i^1$ - link type: cane ($\xi_i^1=1$); body ($\xi_i^1=0$),

$\xi_i^2$ - the branching link ($\xi_i^2=1$),

$m_i$ - mass of the link $\underline{i}$,

$J_{si}$, $J_{Ni}$ - the inertia moments of cane ($\xi_i^1=1$),

$J_{xi}$, $J_{yi}$, $J_{zi}$ - principle moments of inertia ($\xi_i^1=0$),

$\vec{\tilde{e}}_i$ - unit vector of the joint axis,

$\vec{\tilde{r}}_{ii}$ - the vector from the centre of joint $\underline{i}$ to the mass centre of link $\underline{i}$,

$\vec{\tilde{r}}_{ik}$ - the vector from the centre of joint $\underline{k}$ to the mass centre of the link $\underline{i}$ of the kinematic pair $P_{ik}$,

$\vec{\tilde{r}}_{i\ell}$ - the vector from the centre of joint $\underline{\ell}$ to the mass centre of the link $\underline{i}$ of the kinematic pair $P_{i\ell}$,

Note 1: The vectors marked with the tilde ($\tilde{\phantom{x}}$) are given with respect to the local coordinate frame

$\vec{r}_{01}$ - the "home" position of the first link in the absolute coordinate frame,

$\vec{e}_1$ - unit vector of the first joint axis in the absolute coordinate frame,

$q^i$, $\dot{q}^i$ - positions and velocities ($i=1,2,\ldots,NT$),

$H(i,j) = 0$, $h(i) = 0$, $i,j = 1,2,\ldots,NT$.

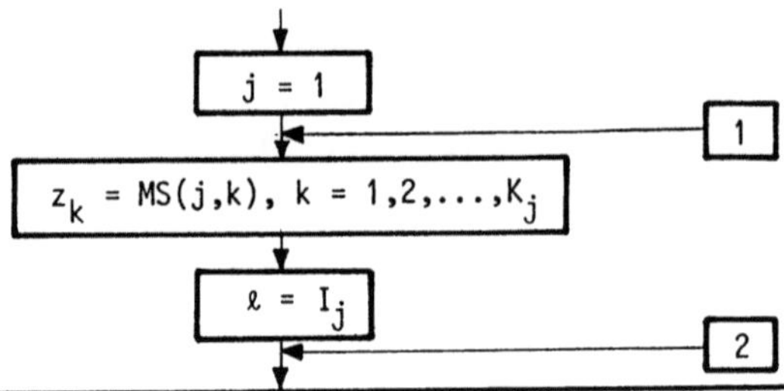



**Mechanism "assembly"**    2

$i = MS(j,\ell)$; $t = MS(j,\ell-1)$

$$\vec{r}_N = \vec{e}_i^{\,0} \times (\vec{r}_{ti}^{\,0} \times \vec{e}_i^{\,0}); \quad \vec{\tilde{r}}_N = \vec{\tilde{e}}_i \times (\vec{\tilde{r}}_{ii} \times \vec{\tilde{e}}_i)$$

$$\vec{a}_i^{\,0} = \frac{-\vec{r}_N}{|\vec{r}_N|} \; ; \quad \vec{\tilde{a}}_i = \frac{\vec{\tilde{r}}_N}{|\vec{\tilde{r}}_N|}$$

$$\vec{b}_i^{\,0} = \vec{e}_i^{\,0} \times \vec{a}_i^{\,0}; \quad \vec{\tilde{b}}_i = \vec{\tilde{e}}_i \times \vec{\tilde{a}}_i$$

$$Q_i^0 = [\vec{q}_{i1}^{\,0} \; \vec{q}_{i2}^{\,0} \; \vec{q}_{i3}^{\,0}] = [\vec{e}_i^{\,0} \; \vec{a}_i^{\,0} \; \vec{b}_i^{\,0}][\vec{\tilde{e}}_i \; \vec{\tilde{a}}_i \; \vec{\tilde{b}}_i]^T$$

($Q_i^0$ - the matrix of transformation before the rotation)

**Determination of the mechanism position**    3

$i = MS(j,\ell)$; $k = MS(j,\ell+1)$

$$\vec{r}_{ii}^{\,0} = Q_i^0 \vec{\tilde{r}}_{ii}; \quad \vec{r}_{ik}^{\,0} = Q_i^0 \vec{\tilde{r}}_{ik}; \quad \vec{e}_k^{\,0} = Q_i^0 \vec{\tilde{e}}_k$$

$$\vec{q}_{ij} = \vec{q}_{ij}^{\,0} \cos q^i + (1-\cos q^i)(\vec{e}_i \cdot \vec{q}_{ij}^{\,0}) \vec{e}_i + \vec{e}_i \vec{q}_{ij}^{\,0} \sin q^i$$

$Q_i = [\vec{q}_{i1} \; \vec{q}_{i2} \; \vec{q}_{i3}]$ - the matrix of transformation after the rotation

$$\vec{r}_{ii} = Q_i \vec{\tilde{r}}_{ii}; \quad \vec{r}_{ik} = Q_i \vec{\tilde{r}}_{ik}; \quad \vec{e}_k = Q_i \vec{\tilde{e}}_k$$

$$\vec{r}_{i,z_n} = \vec{r}_{k,z_n} - \vec{r}_{ki} + \vec{r}_{ii}; \quad n = 1,2,\ldots,\ell-1$$

Note 2: $\vec{r}_{i,z_n}$ are the position vectors of the mass centre of the i-th link with respect to the centre of joint $z_n$ in the series of "+" joints, for example,

$$\vec{r}_{31} = \vec{r}_{21} - \vec{r}_{23} + \vec{r}_{33} \text{ where } \vec{r}_{21} = \vec{r}_{11} - \vec{r}_{12} + \vec{r}_{22}, \; \vec{r}_{11} = \vec{r}_{11}$$

**Angular velocities and accelerations** 4

$$i = MS(j,\ell); \quad k = MS(j,\ell-1)$$

$$\vec{\omega}_i = \vec{\omega}_k + \dot{q}^i \vec{e}_i,$$

$$\vec{\alpha}_{i,z_n} = \vec{\alpha}_{k,z_n}, \quad z_n \neq i$$

$$\vec{\alpha}_i^0 = \vec{\alpha}_k^0 + \dot{q}^i (\vec{\omega}_k \times \vec{e}_i)$$

$$\vec{\alpha}_{ii} = \vec{e}_i$$

**Linear velocities and accelerations** 5

$$i = MS(j,\ell); \quad k = MS(j,\ell-1)$$

$$\vec{v}_i = \vec{v}_k - \vec{\omega}_k \times \vec{r}_{ki} + \vec{\omega}_i \times \vec{r}_{ii}$$

$$\vec{\beta}_{i,z_n} = \vec{\beta}_{k,z_n} + \vec{\alpha}_{k,z_n} \times (\vec{r}_{ii} - \vec{r}_{ki}), \quad z_n \neq i$$

$$\vec{\beta}_{ii} = \vec{e}_i \times \vec{r}_{ii}$$

$$\vec{\beta}_i^0 = \vec{\beta}_k^0 + \vec{\alpha}_k^0 \times (\vec{r}_{ii} - \vec{r}_{ki}) + \dot{q}^i (\vec{\omega}_k \times \vec{e}_k) \times \vec{r}_{ii} - \vec{\omega}_k \times (\vec{\omega}_k \times \vec{r}_{ki}) + \vec{\omega}_i \times (\vec{\omega}_i \times \vec{r}_{ii})$$

**Inertial forces** 6

$$i = MS(j,\ell)$$

$$\vec{F}_i = m_i \vec{w}_i = \sum_{n=1}^{\ell} \vec{a}_{i,z_n} \ddot{q}^{z_n} + \vec{a}_i^0; \quad \vec{a}_{i,z_n} = -m_i \vec{\beta}_{i,z_n}; \quad \vec{a}_i^0 = -m_i \vec{\beta}_i^0$$

$$\xi_i^1 = 1 \qquad \text{NO}$$

$$\text{YES}$$

**Moments of inertial forces - cane** 7

$$i = MS(j,\ell)$$

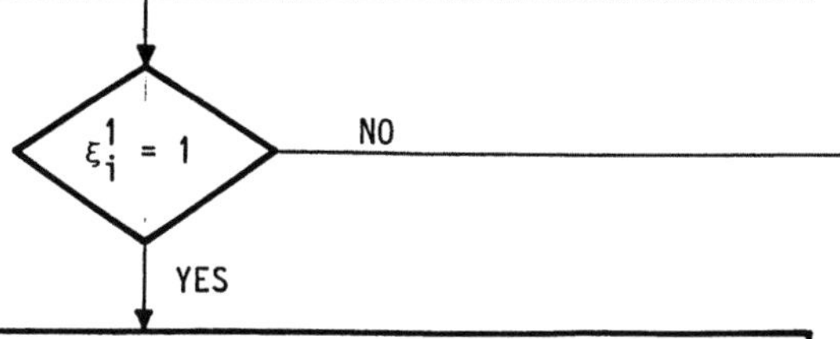

$$\vec{M}_i = \sum_{n=1}^{\ell} \vec{b}_{i,z_n} \ddot{q}^{z_n} + \vec{b}_i^0; \quad \vec{b}_{i,z_n} = -J_{Ni}(\vec{s}_i \times \vec{\alpha}_{i,z_n}) \times \vec{s}_i - J_{si}(\vec{\alpha}_{i,z_n} \vec{s}_i) \vec{s}_i$$

$$b_i^o = -J_{Ni}[(\vec{s}_i \times \vec{\alpha}_i^o) \times \vec{s}_i + \vec{\tau}_i] - J_{si}(\vec{\alpha}_i^o \vec{s}_i)\vec{s}_i$$

$$\vec{\tau}_i = (\vec{\omega}_i \vec{s}_i)(\vec{s}_i \times \vec{\omega}_i); \quad \vec{s}_i = \frac{\vec{r}_{ii}}{|\vec{r}_{ii}|} - \text{unit vector along the "cane" axis}$$

Moments of inertial forces - body      8

$$i = MS(j,\ell)$$

$$\vec{M}_i = \sum_{n=1}^{\ell} \vec{b}_{i,z_n} \ddot{q}^{z_n} + \vec{b}_i^o; \quad \vec{b}_{i,z_n} = -T_i \vec{\alpha}_{i,z_n}; \quad \vec{b}_i^o = -T_i \vec{\alpha}_i^o + \vec{\lambda}_i$$

$$T_i^{nk} = \sum_{m=1}^{3} q_{im}^k q_{im}^n J_{im}; \quad T_i = [T_i^{nk}]; \quad n,k = 1,2,3$$

$$\vec{\lambda}_i = Q_i \begin{bmatrix} (J_{i2}-J_{i3})(\vec{\omega}_i \vec{q}_{i2})(\vec{\omega}_i \vec{q}_{i3}) \\ (J_{i3}-J_{i1})(\vec{\omega}_i \vec{q}_{i3})(\vec{\omega}_i \vec{q}_{i1}) \\ (J_{i1}-J_{i2})(\vec{\omega}_i \vec{q}_{i1})(\vec{\omega}_i \vec{q}_{i2}) \end{bmatrix}; \quad Q_i = \begin{bmatrix} q_{i1}^1 & q_{i2}^1 & q_{i3}^1 \\ q_{i1}^2 & q_{i2}^2 & q_{i3}^2 \\ q_{i1}^3 & q_{i2}^3 & q_{i3}^3 \end{bmatrix}$$

$$\xi_i^2 = 1 \quad \text{NO}$$

YES

save matrix $Q_i$: $Q_p = Q_i$

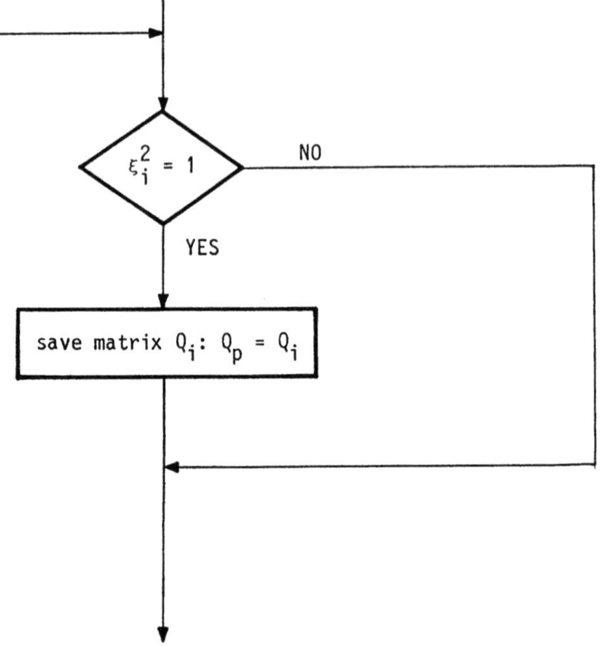

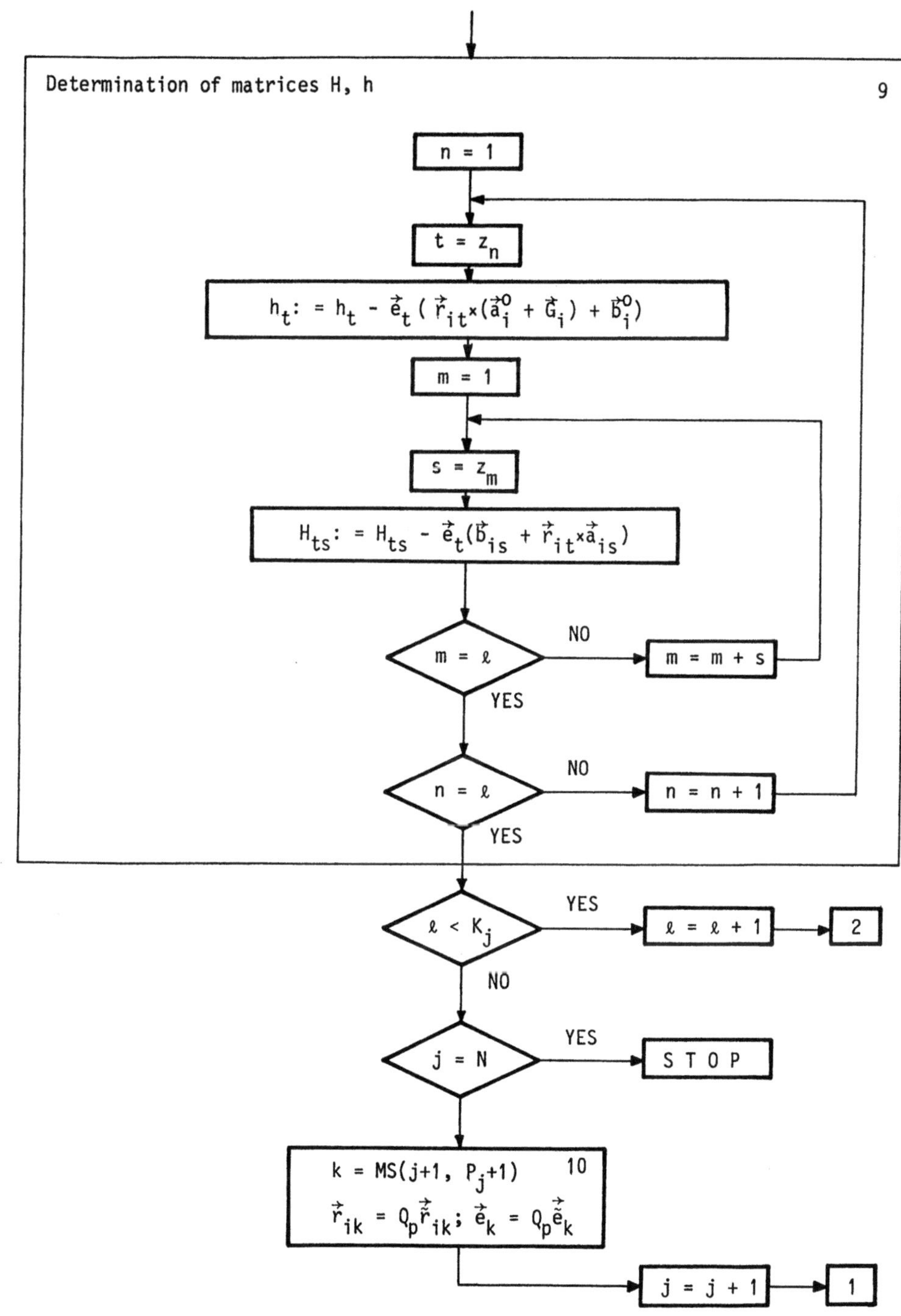


Fig. 2.9. Flow-chart of the algorithm for forming the mathematical model of a complex kinematic chain

kinematic and $\{q^*\}$ a set of dynamic actuators coordinates. Let the numbers of elements of $\{q^*\}$ and $\{\bar{q}\}$ be $n_q*$ and $n_{\bar{q}}$, respectively.

Consider now the generation of equations for dynamic connections of the anthropomorphic mechanism. Let the ZMP coordinates in the fixed coordinate frame be given as vector $\vec{R}_z$. Let $\vec{R}_i$ be the radius-vector of the centre of mass of the i-th link, and $\vec{E}$ the unit vector with respect to which the moments are summed. Then,

$$\vec{E} \cdot \sum_{i=1}^{NT} [\, (\vec{R}_i - \vec{R}_z) \times \vec{F}_i^u + \vec{M}_i^u ] = 0 \qquad (2.4.3)$$

By introducing into this expression the relations for $\vec{F}_i^u$ and $\vec{M}_i^u$ (2.2.34), one obtains,

$$\vec{E} \cdot \sum_{i=1}^{NT} [\, \sum_{k=1}^{i} ((\vec{R}_i - \vec{R}_z) \times \vec{a}_{ik} + \vec{b}_{ik}) \ddot{q}^k + (\vec{R}_i - \vec{R}_z) \times \vec{a}_i^o + \vec{b}_i^o +$$

$$+ (\vec{R}_i - \vec{R}_z) \times \vec{G}_i + \vec{M}_i^G ] = 0 \qquad (2.4.4)$$

Let us separate the subsets of $q^*$ and $\bar{q}$ coordinates and mark appropriate coefficients by upper indices $q^*$ and $\bar{q}$. As a result we get the dynamic connections in the form:

$$\Pi \ddot{q}^* + \eta = 0 \qquad (2.4.5)$$

where the elements of the matrix $\Pi$ and vector $\eta$ are

$$\Pi_{kj} = \sum_{i=1}^{n_q*} [\, (\vec{R}_i - \vec{R}_z) \times \vec{a}_{ik}^{q^*} + \vec{b}_{ik}^{q^*} ] \vec{E}_j \qquad (2.4.6)$$

$$\vec{\eta}_j = \vec{E}_j [\, \sum_{i=1}^{NT} \sum_{k=1}^{n_{\bar{q}}} ((\vec{R}_i - \vec{R}_z) \times \vec{a}_{ik}^{\bar{q}} + \vec{b}_{ik}^{\bar{q}}) \ddot{q}^k +$$

$$+ (\vec{R}_i - \vec{R}_z) \times \vec{a}_i^o + \vec{b}_i^o + (\vec{R}_i - \vec{R}_z) \times \vec{G}_i + \vec{M}_i^G ] \qquad (2.4.7)$$

The vector $\vec{R}_i$ can be calculated in the following way. Let $\vec{R}^*$ be the radius-vector of the basic link mass centre of that chain which contains the i-th link in the series of "+" joints. Let the ordinal number of this link in the considered chain be $\ell$. Then,

$$\vec{R}_i = \vec{R}^* + \sum_{j=1}^{\ell-1} (\vec{r}_{jj} - \vec{r}_{j,j+1}) + \vec{r}_{\ell\ell} - \vec{r}_{o\ell} \qquad (2.4.8)$$

In some cases, a moment different from zero with respect to the $\vec{E}$ axis, may be required. For example, this may appear if one takes into account

the friction moments with respect to the vertical axis. Then, the right-
-hand sides of equations (2.4.3) and (2.4.4) should not be equal to
zero, but to the corresponding values of the moment. The procedure is
further unchanged. The flow-chart for forming the dynamic connections
is given in Fig. 2.10, where the parameter $\xi_i^3$ has the following proper-
ty: if $\xi_i^3 = 1$, then the actuator at the i-th joint is "kinematic",
whereas if $\xi_i^3 = 0$, the joint is "dynamic" (powered).

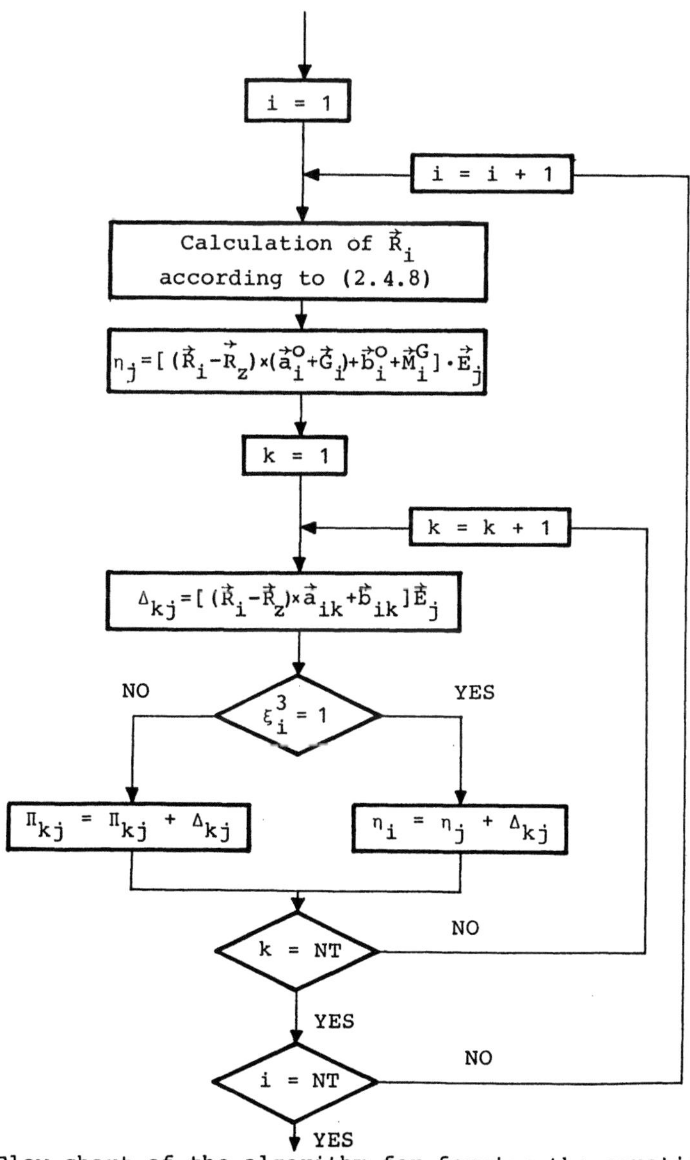


Fig. 2.10. Flow-chart of the algorithm for forming the equations
of dynamic connection

## 2.5. Closed Kinematic Chains

Let us adopt an open kinematic chain equivalent to the considered closed mechanism structure. The equivalent chain is illustrated in Fig.
2.11. It consists of 15 links and 15 revolute joints. Let the links 1-
-14 and all 15 joints have the same parameters and geometry as the corresponding links and joints of the initial closed chain structure. Let
the terminal, 15-th link be fictitious. Let $\vec{Q}_1$, $\vec{Q}_2$, $\vec{Q}_3$ and $\vec{Q}_1'$, $\vec{Q}_2'$, $\vec{Q}_3'$
be the axes of the coordinate frames connected to the terminal link
and the support (base), respectively. Let us determine the terminal
link position in such a way that the coincidence of the equivalent and
closed chain structures implies coincidence of the coordinate frames Q
and $\vec{Q}'$, too.

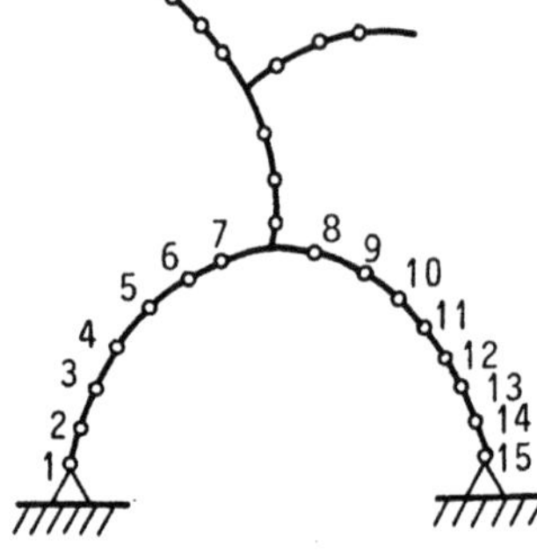

Fig. 2.11. The anthropomorphic mechanism in double-support position

Suppose the terminal link of the equivalent chain does not coincide
with the basic link (Fig. 2.12). Then, it is possible to determine the
translational ($\Delta\vec{\rho}$) and angular ($\Delta\vec{\sigma}$) displacement which will yield the
coincidence of the two coordinate systems. The necessary translational
displacement will be [6, 8]:

$$\Delta\vec{\rho} = \vec{\rho}' - \vec{\rho} \tag{2.5.1}$$

The angular displacement is determined from the condition that the ro-
tation of "Q" about $\Delta\vec{\sigma}$ for the angle $\theta = |\Delta\vec{\sigma}|$ yields the coincidence
of the vectors $\vec{Q}_i$ and $\vec{Q}_i'$. Define $\Delta\vec{d}_i$ as

$$\Delta\vec{d}_i = \vec{Q}_i' - \vec{Q}_i \tag{2.5.2}$$

and the conditions for "closing" the equivalent chain as

$$\Delta\vec{\rho} = 0, \quad \Delta\vec{d}_i = 0, \quad i = 1, 2, 3 \tag{2.5.3}$$

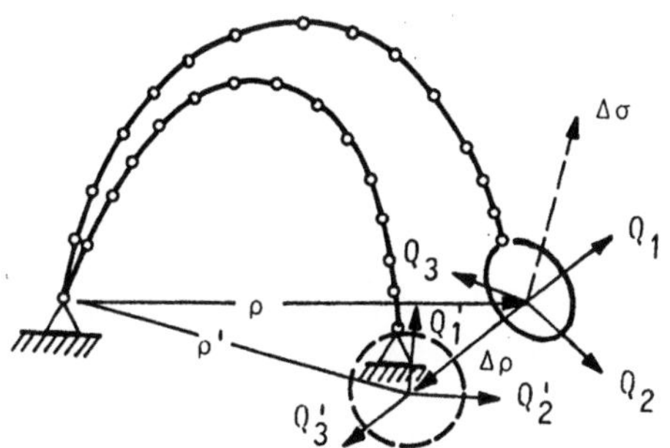

Fig. 2.12. Terminal link of the equivalent chain

Suppose the position of the equivalent chain is set up quite near to the closed chain structure, i.e., $\Delta\rho$ and $\Delta\vec{d}_i$ are sufficiently small. Then, it can be written [6, 8]:

$$\Delta\vec{d}_\ell = \Delta\vec{\sigma} \times \vec{Q}_\ell; \quad \Delta\vec{d}_r = \Delta\vec{\sigma} \times \vec{Q}_r \tag{2.5.4}$$

where $\Delta\vec{d}_\ell$ and $\Delta\vec{d}_r$ are two arbitrary nonzero increments from the set $\Delta\vec{d}_i$ (i=1,2,3); $\vec{Q}_\ell$, $\vec{Q}_r$ are two unit vectors along the corresponding axes $\vec{Q}_i$ (i=1,2,3). If $\vec{Q}_i$ does not coincide with $\vec{Q}'_i$ then at least two increments are different from zero.

According to (2.5.4) the vector $\Delta\vec{\sigma}$ is perpendicular to $\Delta\vec{d}_\ell$ and $\Delta\vec{d}_r$ and, therefore, colinear with their vector product. Let $\vec{C}$ be the unit vector along $\Delta\vec{\sigma}$

$$\Delta\vec{\sigma} = \theta\vec{C}; \quad \theta = |\Delta\vec{\sigma}| \tag{2.5.5}$$

Then

$$\vec{C} = \alpha \frac{\Delta\vec{d}_\ell \times \Delta\vec{d}_r}{|\Delta\vec{d}_\ell \times \Delta\vec{d}_r|} \tag{2.5.6}$$

where $\alpha$ may be either +1 or -1. The factor $\alpha$ is determined on the basis of the following. Let us denote the unit vector of the normal on $\Delta\vec{d}_i$ by $\vec{d}_{ni}$. The vector $\Delta\vec{\sigma}$ should have such direction that the rotation of all vectors $\vec{d}_{n\ell}$ is counterclockwise if viewed from the end of $\Delta\vec{\sigma}$. Therefore the necessary condition $(\vec{d}_{n\ell} \times \Delta\vec{d}_\ell) \cdot \vec{C} > 0$ should be fulfilled. By introducing $\vec{C}$ from (2.5.6) one obtains:

$$\alpha (\vec{d}_{n\ell} \times \Delta\vec{d}_\ell)(\Delta\vec{d}_\ell \times \Delta\vec{d}_r) > 0$$

from which follows that

$$\alpha = \text{sign}[(\vec{d}_{n\ell} \times \Delta\vec{d}_\ell)(\Delta\vec{d}_\ell \times \Delta\vec{d}_r)] \qquad (2.5.7)$$

Relation (2.5.6) enables one to calculate C in all cases except when $\Delta\vec{d}_\ell || \Delta\vec{d}_r$. These vectors may be parallel only if the vector $\Delta\vec{\sigma}$ lies in the plane defined by the vectors $\vec{d}_{n\ell}$ and $\vec{d}_{nr}$; then, both increments are perpendicular to that plane. In this case, however, $\Delta\vec{\sigma}$ is perpendicular to the third ort denoted by $\vec{d}_{ns}$, and thus, the increment $\Delta\vec{d}_s$ should be, by its module, larger than $\Delta\vec{d}_\ell$ and $\Delta d_r^{*)}$. Obviously, an opposite statement is also justifiable: if there be a maximal increment between $\Delta\vec{d}_\ell$ and $\Delta\vec{d}_r$ (with respect to the module), the vectors $\Delta\vec{d}_\ell$ and $\Delta\vec{d}_r$ could not be parallel. Therefore, relation (2.5.6) can be always used, provided that one of its increments (e.g., $\Delta\vec{d}_\ell$) has the maximal intensity

$$|\Delta\vec{d}_\ell| = \max_{(j)}(|\Delta\vec{d}_j|), \qquad j = \ell, r, s$$

By introducing (2.5.5) into (2.5.4) we determine the factor

$$\theta = \frac{|\Delta\vec{d}_\ell|}{|\vec{C} \times \vec{Q}_\ell|} \qquad (2.5.8)$$

Finally, by introducing (2.5.6), (2.5.7), and (2.5.8) into (2.5.5) we obtain the necessary angular displacement

$$\Delta\vec{\sigma} = \frac{|\Delta\vec{d}_\ell|}{|\vec{C} \times \vec{Q}_\ell|} \cdot \frac{\Delta\vec{d}_\ell \times \Delta\vec{d}_r}{|\Delta\vec{d}_\ell \times \Delta\vec{d}_r|} \, \text{sign}[(\vec{d}_{n\ell} \times \Delta\vec{d}_\ell)(\Delta\vec{d}_\ell \times \Delta\vec{d}_r)] \qquad (2.5.9)$$

As a result of closing, the chain (Fig. 2.11) loses 6 d.o.f. Because of that, the relative displacements cannot be prescribed for all joints, but only for 9 of them. These 9 joints and the corresponding coordinates we shall call "basic" and the remaining joints "additional". Let us denote the basic coordinates of the anthropomorphic mechanism by u and the additional ones by S. In this way, the generalized coordinates q for the first chain are divided in two groups of coordinates u and S.

Let us consider now the relation between the basic (u) and additional (S) coordinates in the case of small displacements $\Delta\vec{\rho}$ and $\Delta\vec{\sigma}$ of the equivalent chain.

------

$^{*)}$ The vector $\Delta\vec{\sigma}$ cannot be perpendicular to $\vec{d}_{ns}$ and at the same time to one of the vectors $\vec{d}_{n\ell}$ or $\vec{d}_{nr}$, since one of the increments would be then equal to zero, which contradicts the conditions for $\Delta\vec{d}_\ell$ and $\Delta\vec{d}_r$.

Let small displacements of all joints of the equivalent chain, $\Delta q$, be introduced. As a result, the terminal link undergoes certain translational and angular displacements, so that it can be written

$$\sum_{i=1}^{m} (\vec{e}_i \times \vec{r}_{mi})\, \Delta q^i = \Delta \vec{\rho}; \qquad \sum_{i=1}^{m} \vec{e}_i\, \Delta q^i = \Delta \vec{\sigma} \qquad (2.5.10)$$

where m is the number of elements of the equivalent chain (for the model considered m = 15).

Let us write these equations in the frame Q'. Before that, we should separate the basic and additional coordinates. Denote the basic coordinates by $u_i$ (i = 1,...,9) and the additional ones by $S_i$ (i=1,...,6). Then, the equations (2.5.10) can be written as

$$\sum_{i=1}^{n-n_u} (\vec{e}_k \times \vec{r}_{m\,k})\, \vec{Q}_j' \Delta S_i = - \sum_{i=1}^{n_u} (\vec{e}_i \times \vec{r}_{m\,i}) \cdot \vec{Q}_j' \Delta u_i + \Delta \vec{\rho}\vec{Q}_j'$$

$$\sum_{i=1}^{n-n_u} \vec{e}_k \vec{Q}_j' \Delta S_i = - \sum_{i=1}^{n_u} \vec{e}_i \vec{Q}_j' \Delta u_i + \Delta \vec{\sigma}\vec{Q}_j' \qquad (2.5.11)$$

where $k = n_u + i$; j=1, 2, 3; $n_u$ is the number of basic coordinates (in this case $n_u = 9$).

Let us form the following matrices

$$A = [A_{jk}] = [(\vec{e}_k \times \vec{r}_{m\,k})\, \vec{Q}_j'];\quad \tilde{A} = [\tilde{A}_{jk}] = [\vec{e}_k \cdot \vec{Q}_j']$$

$$B = [B_{ji}] = -[(\vec{e}_i \times \vec{r}_{m\,i}) \cdot \vec{Q}_j'];\quad \tilde{B} = [\tilde{B}_{ji}] = -[\vec{e}_i \cdot Q_j']$$

$$C = \begin{bmatrix} \vec{i}\cdot\vec{Q}_1' & \vec{j}\cdot\vec{Q}_1' & \vec{k}\cdot\vec{Q}_1' \\ \vec{i}\cdot\vec{Q}_2' & \vec{j}\cdot\vec{Q}_2' & \vec{k}\cdot\vec{Q}_2' \\ \vec{i}\cdot\vec{Q}_3' & \vec{j}\cdot\vec{Q}_3' & \vec{k}\cdot\vec{Q}_3' \end{bmatrix} \qquad (2.5.12)$$

Then, equations (2.5.11) can be written in the form [6, 8]:

$$\left[\frac{A}{\tilde{A}}\right] \Delta S = \left[\frac{B}{\tilde{B}}\right] \Delta u + \left[\frac{C}{0}\right] \Delta \rho + \left[\frac{0}{C}\right] \Delta \sigma \qquad (2.5.13)$$

where $\Delta S = (\Delta S_1, \Delta S_2, ..., \Delta S_{n_s})^T$; $\Delta u = (\Delta u_1, \Delta u_2, ..., \Delta u_{n_u})^T$.

Solving last equation with respect to $\Delta S$ gives the equation of connection in the form

$$\Delta S = \Lambda \Delta u + \Lambda_1 \Delta \rho + \Lambda_2 \Delta \sigma \qquad\qquad (2.5.14)$$

The connection matrices $\Lambda$, $\Lambda_1$, $\Lambda_2$ have dimensions $(n_s \times n_u)$, $(n_s \times 3)$, and $(n_s \times 3)$, respectively; $n_s$ being the number of additional coordinates.

In the case when the coordinate frame $Q'$ has been chosen in such a way that its orts are colinear with the orts $\vec{i}$, $\vec{j}$, $\vec{k}$ of the fixed coordinate frame, the matrix C is reduced to the unit matrix and equations (2.5.12), (2.5.13) are simplified.


*Determining position of the closed chain*


To solve this problem, we shall consider an equivalent open chain and, by means of successive iterations, satisfy the conditions for its closing.

Suppose the values of basic coordinates and vectors $\vec{\rho}\,'$, $\vec{Q}_j'$ $(j=1,2,3)$, that characterize the position of the closed chain, are known. As for the additional coordinates, let only their approximate values, ensuring sufficiently small displacements of $\Delta\vec{\rho}$ and $\Delta\vec{\sigma}$, be known. Denote by $u_o$ and $S_o$ the set of all coordinates corresponding to the "home" position. Determination of the chain position (i.e., calculation of the vectors $\vec{e}_i$, $\vec{r}_{ij}$ and the values of the additional coordinates $\Delta\vec{S}$) will be executed according to the sequence:

1. Forming the equivalent chain and determining its position which corresponds to the coordinates $u_o$, $S_o$;

2. Calculating $\Delta\vec{\rho}$ and $\Delta\vec{d}^j$ for the equivalent chain according to (2.5.1) and (2.5.2);

3. Choosing $\Delta\vec{d}_\ell$ and $\Delta\vec{d}_r$ which differ from zero, and calculating $\Delta\vec{\sigma}$ according to (2.5.9);

4. Calculating the connection matrices according to (2.5.12)-(2.5.14);

5. Determining the values of additional coordinates according to:

$$\Delta S = \Lambda_1 \Delta \rho + \Lambda_2 \Delta \sigma \qquad (2.5.15)$$

Expression (2.5.15) does not include $\Delta u$, since the values of the basic coordinates are known, and $\Delta u \equiv 0$.

Now, all generalized coordinates $q_o$ are known. However, the values of $S_o$ are determined in the first approximation. Because of that, after displacement of the equivalent chain to the position corresponding to $q_o$, the closing conditions (2.5.3) may appear unsatisfactory in regard to the prescribed accuracy. In that case, it is necessary to repeat the whole procedure 1-5. The flow-chart of the iterative procedure described is given in Fig. 2.13.

*Velocities and accelerations*

Equations (2.5.14) have been written for small translational and angular displacements. The analogous relations are justifiable for velocities, but these should not necessarily be small.

The translational velocity of the mass centre of the equivalent chain terminal link is related to the relative velocity of the j-th joint by:

$$(\vec{v}_m)_j = \dot{q}^j \vec{e}_j \times \vec{r}_{mj}$$

On the basis of (2.3.6) the translational velocity may be written in the form

$$\sum_{i=1}^{m} (\vec{e}_i \times \vec{r}_{m\,i})\, \dot{q}^i = \vec{v}_m$$

Analogously, for the angular velocity of the terminal link of the equivalent chain it may be written

$$\sum_{i=1}^{m} \vec{e}_i \dot{q}^i = \vec{\omega}_m$$

Therefore, it follows that the velocity equations analogous to (2.5.14) are justifiable, that is [6, 8]:

$$\dot{S} = \Lambda \dot{u} + \Lambda_1 v_m + \Lambda_2 \omega_m \qquad (2.5.16)$$

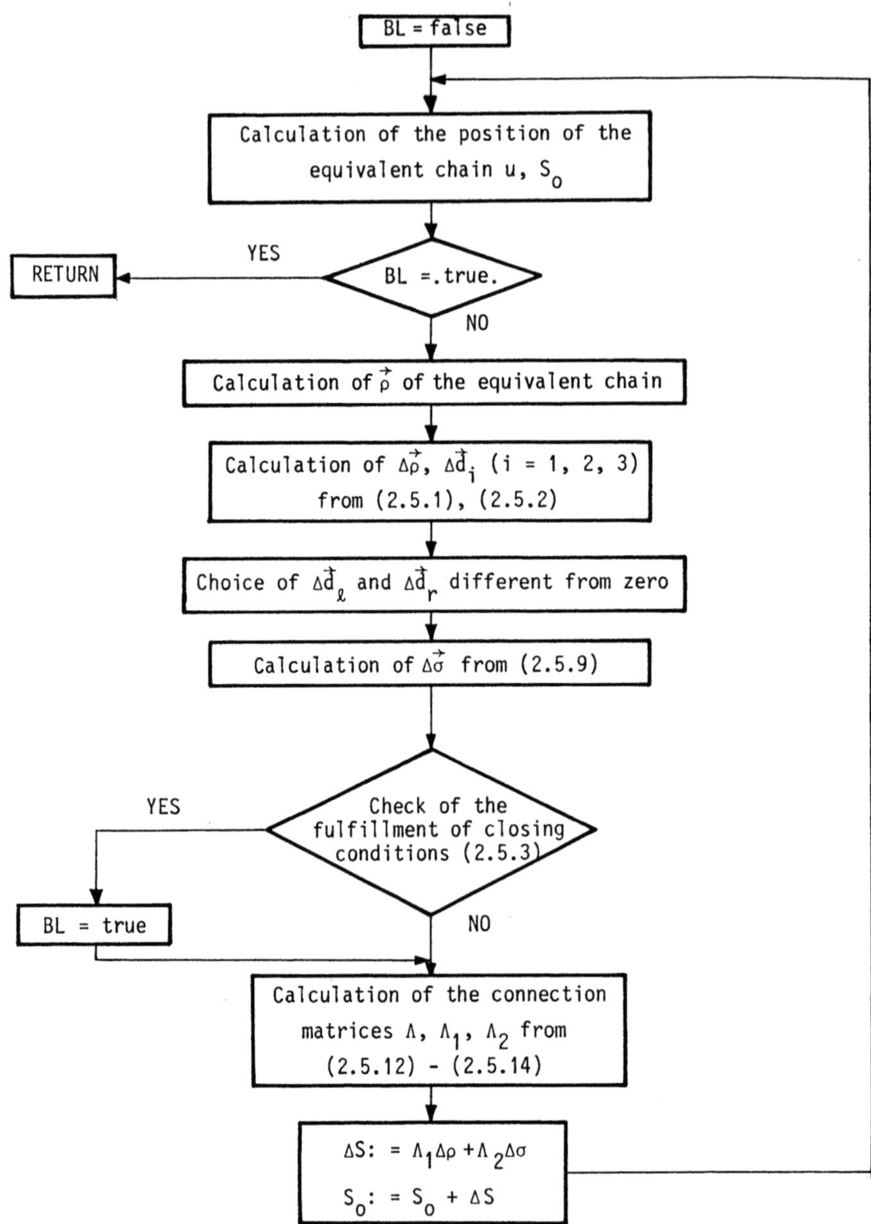



Fig. 2.13. Flow-chart of the algorithm for calculating the position of the closed chain

where $\Lambda$, $\Lambda_1$, $\Lambda_2$ are the connection matrices; $\vec{v}_m$, $\vec{\omega}_m$ are the translational and angular velocity of the terminal link of the equivalent chain. After determining the velocities $\dot{S}$ of all additional joints according to (2.5.16), the next step is to solve the problem of the open equivalent chain. In determining accelerations it is assumed that the relative accelerations of the basic joints, $\ddot{u}$, are known. The necessary $\ddot{S}$ values should be determined from the conditions of closing the chain. For this purpose, let us write according to (2.3.6), the accelerations $\vec{\varepsilon}_m$ and $\vec{w}_m$ for the terminal link of the equivalent chain,

$$\vec{\varepsilon}_m = \sum_{i=1}^{m} \vec{\alpha}_{m\,i}\,\ddot{q}^{\,i} + \vec{\alpha}_m^{\,O}; \qquad \vec{w}_m = \sum_{i=1}^{m} \vec{\beta}_{m\,i}\,\ddot{q}^{\,i} + \vec{\beta}_m^{\,O} \tag{2.5.17}$$

The vectors $\vec{\alpha}_m^{\,O}$ and $\vec{\beta}_m^{\,O}$ can be calculated since they are dependent on the relative velocities $\dot{q}^{\,i}$, determined before. Further, according to (2.3.8) it can be written

$$\vec{\alpha}_{m\,i} = \vec{e}_i; \qquad \vec{\beta}_{m\,i} = \vec{e}_i \times \vec{r}_{m\,i} \tag{2.5.18}$$

After introducing (2.5.18) into (2.5.17) one obtains

$$\sum_{i=1}^{m} \vec{e}_i\,\ddot{q}^{\,i} = -\vec{\alpha}_m^{\,O}; \qquad \sum_{i=1}^{m} (\vec{e}_i \times \vec{r}_{m\,i})\,\ddot{q}^{\,i} = -\vec{\beta}_m^{\,O} \tag{2.5.19}$$

A comparison of (2.5.19) and (2.5.10) reveals that only their right-hand sides are different, where

$$\Delta\vec{\rho} \text{ is replaced by } -\vec{\alpha}_m^{\,O}$$

$$\Delta\vec{\sigma} \text{ is replaced by } -\vec{\beta}_m^{\,O} \tag{2.5.20}$$

Equations (2.5.10) are the initial equations in deriving the connection equations (2.5.14); the values $\Delta\vec{\rho}$ and $\Delta\vec{\sigma}$ in them being linear. Therefore, if we put $\ddot{S}$ and $\ddot{u}$ instead $\Delta S$ and $\Delta u$ and the corresponding expressions (2.5.20) instead of $\Delta\vec{\rho}$ and $\Delta\vec{\sigma}$, respectively, into the connection equations (2.5.14), they will still hold [6, 8]:

$$\ddot{S} = \Lambda\ddot{u} + \Lambda_1(-\vec{\beta}_m^{\,O}) + \Lambda_2(-\vec{\alpha}_m^{\,O}) \tag{2.5.21}$$

By means of these equations it is possible to determine all accelerations of the additional joints, $\ddot{S}$. After that, the accelerations can be determined by considering the open equivalent chain.

Further, the accelerations are defined by considering the open equivalent chain using the algorithm developed for simple chains. The flow
-chart of the algorithm for calculation of the position, velocity, and
acceleration of the open and closed kinematic chain is presented in
Fig. 2.14.

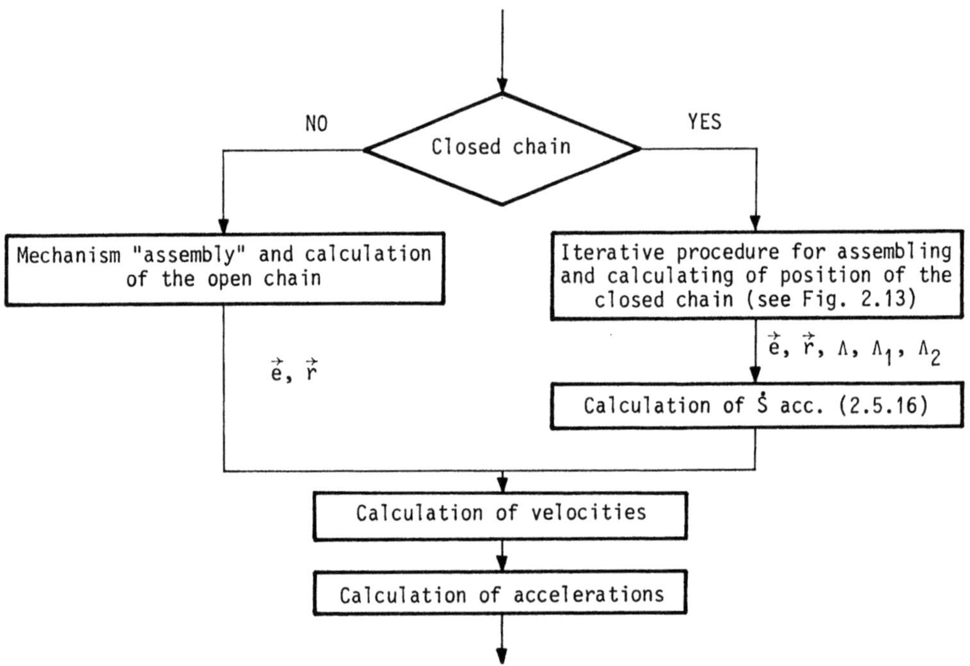


Fig. 2.14. Calculation of the position, velocity, and acceleration
of the open and closed kinematic chain

*Closed chain dynamics*

In the analysis of dynamics we shall use the same procedure as before,
i.e., to consider first the open chain, and then, add the corresponding
equations for additional coordinates. If, instead of the first kinematic chain presented in Fig. 2.11, we consider an equivalent chain, we
shall obtain a mechanism having only open chains. In forming the differential equations we shall assume that an unknown force $\vec{R}^*$ and a
moment $\vec{M}^*$ act upon the terminal link of the equivalent chain.

The quantities $\vec{R}^*$ and $\vec{M}^*$ are the ground reactions of the closed chain.
These two vectors have 6 unknown components which can be described by
the vector

$$\vec{\sigma} = (\sigma_1, \sigma_2, \ldots, \sigma_6)^T = (R_x^*, R_y^*, R_z^*, M_x^*, M_y^*, M_z^*)^T \qquad (2.5.22)$$

The differential equations of motion for the case of an open equivalent chain differ from those for the conventional open structures, and they are of the following form [6, 8]:

$$H\ddot{q} + B\sigma + h = P \qquad (2.5.23)$$

where

$$-H_{ik} = \vec{e}_i \sum_{(j)} (\vec{b}_{jk} + \vec{r}_{ji} \times \vec{a}_{jk}), \quad -h_i = \vec{e}_i \sum_{(j)} (\vec{r}_{ji} \times \vec{a}_j^O + \vec{b}_j^O + M_j^G + \vec{r}_{ji} \times \vec{G}_j)$$

$$H = [H_{ik}]; \quad h = [h_1, \ldots, h_n]^T; \quad P = [P_1, \ldots, P_{n-1}, 0]^T \qquad (2.5.24)$$

Let us notice that one of driving torques is equal to zero because of the presence of the "zero" drive in the equivalent open chain, and that the relationship between the reaction moment (load) and driving torque is given by (2.3.18).

The elements of matrices are composed of the coefficients already used in (2.3.21) for forming differential equations of motion of the open kinematic chains. Expression (2.5.23) corresponds to a system of 26 equations, containing 26 unknown accelerations $\ddot{q}^i$ and 6 unknown $\sigma_j$. The additional matrix equation to be added to (2.5.23) is:

$$\ddot{S} = \Lambda\ddot{u} - \Lambda_1\beta_m^O - \Lambda_2\alpha_m^O \qquad (2.5.25)$$

Finally, 32 equations are obtained from which it is possible to determine the relative accelerations $\ddot{q}^i$ and components $\sigma_j$ in the case when the driving torques $P_i$ are known.

When an inverse problem is considered (with the known accelerations $\ddot{q}^i$ and unknown torques $P_i$), then, the system of 26 equations (2.5.23) contains 32 unknown quantities (torques P and vector $\vec{\sigma}$). In that case, equations (2.5.25) cannot be uniquely solved.

One can prescribe 6 arbitrary torques of the first chain joints and, then, determine the other moments and $\vec{\sigma}$ from (2.5.23). This can be easily explained. Let us consider the simplest four-linked mechanism (Fig. 2.15).

It is straightforward that the mechanism considered can be in equilibrium state for different torques values, provided $P_1 = -P_2$. Hence, only one value of the torque should be prescribed. The analogous case also arises with the more complex kinematic chains, where it is necessary to prescribe (m-n$^*$) torques; m and n$^*$ being the numbers of d.o.f. of the equivalent and closed chain, respectively.

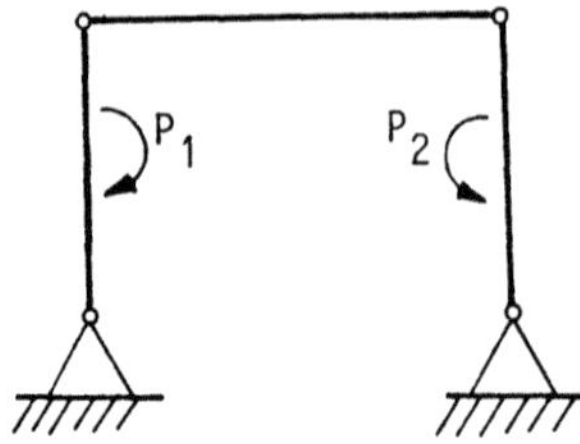


Fig. 2.15. The four-linked mechanism

Since in the dynamics analysis we actually consider an open chain which is equivalent to the closed chain, it is possible to use basically the same procedure as for open chains, but with certain modifications. Thus, the algorithm should be supplied with an iterative procedure for calculation of the position (see the flow-chart in Fig. 2.13) and additional velocities $\dot{S}$ of the first chain. After these alterations, the flow-chart takes the form as presented in Fig. 2.14. The equations for dynamic connections are formed in the same way for both single and double-support phase, the only difference being that in the latter instance an equivalent kinematic chain of the mechanism is considered. Because of that, the algorithm in Fig. 2.9. remains unchanged in its part where these equations are formed. However, in the part dealing with the formation of differential equations it is necessary to introduce two modifications: a) separate the coefficients multiplying the variables $\ddot{\sigma}$ (i.e. the matrices B in (2.5.23)) and b) add equations (2.5.25), where the matrices $\Lambda$, $\Lambda_1$, $\Lambda_2$ and vectors $\ddot{\beta}_m^O$, $\ddot{\alpha}_m^O$ are calculated according to the algorithm given in Fig. 2.9. Since these alterations are not essential, the corresponding flow-chart is not given separately.


## 2.6. Synthesis of Functional Movements

The methods for forming mathematical models of active spatial mechanisms by means of digital computers, described in the preceding section, will be used now for the synthesis of the nominal dynamics [14-20]. The only acceptable solutions in the synthesis procedure are those yielding the functional movements. In the rehabilitation robotics, for example, it is necessary to realize the artificial locomotion of an anthropomorphic system. Therefore, only these solutions are of interest which ensure similarity to regular human gait. Having in mind that the nominal trajectory and the corresponding nominal dynamics synthesized on the basis of a certain optimization criterion, does not a priori ensure functional movements, an engineering suboptimal procedure has been proposed for the synthesis of functional movements [11, 12]. This procedure, which was afterwards named semi-inverse method [8] consists, most succintly, in the following.

In order to ensure functional movements, the dynamics for one part of the system should be given in advance. In this way the system's functionality is achieved without any simplification of the original system, and the synthesis of functional movements is carried out using a mathematical model of substantially reduced dimensionality. The remaining part of the system should be solved in such a way that the "open" (unprescribed) dynamics maintains the system in dynamic equilibrium for the prescribed boundary conditions and the corresponding dynamic connections in dependence of the class of the task. The dynamics thus calculated will be called the compensating dynamics, or compensating synergy, in contrast to the dynamics defined in advance which we call the prescribed synergy. The proposed synergy can also be used for other classes of complex dynamic systems. However, we shall concentrate here only on the application of this procedure for solving the problems of the artificial gait which, as a problem of dynamics, possesses certain specificities that should be taken into account in order to obtain the required functional solutions.

The prescribed part of dynamics and the law of ZMP displacement for the synthesis of an anthropomorphic gait is obtained by studying the human locomotion. The remaining part of dynamics $q^*$ must satisfy the repeatibility conditions, inherent to the nature of human gait. Thus, the synthesis of functional movements reduces to solving the system of differential equations (2.4.5) for the prescribed boundary conditions of repeatibility and the adopted law of ZMP displacement. It is advisable to choose that the prescribed part of dynamics $\bar{q}$ is referred to the displacement of the robot's "legs", which does not mean that the coordinates $\bar{q}$ cannot represent movements of some other parts of the system. Such a choice of the prescribed synergy is dictated by, at least, two essential characteristics of the locomotion of anthropomorphic systems.

First, the gait is mainly determined by legs' movements and, second, these movements are more complex than the movements of other parts of an anthropomorphic system. The compensating synergy, i.e., the coordinates $q^*$ refer to the upper part ("trunk") of the model. Therefore, a simplified scheme of robot's gait would be as follows: the "legs" move according to the algorithm synthesized on the basis of the parameters obtained for human gait, and the "trunk" performs some periodical compensating movements. These movements ensure an appropriate displacement of the ZMP and maintain the system in dynamic equilibrium with respect to the adopted axes $E_j$ according to expression (2.4.6).

If we adopt that $\vec{E}_j$ (j=1,2,3) are the unit vectors of the fixed coordinate frame, then (2.4.5) is a system of three second-order differential equations. In this case, it is necessary to prescribe the synergy for the M-3 d.o.f. The compensating synergy concerns the remaining 3 d.o.f. This synergy (trajectories of the upper part of the system) is obtained by solving the obtained system of three differential equations for the prescribed boundary conditions of repeatibility. The nominal dynamics thus obtained ensures both the adopted displacement of the ZMP and dynamic equilibrium of the overall system with respect to the axes of the fixed coordinate frame.

Using an analogous procedure for forming the dynamic connections for the point of contact with the ground, it is possible to establish dynamic connections for other mechanism's joints. Besides, it can be supposed that an anthropomorphic mechanism can also possess some non-powered joints, as in the case of free arms. The only moments present in these joints are those of friction. Dynamic equations for the axes of these mechanism joints can be considered as part of the general mathematical model (2.3.16) for the boundary conditions of repeatability conditions type. The dimension of the vector $q^*$ is equal to the number of possible connections, whereas the matrices and vectors in (2.4.5) are of the corresponding order.

*Boundary conditions*

In the synthesis of movements of the anthropomorphic gait, all d.o.f. $q$ and their derivatives $\dot{q}$ should at the end of a step be equal to those in its beginning, that is

$$\{q\}_T = \{q\}_O; \quad \{\dot{q}\}_T = \{\dot{q}\}_O \qquad (2.6.1)$$

The prescribed d.o.f. $\{\bar{q}\}$ satisfy these conditions since the data for predetermined part of the system are based on regular anthropomorphic gait. Thus, the corresponding system of differential equations (2.4.5) should be solved for the following boundary conditions

$$\begin{bmatrix} q^* \\ \hline \dot{q}^* \end{bmatrix}_T = \begin{bmatrix} q^* \\ \hline \dot{q}^* \end{bmatrix}_0 \tag{2.6.2}$$

Let assume an approximate solution to the problem be somehow known. A more accurate solution can be obtained by successive corrections by means of a sensitivity matrix, formed on the basis of linearization in the vicinity of the approximate solution. Under these conditions, small changes in the initial conditions cause small corrections in the terminal conditions [8], which may be written as

$$\begin{bmatrix} q^* \\ \hline \dot{q}^* \end{bmatrix}_T + \begin{bmatrix} \Delta q^* \\ \hline \Delta \dot{q}^* \end{bmatrix}_T = \begin{bmatrix} q^* \\ \hline \dot{q}^* \end{bmatrix}_0 + \begin{bmatrix} \Delta q^* \\ \hline \Delta \dot{q}^* \end{bmatrix}_0 \tag{2.6.3}$$

On the other hand, by means of the sensitivity matrix, the change in terminal conditions may be determined from the change of the initial conditions, that is

$$\begin{bmatrix} \Delta q^* \\ \hline \Delta \dot{q}^* \end{bmatrix}_T = [U] \begin{bmatrix} \Delta q^* \\ \hline \Delta \dot{q}^* \end{bmatrix}_0 \tag{2.6.4}$$

where matrix $[U]$ is the sensitivity matrix with the submatrices expressed in form of Jacobians. This matrix is of the form

$$[U] = \begin{bmatrix} \dfrac{\partial(q^{*i})}{\partial(q^{*j})} & \vdots & \dfrac{\partial(q^{*i})}{\partial(\dot{q}^{*j})} \\ \cdots & + & \cdots \\ \dfrac{\partial(\dot{q}^{*i})}{\partial(q^{*j})} & \vdots & \dfrac{\partial(\dot{q}^{*i})}{\partial(\dot{q}^{*j})} \end{bmatrix} \tag{2.6.5}$$

On the basis of equations (2.6.3) and (2.6.4) we obtain the necessary corrections for the initial conditions

$$\begin{bmatrix} \Delta q^* \\ \hline \Delta \dot{q}^* \end{bmatrix}_0 = ([U]-[I])^{-1} \left( \begin{bmatrix} q^* \\ \hline \dot{q}^* \end{bmatrix}_0 - \begin{bmatrix} q^* \\ \hline \dot{q}^* \end{bmatrix}_T \right) \tag{2.6.6}$$

If we take advantage of the symmetry inherent to human gait, the task may be reduced to the analysis of only one half of a step. At the end of a half-step, all movements in the x-z plane are repeated in the following manner: the central compensating link has the same and the leg links the opposite sign with respect to the start. All movements out of the x-z plane are repeated after each half-step, but in the opposite direction. The repeatibility conditions for a half-step become

$$\left[\frac{q^*}{\dot{q}^*}\right]_{T/2} = [W]\left[\frac{q^*}{\dot{q}^*}\right]_0 \qquad (2.6.7)$$

The transformation matrix [W] expresses the described symmetry. An equation analogous to (2.6.6) for the boundary conditions (2.6.7) renders the corrections of the initial conditions in the form

$$\left[\frac{\Delta q^*}{\Delta \dot{q}^*}\right]_0 = ([U]-[W])^{-1}\left([W]\left[\frac{q^*}{\dot{q}^*}\right]_0 - \left[\frac{q^*}{\dot{q}^*}\right]_{T/2}\right) \qquad (2.6.8)$$

The necessary details concerning the structural matrices of the model can be found in Appendix of Chapter 1, in which we described a procedure for forming a mathematical model of the gait using Euler's angles.


*An iterative procedure for synthesis of
nominal dynamics*


We have already mentioned the so-called miscellaneous type of task in which the motion and generalized forces are partly prescribed. In a general case, this task is described by the corresponding system of nonlinear differential equations with variable coefficients. Solving this system for the boundary conditions, either (2.6.2) or (2.6.7), gives the compensating synergy of the system. The next step is to form an algorithm for obtaining the solution by means of computer [7].

The boundary conditions (2.6.2) and (2.6.7) can be, in the general case, represented by the functional relation

$$Y^T = \chi(Y^O) \qquad (2.6.9)$$

where

$$Y^T = \left[\frac{q^*}{\dot{q}^*}\right]_T \, , \qquad Y^O = \left[\frac{q^*}{\dot{q}^*}\right]_0$$

Let $\tilde{Y}(t)$ be the solution to the above problem which does not satisfy relation (2.6.9). Take that $\tilde{Y}(0)=\tilde{Y}^O$ and $Y(T)=\tilde{Y}^T$. Let us introduce the index of performance to fulfil the general condition (2.6.9) in the form

$$J = ||\tilde{Y}^T - \chi(\tilde{Y}^O)|| \tag{2.6.10}$$

where $||\cdot||$ denotes Euclid's vector norm. The values $\tilde{Y}^T$ and $\tilde{Y}^O$ are connected by differential equations of dynamic equilibrium, which means that the index of performance J is only a function of $\tilde{Y}^O$, that is

$$J = J(\tilde{Y}^O) \tag{2.6.11}$$

According to (2.6.10), the condition (2.6.9) will be satisfied provided the function (2.6.10) is minimized

$$J(\bar{Y}^O) = \min_{\tilde{Y}^O} J(\tilde{Y}^O) = 0 \tag{2.6.12}$$

where $\bar{Y}^O$ is the nominal solution satisfying relation (2.6.9).

To solve this optimization problem, the gradient method may be used in the form of the iterative process [8]

$$\tilde{Y}^O_{i+1} = \tilde{Y}^O_i - \varepsilon \nabla J \tag{2.6.13}$$

where $\varepsilon$ is a positive real constant, $\nabla J$ is the gradient of the function J with respect to $\tilde{Y}^O$, and i is the ordinal number of the iteration.

When the state vector $\tilde{Y}^O$ is sufficiently close to its nominal value $\bar{Y}^O$, then a local method may be introduced by the following iterative procedure

$$\tilde{Y}^O_{i+1} = \tilde{Y}^O_i + \varepsilon \phi(\tilde{Y}^O_i) \tag{2.6.14}$$

where the increment $\phi(\tilde{Y}^O_i)$ is calculated either from (2.6.6) or (2.6.8).

*The algorithm for synthesis of nominal dynamics*

Synthesis of nominal dynamics of an anthropomorphic mechanism for artificial biped gait generation is based on the procedure for automatic forming the differential equations of motion of these mechanisms (see Fig. 2.9). In Fig. 2.16. is presented a global algorithm for determi-

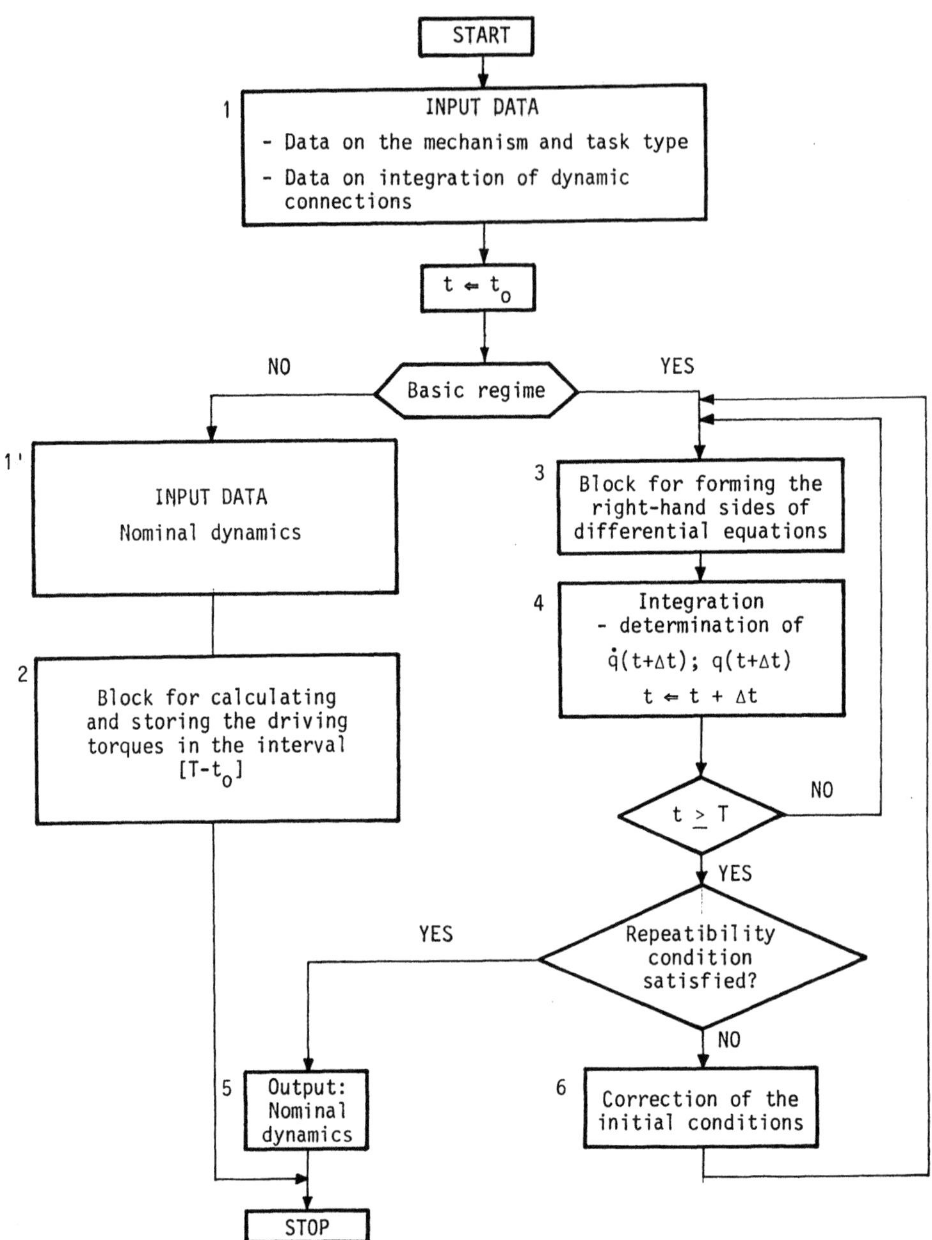


Fig. 2.16. Flow-chart of the algorithm for determination
of the nominal dynamics

nation of nominal dynamics. Two groups of input data (block 1) can be distinguished: the input data referring to the mechanical structure and task type, and the data on integration of dynamic connections of the mechanism. The branch for "basic regime" is responsible for the synthesis of nominal dynamics. Block 3 is the stage in which the right-hand sides of differential equations for the non--prescribed part of dynamics are formed. This block, worked out in detail, is presented in Fig. 2.17. Integration of the system of equations

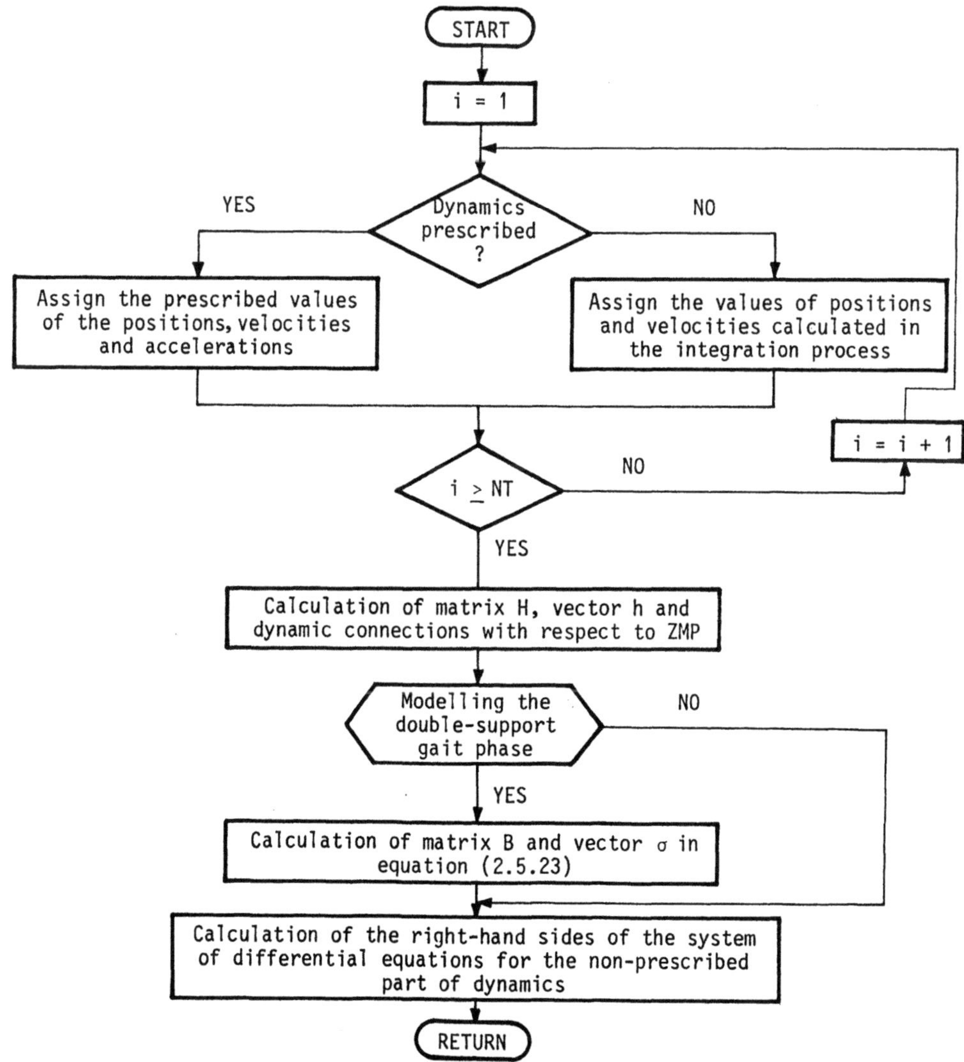


Fig. 2.17. General flow-chart of the algorithm for forming the right--hand sides of differential equations for the non-prescribed part of dynamics

on the interval $[T-t_o]$ is carried out in block 4. If the repeatibility
conditions are not satisfied after the first correction of the initial
conditions, the procedure is repeated. If, however, the repeatibility
conditions are satisfied, the output is the nominal dynamics. Now, when
the complete nominal dynamics is known, the driving torques can be
calculated in block 2.

*Modelling the two-linked foot*

In the course of single-support phase of a chosen type of gait, the movement of the foot that is being in contact with the ground is accomplished in three stages, as shown in Fig. 2.18.

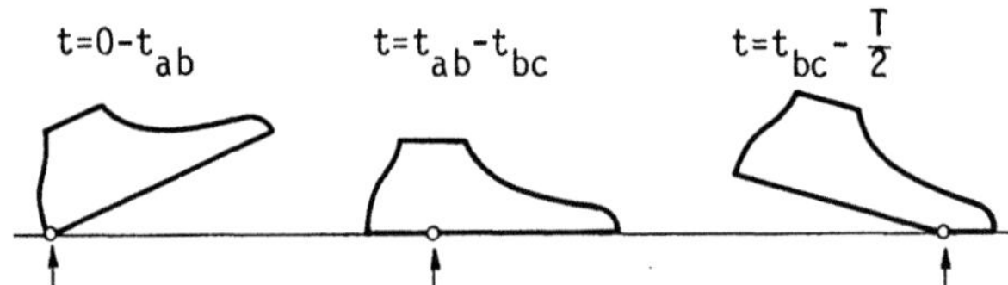


Fig. 2.18. Stages of foot contact with the ground

In stage I the contact with the ground is realized through the heel, in
stage II through the entire foot, whereas in stage III the support is
on the toes. Correspondingly, the position of the ZMP is under the heel,
under the foot and under the toes. After stage III, the ZMP jumps over
to a position under the other foot, at the very moment this foot touches the ground. Afterwards, the motion sequences are repeated.

In the first two stages of its contact with the ground, the foot is
described by one link. However, in the third stage the foot is better
modelled as if composed of two links. The first link corresponds to the
toes and the second to the rest of the foot (Fig. 2.19).

This kind of the foot modelling and displacement of the ZMP is executed
in the programme package within the subroutine called SUPP in the following way.

Let us consider the foot which is in the support phase. It always belongs to the 1-st kinematic chain which is represented by the 1-st row
in the matrix MS. Let the foot be in phase III when two-link model is

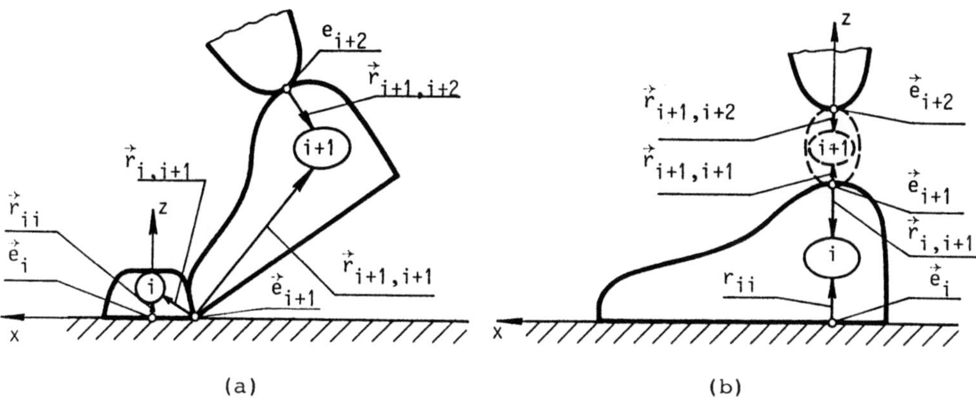

(a)           (b)

Fig. 2.19. Modelling the two-link foot of suppoting leg

required. Then, the link "i" corresponds to the toes and "i+1" to the
rest of the foot (Fig. 2.19.(a)). The law of the change of the angle of
the joint "i+1", connecting these two links, is prescribed. When the
foot is in the stage I or II, it is modelled as a single link (Fig.
2.19.(b)) according to the following procedure. The data related to the
complete foot, are associated with the i-th link, while the "i+1"-st
link is considered as fictious (its mass, inertia moments and length
are zero) which is in Fig. 2.19(b) represented by the dashed line. Now,
the angle $q^{i+1}$ is constant during the phases I and II. The flow chart
of this procedure is presented in Fig. 2.20.

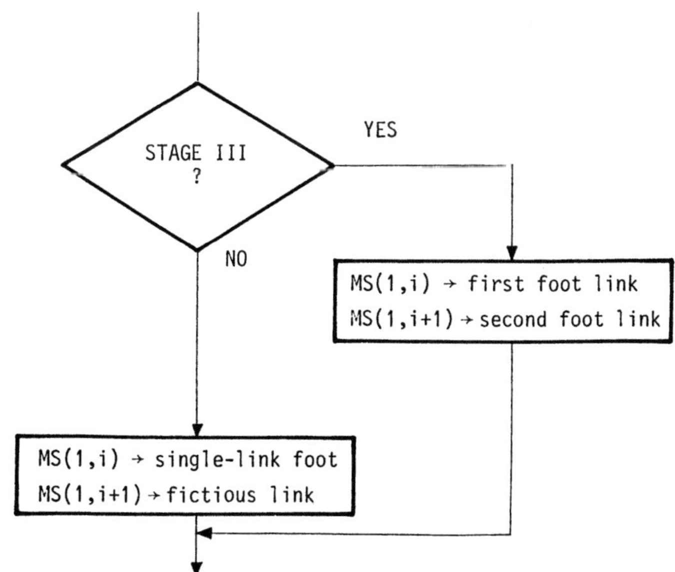


Fig. 2.20. Flow-chart for modelling the two-link foot

## 2.7. The Prescribed Part of Dynamics

The prescribed part of dynamics comprises data on the synergy of lower extremities of an anthropomorphic robot and the law of the ZMP displacement. The data on movements of legs are obtained from biometric measurements of a normal human gait. These data are referred to the angle trajectories of the hip, knee, and ankle joint as a function of time. The law of the ZMP displacement is determined on the basis of measurement of the vertical components of reaction forces at the human foot.

*Synthesis of the prescribed synergy*
*for the leg motion*

The motion of "legs" is adopted as the prescribed part of dynamics. The necessary data are obtained on the basis of biometric measurements of a normal human gait. Time trajectories of the change of angles of the hip, knee and ankle joint in external coordinates are measured in the course of a full-step and for a single-support phase gait. After some training, a healthy person is able to perform a symmetric single-support gait accompanied by minimal twisting movements of the pelvis. A trained person can learn how to move, either with the normally bent,or with parallel feet. The person which is being studied should keep the upper part of the body in vertical position during recording the gait parameters. A period of the stationary gait is recorded for a sufficiently long time, so that any suitable part of a step may be taken for the analysis.

Values of the angles for each joint are measured at a certain number of descrete points. On the basis of the measured values, the prescribed synergy (prescribed trajectory) is synthesized in the following way. The duration time of a full step T is devided into 80 equal intervals, and the values of angles at the discrete points are interpolated using a linear function. The values thus obtained are smoothed by means of a fifth-order polynom using the IBM-SSP subroutine SE 15. This subroutine requires four additional points at the end of each full step. These additional points are determined from the periodicity conditions of the angles prepared for approximation. Furthermore, the double differention is carried out with the aid of the IBM-SSP subroutine DET 3. Thus the values are obtained for the position, velocity and acceleration at all discretization points. From these data are further extracted the data for both the supporting and swing phase. In this way,the

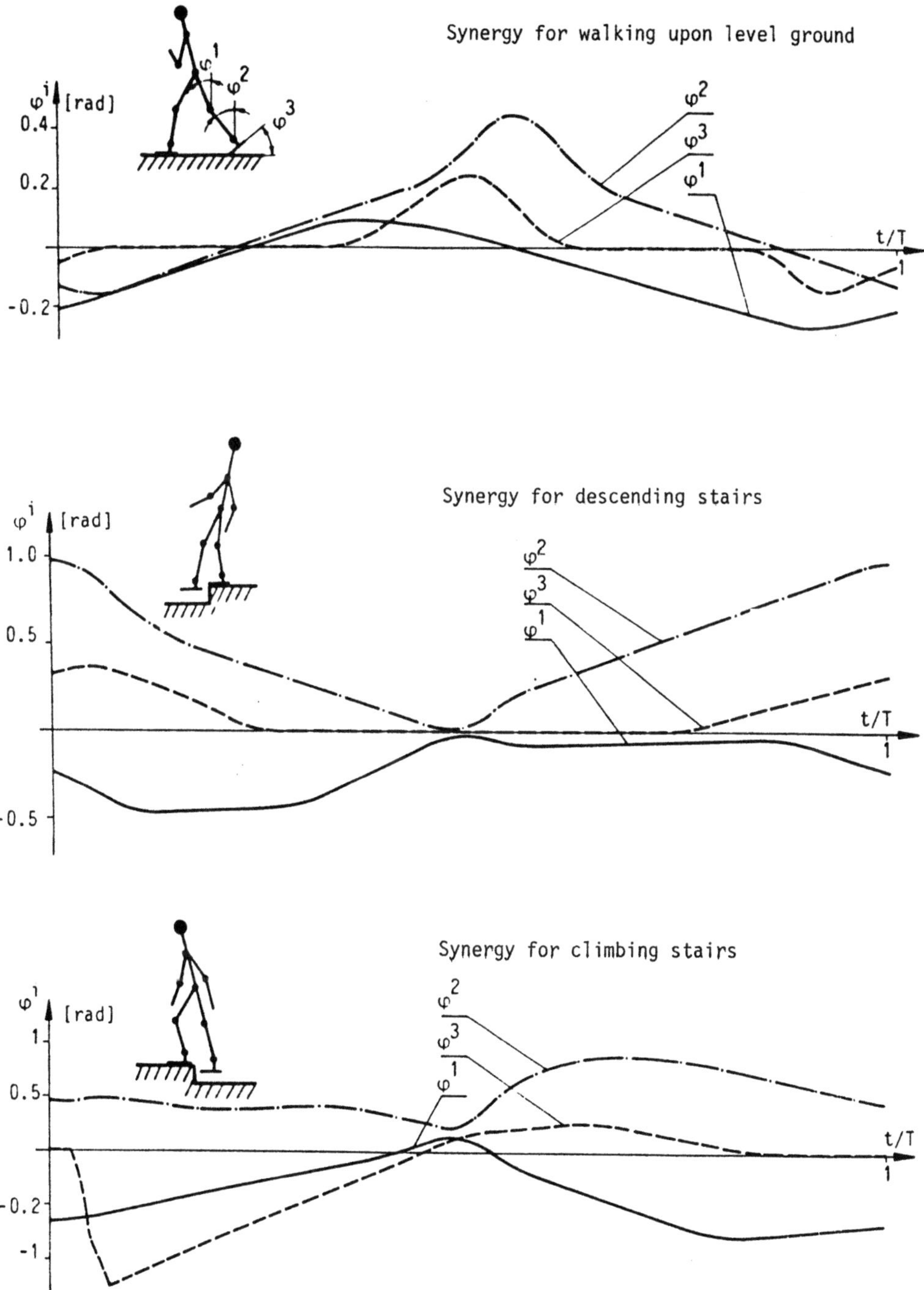



Fig. 2.21. Prescribed synergy for legs motion

prescribed synergy is synthesized for the left leg as the "supporting"
and the right leg as the "swing" one.

In this way we obtain the prescribed synergy in external coordinates
for the hip, knee, and ankle joint. All the obtained values are conver-
ted into the internal coordinates, and they are given in Appendix.

Fig. 2.21. shows the smoothed diagrams for the change of the legs angles in external coordinates. The graphs show changes in the angle of
the hip, knee, and ankle joint for walking: upon level ground, climbing
stairs, and descending stairs. The chosen gait types are characterized
by a very "smooth" pelvic motion. This supposition is of purely practical nature, because the applicability of these results to exoskeleton
type biped robots is kept in mind.

*Law of the ZMP displacement*

As mentioned above (see Fig. 2.18), the supporting point may be either
under the heel (stage I), under the foot (stage II), or under the toes
(stage III). The transitions from stage I to stage II, and from stage
II to stage III occur at the time $t=t_{ab}$ and $t=t_{bc}$, respectively. Thus,
the law of the ZMP displacement may be interpreted as follows: in the
beginning of stage I the ZMP is under the heel; in the end of stage I
it "jumps" to the foot "centre"; in the end of stage II it is shifed
under the toes; in the end of the half-step the ZMP "jumps over" under
the other foot which is now being in contact with the ground. The ZMP
trajectory thus described corresponds to the single-support phase of
gait. Five different laws of the ZMP displacement in the single-support gait phase, corresponding to a half-step (T denotes the duration
of a full step) are presented in Fig. 2.22.

In Fig. 2.23. are presented three different laws of the ZMP displacement for a double-support gait phase. The parameter p (expressed as %)
denotes the ratio of the duration period of a double-support phase
and of a full step, while S denotes the length of a step and 2d its
width.

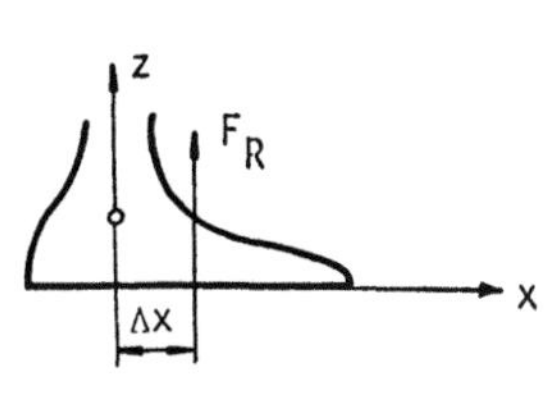



ZMP displacement

| CASE | t(sec) | Δx(m) |
|------|--------|-------|
| I | 0 ÷ T/2 | 0.0 |
| II | 0 ÷ 0.3<br>0.3 ÷ T/2 | 0.0<br>0.035 |
| III | 0 ÷ 0.5<br>0.5 ÷ T/2 | 0.0<br>0.035 |
| IV | 0 ÷ 0.2<br>0.2 ÷ 1.0<br>1.0 ÷ T/2 | -0.02<br>0.0<br>0.0 |
| V | 0 ÷ 0.2<br>0.2 ÷ 0.6<br>0.6 ÷ T/2 | -0.02<br>0.0<br>0.035 |

Fig. 2.22. Set of ZMP trajectories for single-support gait phase

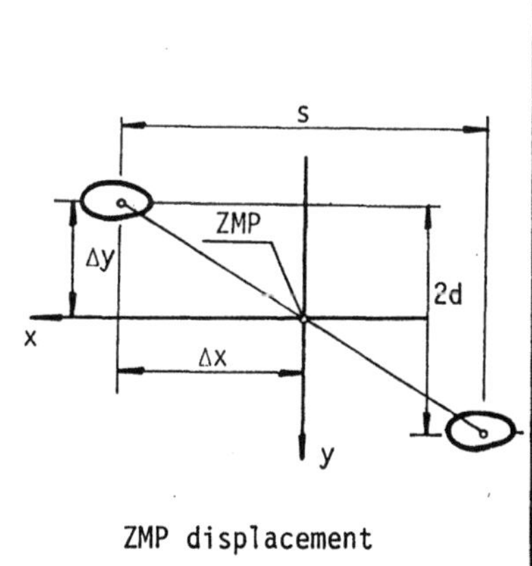



ZMP displacement

| CASE | Δx(m) | Δy(m) | t(sec) |
|------|-------|-------|--------|
| 1 | 0.5 S | -d | 0 ÷ τ (τ=T·p/400) |
| | 0 | 0 | τ ÷ (T/2-τ) |
| | -0.5 S | -d | (T/2-τ) ÷ T/2 |
| 2 | 0.75 S | -1.5 d | 0 ÷ τ/2 |
| | 0.25 S | -0.5 d | τ/2 ÷ τ |
| | 0 | 0 | τ ÷ (T/2-τ) |
| | -0.25 S | -0.5 d | (T/2-τ) ÷ (T-τ)/2 |
| | -0.75 S | -1.5 d | (T-τ)/2 ÷ T/2 |
| 3 | 0.83 S | -1.66 d | 0 ÷ τ/3 |
| | 0.50 S | -d | τ/3 ÷ 2τ/3 |
| | 0.17 S | -0.34 d | 2τ/3 ÷ τ |
| | 0 | 0 | τ ÷ (T/2-τ) |
| | -0.17 S | -0.34 d | (T/2-τ) ÷ (T/2-2τ/3) |
| | -0.50 S | -d | (T/2-2τ/3) ÷ (T/2-τ/3) |
| | -0.83 S | -1.66 d | (T/2-τ/3) ÷ T/2 |

Fig. 2.23. Set of ZMP trajectories for double-support gait phase

*Gait parameters*

On the basis of the above consideration, the following parameters of symmetric gait types may be distinguished: the step length S, the duration of a full step T, the law of ZMP displacement, and the duration time of a double-support phase, p. The parameters T and S control the walking speed for a fixed gait type. To accomplish this for the given d.o.f. at the hip, knee, and ankle joint, the values of positions are multiplied by $\alpha_1$, of velocities by $\alpha_2$, and of accelerations by $\alpha_3$, where $\alpha_1 = S$, $\alpha_2 = \alpha_1/T$, $\alpha_3 = \alpha_2/T$.

## 2.8. Numerical Results

The algorithms described above were the basis for writting a computer programme in FORTRAN to synthesize the functional movements of spatial mechanisms for realization of the artificial anthropomorphic gait. The programme listing is given in Appendix. In this section we shall present some of the results on the synthesis of functional movements using this programme.

Example 1. The adopted structure of the anthropomorphic mechanism is shown in Fig. 2.24. The mechanism consists of 14 links and 14 revolute joints of the 5-th class. Links 5 and 10 are the branching links. This structure can be split into three kinematic chains containing the joints 1-8, 9-12 and 13-14, in the first, the second, and the third chain, respectively. The topological structure of the complex kinematic chain can be represented by a series of "+" joints in a matrix form

$$MS = \begin{bmatrix} 1 & 2 & 3 & 4 & 5 & 6 & 7 & 8 & 0 \\ 1 & 2 & 3 & 4 & 5 & 9 & 10 & 11 & 12 \\ 1 & 2 & 3 & 4 & 5 & 9 & 10 & 13 & 14 \end{bmatrix} \qquad (2.8.1)$$

Table 2.1. contains numerical values for the mechanical part of the system presented in Fig. 2.24. The vectors of column 6 of the table, for the links 4 and 5, are shown in Fig. 2.25; the notation for other links being analogous. The prescribed part of dynamics is adopted for the pairs 1-8, i.e., for the first chain. This part of synergy (prescribed synergy) is defined based on measurement of the human gait (Fig. 2.21). As seen from Fig. 2.24, the mechanism's arms are fixed during

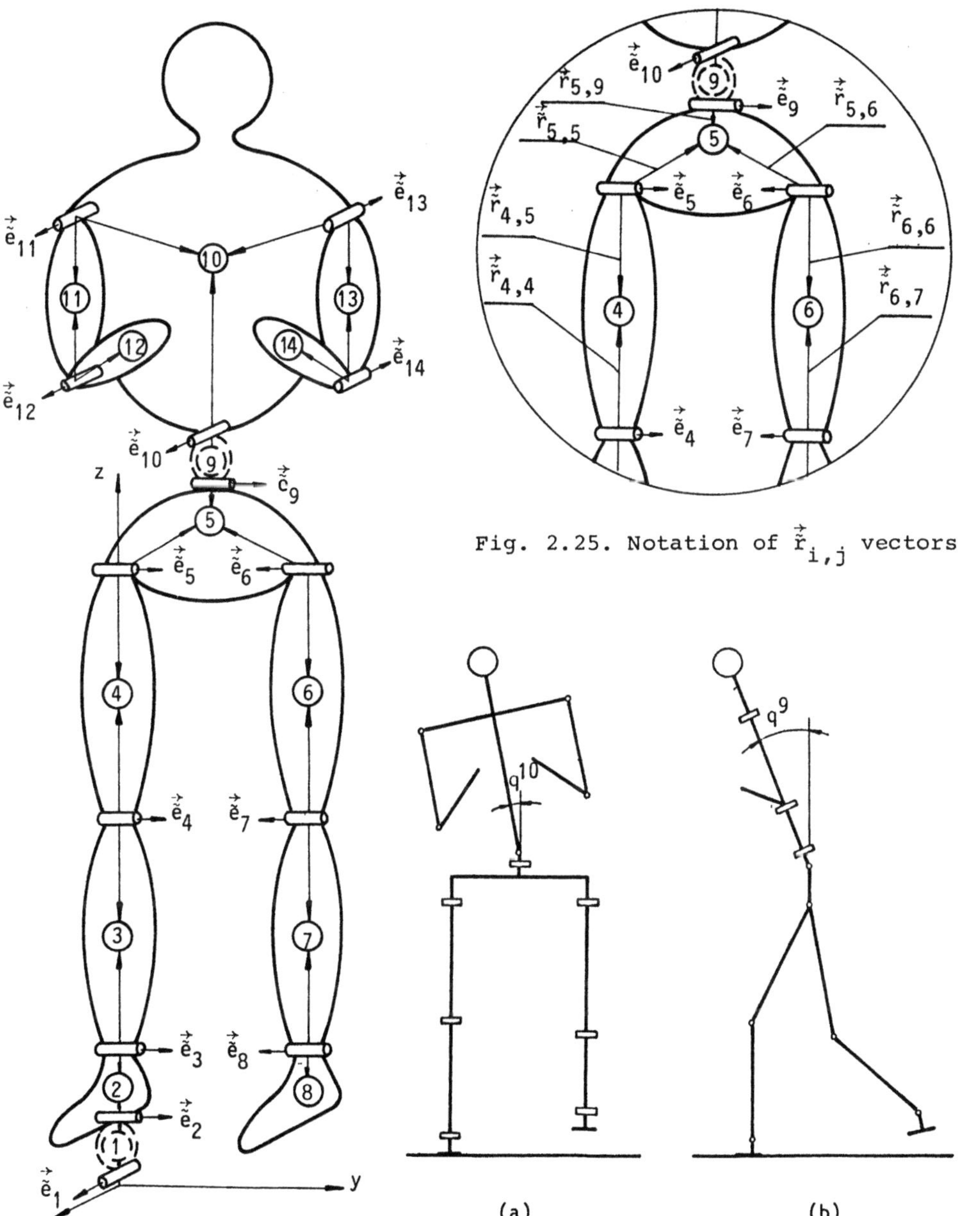



Fig. 2.25. Notation of $\vec{\tilde{r}}_{i,j}$ vectors



Fig. 2.24. Mechanical scheme of
the anthropomorphic
mechanism with fixed
arms

Fig. 2.26. Compensating d.o.f. in
the frontal (a) and
sagittal (b) plane

the gait. This is achieved by ascribing constant values to the positions $q^{11}$-$q^{14}$. They are: $q^{11}=q^{13}=1.862253$ [rad]; $q^{12}=q^{14}=1.5707$ [rad].

A set of prescribed ZMP trajectories for the single-support gait phase is given in Fig. 2.22. We shall additionally assume that $M_x=M_y=0$, where M stands for the moment expressed with respect to the ZMP.

The remaining part of the system, $q^9$ and $q^{10}$, is given as the compensating part of dynamics, which has to be determined in such a way to ensure the mechanism's stability during the motion, both in the sagittal and frontal plane. This is illustrated in Fig. 2.26, where $q^9$ is compensation in the sagittal (Fig. 2.26.b) and $q^{10}$ in frontal (Fig. 2.26. a) plane.

The data on mechanism structure are given by matrix (2.8.1), the mechanism parameters in Table 2.1. and by all above data on the prescribed part of dynamics. The numerical values for other input data, corresponding to the notation in the flow-chart in Fig. 2.9, are: $N=3$; $I_1=1$, $I_2=5$, $I_3=7$; $K_1=8$, $K_2=K_3=9$; $NT=14$; $\xi_i^1=0$, $i=1,\ldots,14$; $\xi_5^2=1$, $\xi_{10}^2=1$; the law of motion of kinematic pairs referring to the compensating dynamics $\xi_9^3=1$, $\xi_{10}^3=1$, and the dynamics of other kinematic pairs prescribed; $\xi_i^3=0$, $i=1,\ldots,8$, and $\xi_i^3=0$, $i=11$, 12, 13, 14; $\vec{r}_{o1}=(0,\ 0,\ -0.0001)^T$; $\vec{e}_1=(1,\ 0,\ 0)^T$; the initial values of positions ($q^9(0)$ and $q^{10}(0)$), and velocities ($\dot{q}^9(0)$ and $\dot{q}^{10}(0)$) of the open part of the system; the matrix of transformation W for repeatibility conditions of the form

$$W = \begin{bmatrix} 1 & 0 & 0 & 0 \\ 0 & -1 & 0 & 0 \\ 0 & 0 & 1 & 0 \\ 0 & 0 & 0 & -1 \end{bmatrix} \qquad (2.8.2)$$

On the basis of all input data, the programme automatically forms two second-order differential equations of the form (2.4.5), where $E_1 = (1,\ 0,\ 0)^T$ and $E_2 = (0,\ 1,\ 0)^T$. Then the system of equations is solved for the prescribed repeatibility conditions using an iterative procedure. If the iterative procedure converges, the obtained solution to the problem is the compensating dynamics, which, together with the prescribed dynamics makes the overall nominal dynamics of the system.

Some of the output results of this programme for different values of

Table 2.1. Kinematic and dynamic parameters of the mechanism

| Link | Mass [kg] | Moment of inertia [$kgm^2$] | | | Distance of the axes centres of joints from the link centre [m] | Joint unit axes |
|---|---|---|---|---|---|---|
| | | $J_x$ | $J_y$ | $J_z$ | | |
| 1 | 0.0 | 0.0 | 0.0 | 0.0 | $\vec{\tilde{r}}_{1,1} = (0, 0, 0.0001)^T$; $\vec{\tilde{r}}_{1,2} = (0, 0, -0.0001)^T$ | $\vec{\tilde{e}}_1 = (1, 0, 0)^T$ |
| 2 | 1.53 | 0.00006 | 0.00055 | 0.00045 | $\vec{\tilde{r}}_{2,2} = (0, 0, 0.030)^T$; $\vec{\tilde{r}}_{2,3} = (0, 0, -0.070)^T$ | $\vec{\tilde{e}}_2 = (0, 1, 0)^T$ |
| 3 | 3.21 | 0.00393 | 0.00393 | 0.00038 | $\vec{\tilde{r}}_{3,3} = (0, 0, 0.210)^T$; $\vec{\tilde{r}}_{3,4} = (0, 0, -0.210)^T$ | $\vec{\tilde{e}}_3 = (0, 1, 0)^T$ |
| 4 | 8.41 | 0.01120 | 0.01200 | 0.00300 | $\vec{\tilde{r}}_{4,4} = (0, 0, 0.220)^T$; $\vec{\tilde{r}}_{4,5} = (0, 0, -0.220)^T$ | $\vec{\tilde{e}}_4 = (0, 1, 0)^T$ |
| 5 | 6.96 | 0.00700 | 0.00565 | 0.00627 | $\vec{\tilde{r}}_{5,5} = (0, 0.135, 0.1)^T$; $\vec{\tilde{r}}_{5,6} = (0, -0.135, 0.1)^T$; $\vec{\tilde{r}}_{5,9} = (0, 0, -0.05)^T$ | $\vec{\tilde{e}}_5 = (0, 1, 0)^T$ |
| 6 | 8.41 | 0.01120 | 0.01200 | 0.00300 | $\vec{\tilde{r}}_{6,6} = (0, 0, -0.220)^T$; $\vec{\tilde{r}}_{6,7} = (0, 0, 0.220)^T$ | $\vec{\tilde{e}}_6 = (0, -1, 0)^T$ |
| 7 | 3.21 | 0.00393 | 0.00393 | 0.00038 | $\vec{\tilde{r}}_{7,7} = (0, 0, -0.210)^T$; $\vec{\tilde{r}}_{7,8} = (0, 0, 0.210)^T$ | $\vec{\tilde{e}}_7 = (0, -1, 0)^T$ |
| 8 | 1.53 | 0.00006 | 0.00055 | 0.00045 | $\vec{\tilde{r}}_{8,8} = (0, 0, -0.070)^T$ | $\vec{\tilde{e}}_8 = (0, -1, 0)^T$ |
| 9 | 0.0 | 0.0 | 0.0 | 0.0 | $\vec{\tilde{r}}_{9,9} = (0, 0, 0.0001)^T$; $\vec{\tilde{r}}_{9,10} = (0, 0, -0.0001)^T$ | $\vec{\tilde{e}}_9 = (0, 1, 0)^T$ |
| 10 | 30.85 | 0.15140 | 0.13700 | 0.02830 | $\vec{\tilde{r}}_{10,10} = (0, 0, 0.34)^T$; $\vec{\tilde{r}}_{10,11} = (0, 0.2, -0.06)^T$; $\vec{\tilde{r}}_{10,13} = (0, -0.2, -0.06)^T$ | $\vec{\tilde{e}}_{10} = (1, 0, 0)^T$ |
| 11 | 2.07 | 0.00200 | 0.00200 | 0.00022 | $\vec{\tilde{r}}_{11,11} = (0, 0, -0.154)^T$; $\vec{\tilde{r}}_{11,12} = (0, 0, 0.154)^T$ | $\vec{\tilde{e}}_{11} = (1, 0, 0)^T$ |
| 12 | 1.14 | 0.00250 | 0.00425 | 0.00014 | $\vec{\tilde{r}}_{12,12} = (0, 0, -0.132)^T$ | $\vec{\tilde{e}}_{12} = (1, 0, 0)^T$ |
| 13 | 2.07 | 0.00200 | 0.00200 | 0.00022 | $\vec{\tilde{r}}_{13,13} = (0, 0, -0.154)^T$; $\vec{\tilde{r}}_{13,14} = (0, 0, 0.154)^T$ | $\vec{\tilde{e}}_{13} = (-1, 0, 0)^T$ |
| 14 | 1.14 | 0.00250 | 0.00425 | 0.00014 | $\vec{\tilde{r}}_{14,14} = (0, 0, -0.132)^T$ | $\vec{\tilde{e}}_{14} = (-1, 0, 0)^T$ |

gait parameters are presented in Figs. 2.27. - 2.30. Figs. 2.27. - 2.28. show the compensating movements of the mechanism. In Fig. 2.29. are shown the reaction forces and driving torques, and in Fig. 2.30. is  shown the power at joints.

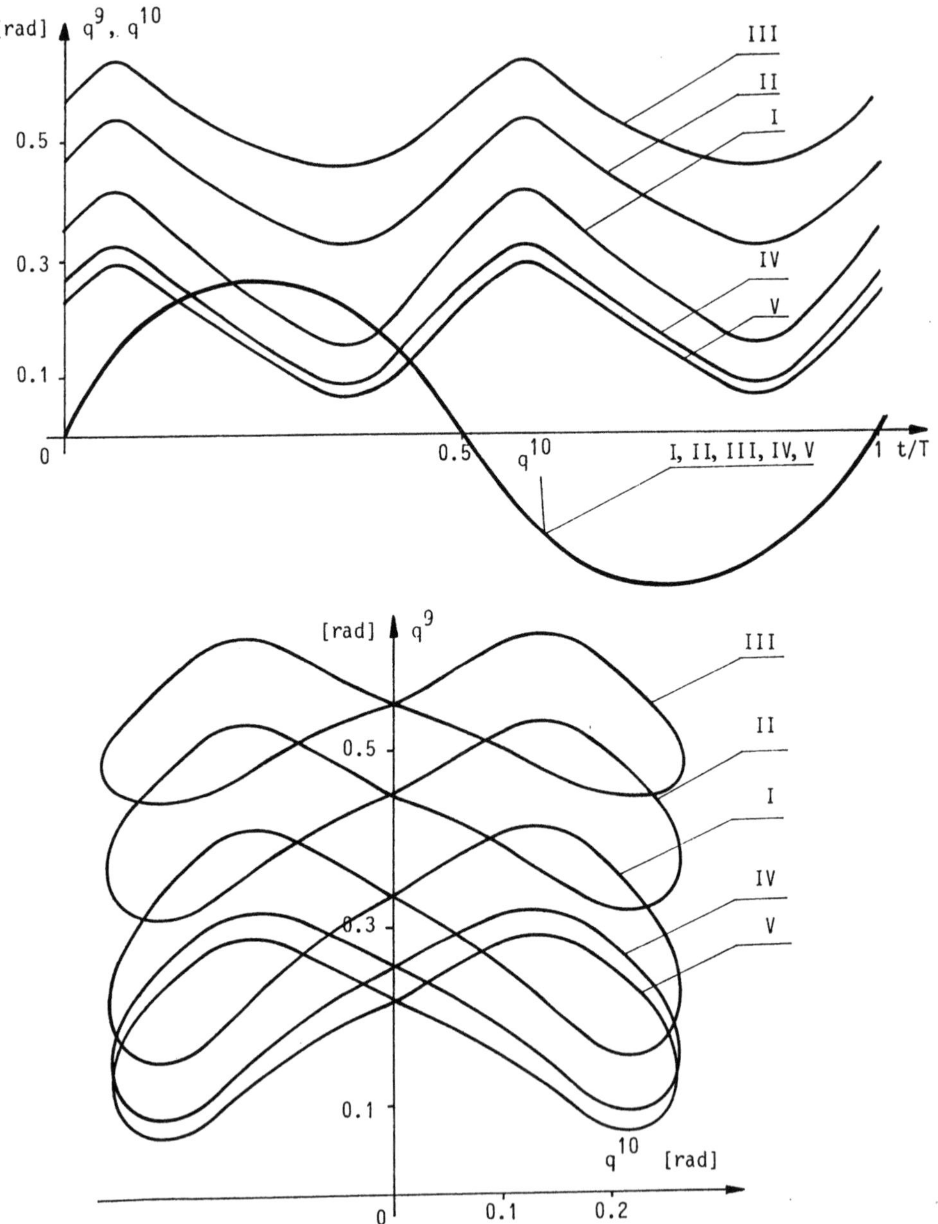


Fig. 2.27. Compensating movements for the single-support gait upon level ground for T=1.5, S=0.6 and ZMP laws from Fig. 2.22.

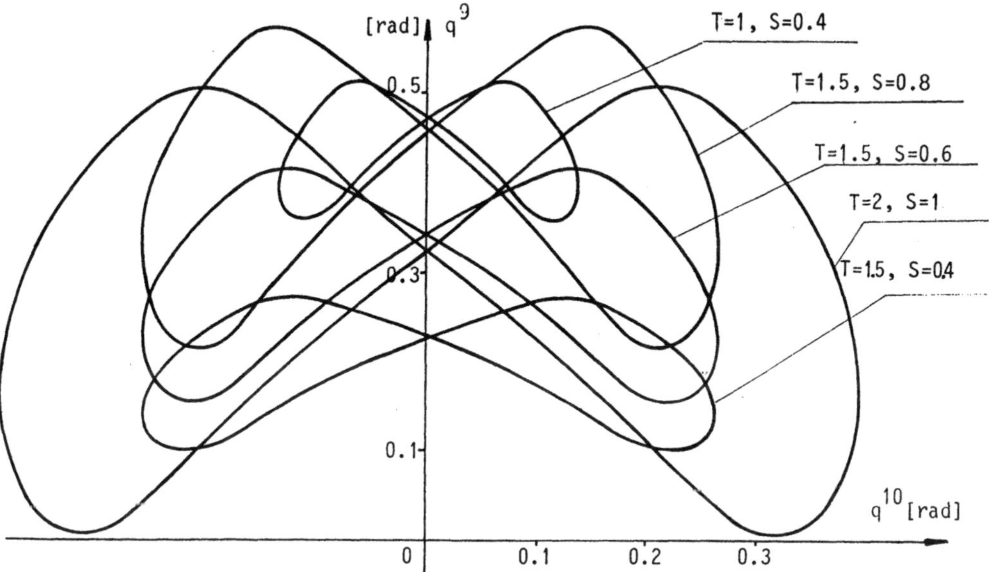


Fig. 2.28. Compensating movements for the single-support gait upon
level ground for ZMP law I and different values of parameters T and S

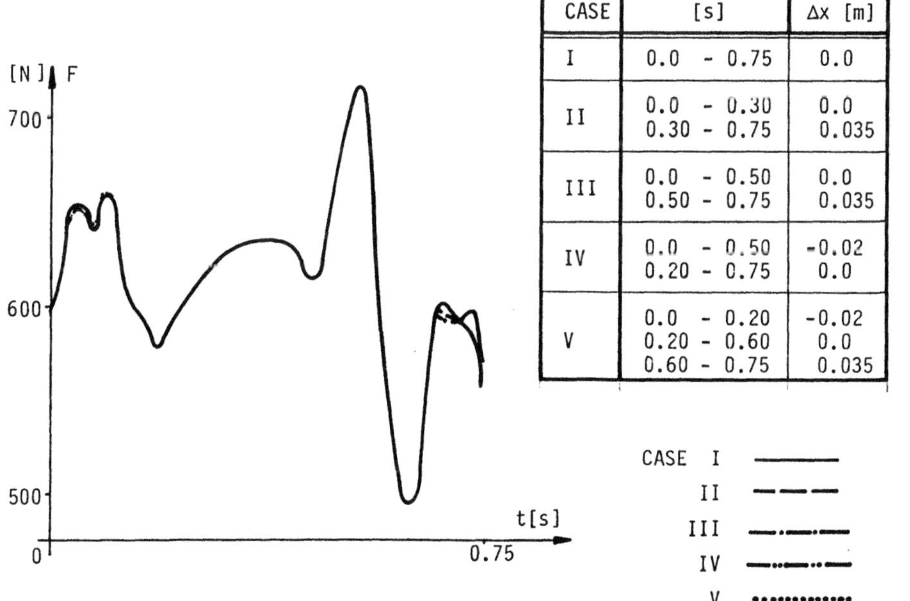


| CASE | [s] | $\Delta x$ [m] |
|---|---|---|
| I | 0.0 - 0.75 | 0.0 |
| II | 0.0 - 0.30<br>0.30 - 0.75 | 0.0<br>0.035 |
| III | 0.0 - 0.50<br>0.50 - 0.75 | 0.0<br>0.035 |
| IV | 0.0 - 0.50<br>0.20 - 0.75 | -0.02<br>0.0 |
| V | 0.0 - 0.20<br>0.20 - 0.60<br>0.60 - 0.75 | -0.02<br>0.0<br>0.035 |



Fig. 2.29.a. Vertical components of ground reaction forces for the
single-support gait upon level ground for T=1.5, S=0.6
and diferent ZMP laws

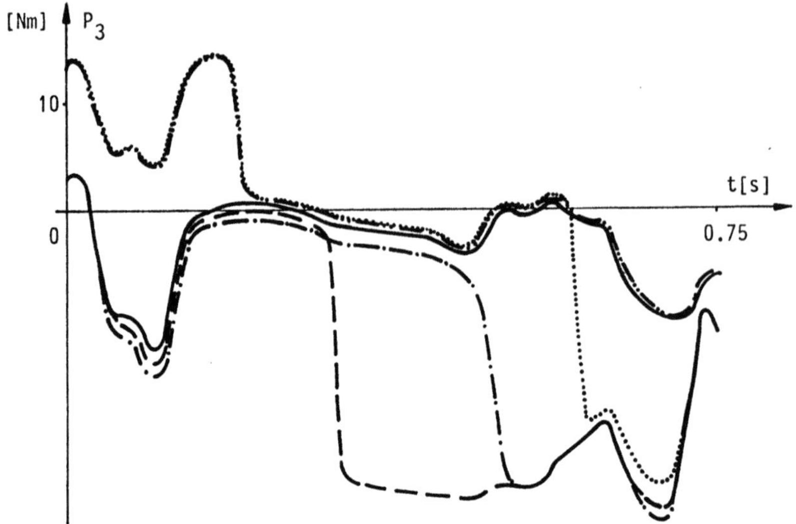

Fig. 2.29.b. Ankle joint torques for the single-support gait upon
level ground for T=1.5, S=0.6 and different ZMP laws
– supporting phase

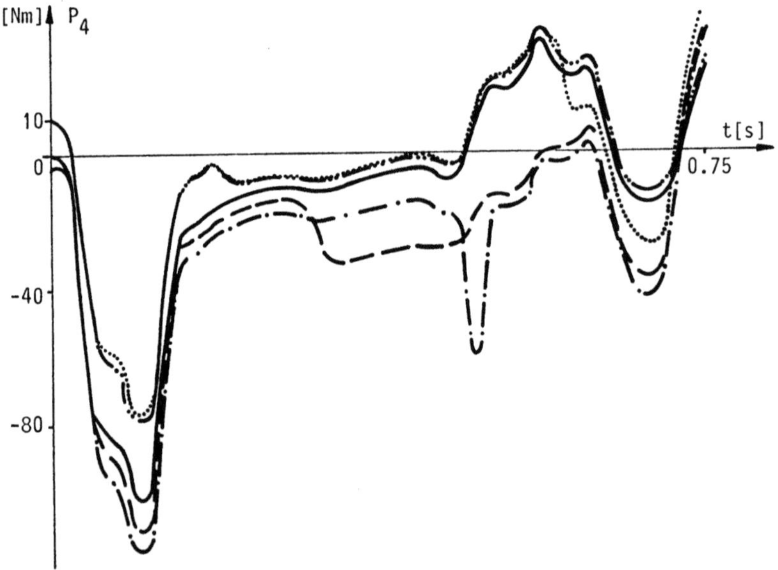

Fig. 2.29.c. Knee joint torques for the single-support gait upon
level ground for T=1.5, S=0.6 and different ZMP laws
– supporting phase

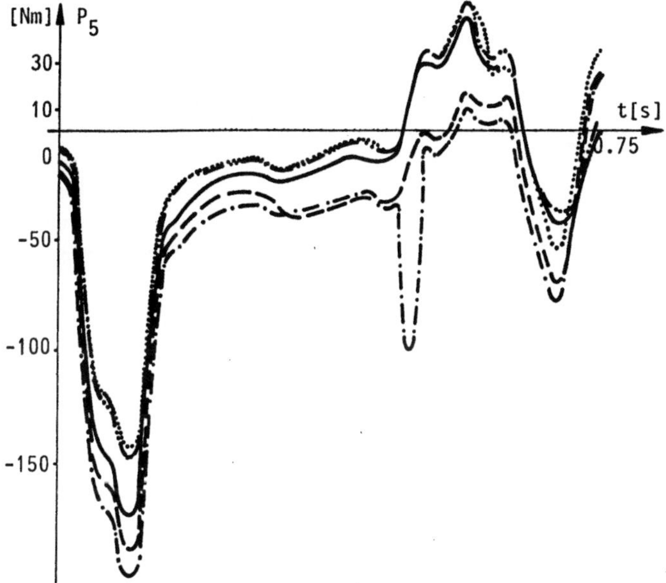

Fig. 2.29.d. Hip joint torques for the single-support gait upon level ground for T=1.5, S=0.6 and different ZMP laws - supporting phase

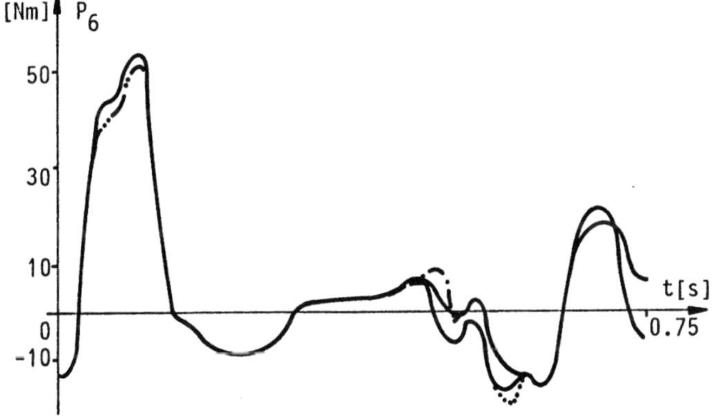

Fig. 2.29.e. Ankle joint torques for the single-support gait upon level ground for T=1.5, S=0.6 and different ZMP laws - swing phase

Fig. 2.29.f. Knee joint torques for the single-support gait upon level ground for T=1.5, S=0.6 and different ZMP laws - swing phase

Fig. 2.29.g. Hip joint torques for the single-support gait upon
            level ground for T=1.5, S=0.6 and different ZMP
            laws - swing phase

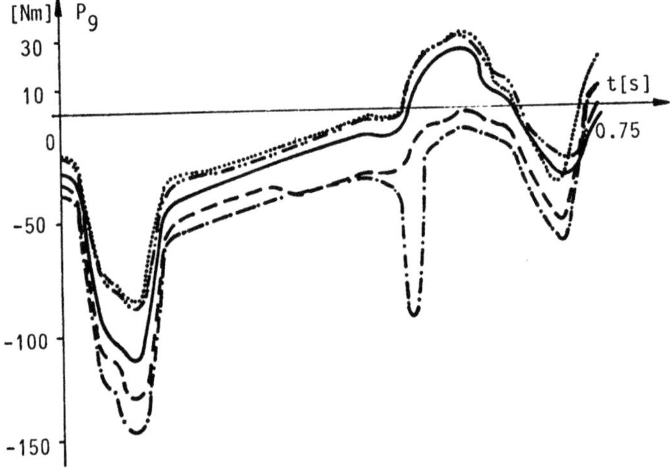

Fig. 2.29.h. Trunk joint torques in the sagittal plane for the
            single-support gait upon level ground for T=1.5,
            S=0.6 and different ZMP laws

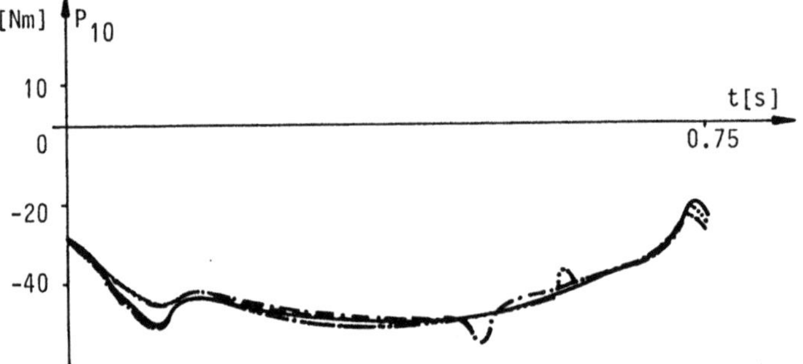

Fig. 2.29.i. Trunk joint torques in the frontal plane for the
            single-support gait upon level ground for T=1.5,
            S=0.6 and different ZMP laws

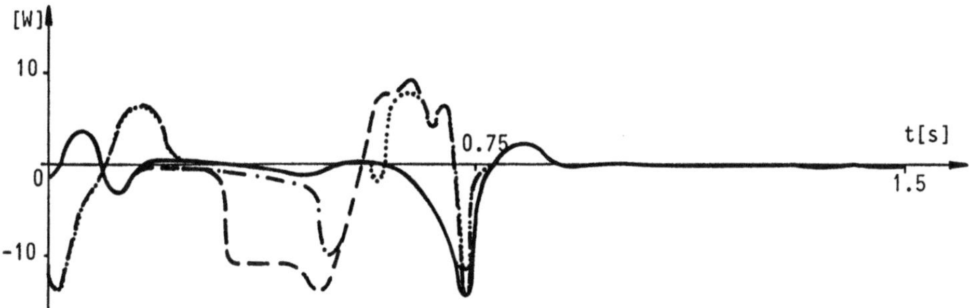

Fig. 2.30.a. Power at ankle joint for the single-support gait upon
level ground for T=1.5, S=0.6 and different ZMP laws

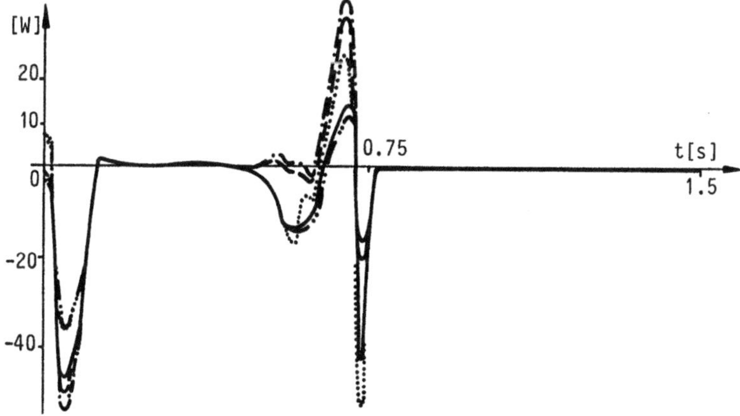

Fig. 2.30.b. Power at knee joint for the single-support gait upon
level ground for T=1.5, S=0.6 and different ZMP laws

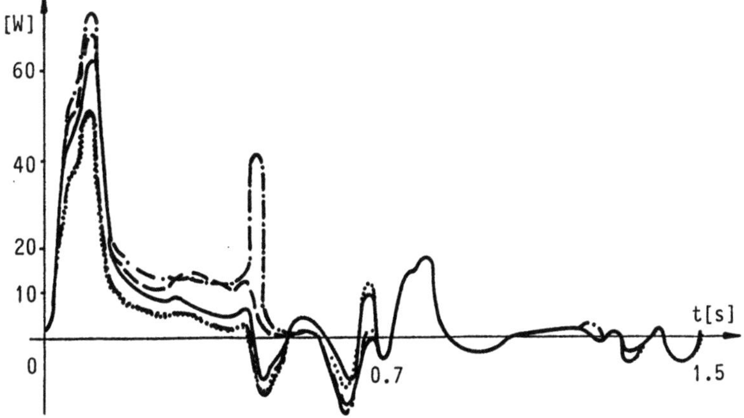

Fig. 2.30.c. Power at hip joint for the single-support gait upon
level ground for T=1.5, S=0.6 and different ZMP laws

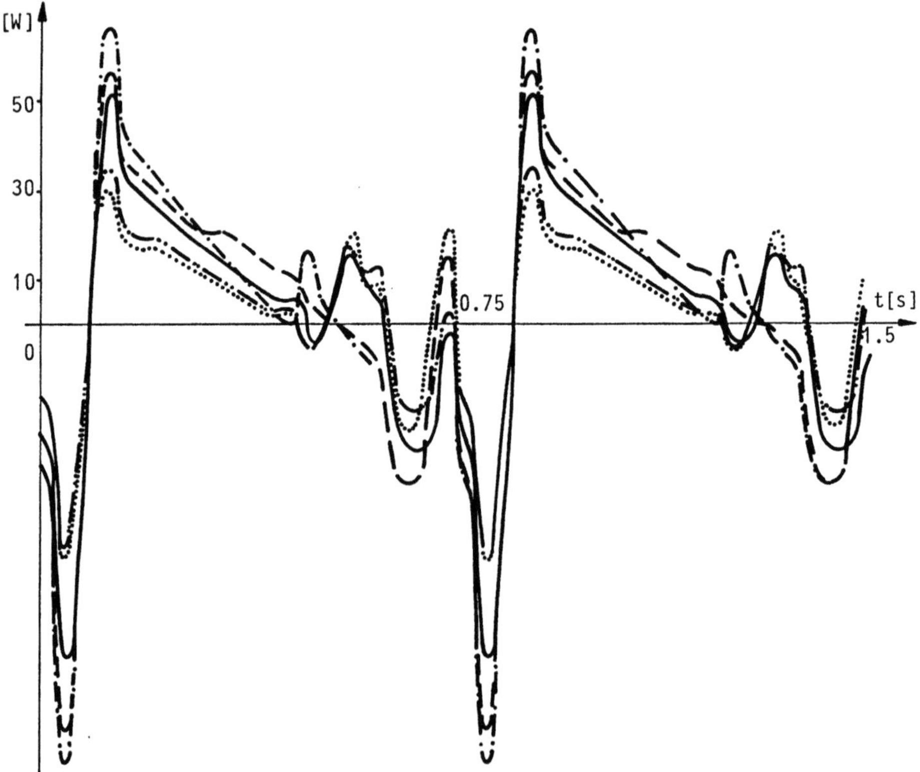

Fig. 2.30.d. Power at trunk joint in the sagittal plane for the
single-support gait upon level ground for T=1.5,
S=0.6 and different ZMP laws

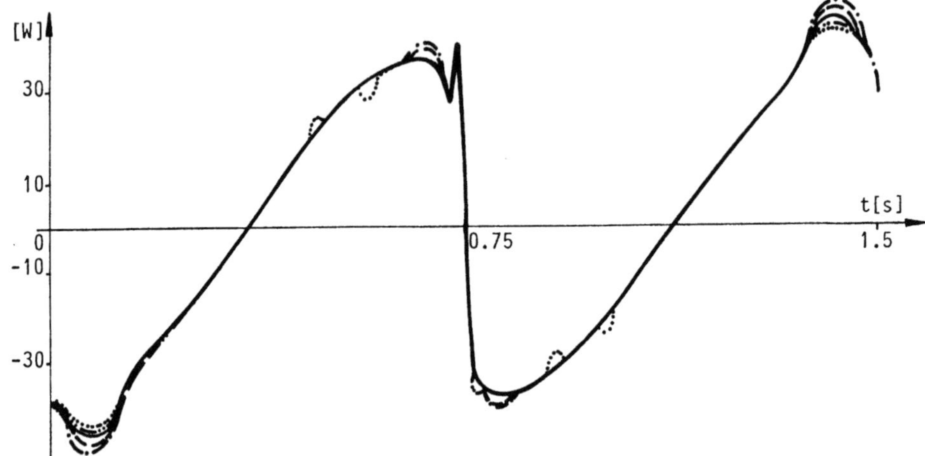

Fig. 2.30.e. Power at trunk joint in the frontal plane for the
single-support gait upon level ground for T=1.5,
S=0.6 and different ZMP laws

<u>Example 2.</u> In this example we consider the anthropomorphic mechanism shown in Fig. 2.31, which consists of 17 links and 17 revolute joints of the 5-th class. The topological structure represented by the series of "+" joints is of the form:

$$MS = \begin{bmatrix} 1 & 2 & 3 & 4 & 5 & 6 & 7 & 8 & 9 & 10 & 11 & 12 \\ 1 & 2 & 3 & 4 & 5 & 6 & 7 & 13 & 14 & 15 & 0 & 0 \\ 1 & 2 & 3 & 4 & 5 & 6 & 7 & 13 & 16 & 17 & 0 & 0 \end{bmatrix} \qquad (2.8.3)$$

The kinematic and dynamic parameters of the mechanism are given in Table 2.2. Fig. 2.32. shows the $q^3$ and $q^{13}$ d.o.f. for compensation in the frontal (2.32a) and sagittal (Fig. 2.32.b.) plane, respectively, under the condition link 7 (pelvis) is parallel to the ground and the projection of the legs are parallel in the frontal plane. This condition is fulfilled by putting $q^7 \rightarrow q^7 - q^3$ and $q^8 \rightarrow q^8 + q^3$. As in Example 1, the mechanism's arms are fixed, i.e., $q^{14} = q^{16} = 1.862253$ [rad] and $q^{15} = q^{17} = 1.5707$ [rad].

The other data are: $N = 3$; $I_1 = 1$, $I_2 = 7$, $I_3 = 8$; $K_1 = 12$, $K_2 = 10$, $K_3 = 10$; $NT = 17$; $\xi_i^1 = 0$, $i = 1, \ldots, 17$; $\xi_7^2 = 1$, $\xi_{13}^2 = 1$; values $\xi_i^3$ is 1 for $i \in \{3, 13\}$ and 0 for other values of index $i$; $\vec{r}_{o1} = (0, 0, -0.0001)^T$; $\vec{e}_{1_3} = (1, 0, 0)^T$; the initial values of the compensating part of the system $(q^3(0), q^{13}(0), \dot{q}^3(0)$ and $\dot{q}^{13}(0))$; the repeatibility conditions are:

$$W = \begin{bmatrix} -1 & 0 & 0 & 0 \\ 0 & 1 & 0 & 0 \\ 0 & 0 & -1 & 0 \\ 0 & 0 & 0 & 1 \end{bmatrix} \qquad (2.8.4)$$

Figs. 2.33 - 2.38. show the compensating movements for the different parameters and gait types.


<u>Example 3.</u> In this case we consider the realization of an artificial anthropomorphic gait by a mechanism with free arms (Fig. 2.39). The mechanism consists of 19 links and 19 revolute joints of the 5-th class. Its topological structure can be represented by a series of "+" joint in the form of the matrix

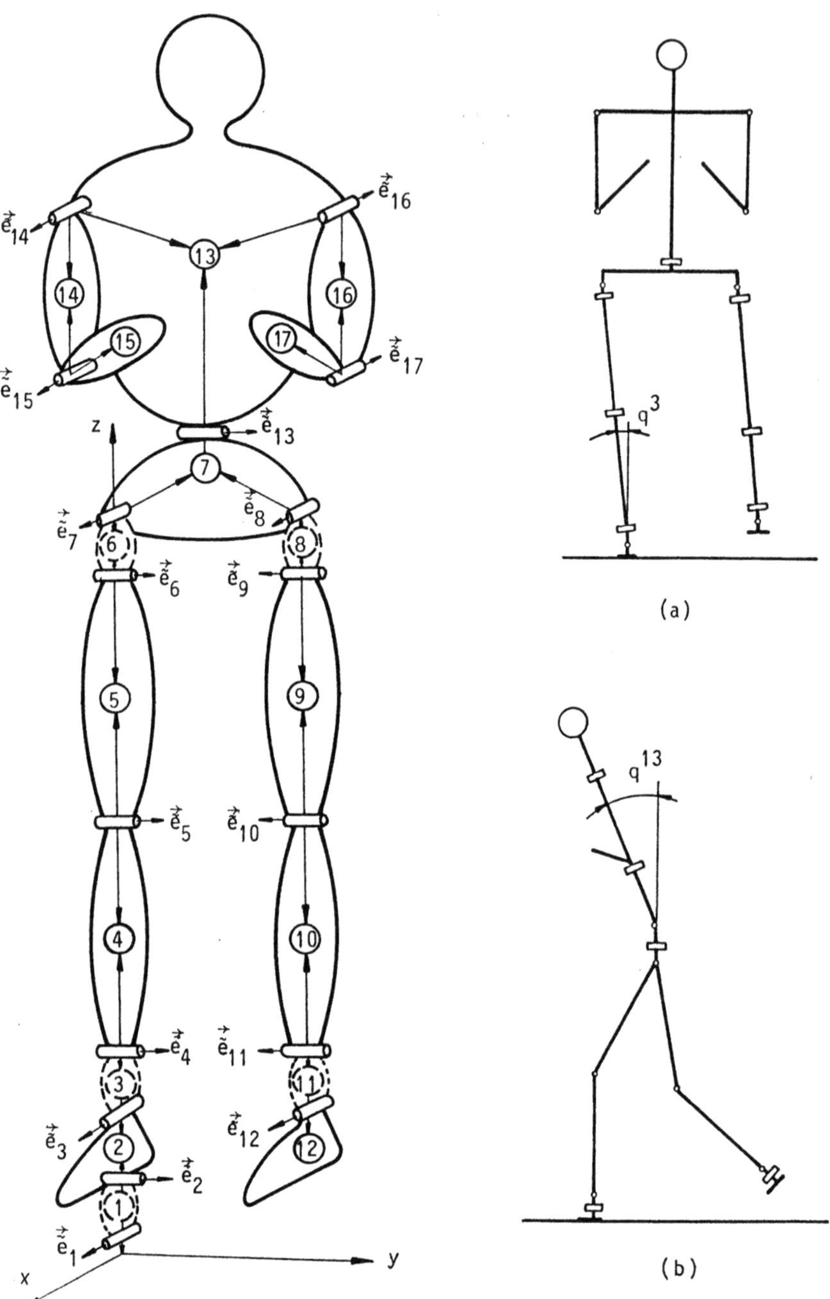



Fig. 2.31. Mechanical scheme of
the antropomorphic
mechanism with fixed
arms

Fig. 2.32. Compensating d.o.f. in
the frontal (a) and
sagittal (b) plane

Table 2.2. Kinematic and dynamic parameters of the mechanism

| Link | Mass [kg] | Moment of inertia [kgm$^2$] | | | Distance of the axes centres of joints from the link centre [m] | Joint unit axes |
| --- | --- | --- | --- | --- | --- | --- |
| | | $J_x$ | $J_y$ | $J_z$ | | |
| 1 | 2 | 3 | 4 | 5 | 6 | 7 |
| 1 | 0.0 | 0.0 | 0.0 | 0.0 | $\vec{\tilde{r}}_{1,1} = (0, 0, 0.0001)^T$; $\quad \vec{\tilde{r}}_{1,2} = (0, 0, -0.0001)^T$ | $\vec{\tilde{e}}_1 = (1, 0, 0)^T$ |
| 2 | 1.53 | 0.00006 | 0.00055 | 0.00045 | $\vec{\tilde{r}}_{2,2} = (0, 0, 0.030)^T$; $\quad \vec{\tilde{r}}_{2,3} = (0, 0, -0.070)^T$ | $\vec{\tilde{e}}_2 = (0, 1, 0)^T$ |
| 3 | 0.0 | 0.0 | 0.0 | 0.0 | $\vec{\tilde{r}}_{3,3} = (0, 0, 0.0001)^T$; $\quad \vec{\tilde{r}}_{3,4} = (0, 0, -0.0001)^T$ | $\vec{\tilde{e}}_3 = (1, 0, 0)^T$ |
| 4 | 3.21 | 0.00393 | 0.00393 | 0.00038 | $\vec{\tilde{r}}_{4,4} = (0, 0, 0.210)^T$; $\quad \vec{\tilde{r}}_{4,5} = (0, 0, -0.210)^T$ | $\vec{\tilde{e}}_4 = (0, 1, 0)^T$ |
| 5 | 8.41 | 0.01120 | 0.01200 | 0.00300 | $\vec{\tilde{r}}_{5,5} = (0, 0, 0.220)^T$; $\quad \vec{\tilde{r}}_{5,6} = (0, 0, -0.220)^T$ | $\vec{\tilde{e}}_5 = (0, 1, 0)^T$ |
| 6 | 0.0 | 0.0 | 0.0 | 0.0 | $\vec{\tilde{r}}_{6,6} = (0, 0, 0.0001)^T$; $\quad \vec{\tilde{r}}_{6,7} = (0, 0, -0.0001)^T$ | $\vec{\tilde{e}}_6 = (0, 1, 0)^T$ |
| 7 | 6.96 | 0.00700 | 0.00565 | 0.00627 | $\vec{\tilde{r}}_{7,7} = (0, 0.135, 0.1)^T$; $\quad \vec{\tilde{r}}_{7,8} = (0, -0.135, 0.1)^T$; <br> $\vec{\tilde{r}}_{7,13} = (0, 0, -0.05)^T$; | $\vec{\tilde{e}}_7 = (1, 0, 0)^T$ |
| 8 | 0.0 | 0.0 | 0.0 | 0.0 | $\vec{\tilde{r}}_{8,8} = (0, 0, -0.0001)^T$; $\quad \vec{\tilde{r}}_{8,9} = (0, 0, 0.0001)^T$ | $\vec{\tilde{e}}_8 = (1, 0, 0)^T$ |
| 9 | 8.41 | 0.01120 | 0.01200 | 0.00300 | $\vec{\tilde{r}}_{9,9} = (0, 0, -0.220)^T$; $\quad \vec{\tilde{r}}_{9,10} = (0, 0, 0.220)^T$ | $\vec{\tilde{e}}_9 = (0, -1, 0)^T$ |
| 10 | 3.21 | 0.00393 | 0.00393 | 0.00038 | $\vec{\tilde{r}}_{10,10} = (0, 0, -0.210)^T$; $\vec{\tilde{r}}_{10,11} = (0, 0, 0.210)^T$ | $\vec{\tilde{e}}_{10} = (0, -1, 0)^T$ |
| 11 | 0.0 | 0.0 | 0.0 | 0.0 | $\vec{\tilde{r}}_{11,11} = (0, 0, -0.0001)^T$; $\vec{\tilde{r}}_{11,12} = (0, 0, 0.0001)^T$ | $\vec{\tilde{e}}_{11} = (0, -1, 0)^T$ |

Table 2.2. Cont.

| 1 | 2 | 3 | 4 | 5 | 6 | 7 |
|---|---|---|---|---|---|---|
| 12 | 1.53 | 0.00006 | 0.00055 | 0.00045 | $\vec{\tilde{r}}_{12,12} = (0, 0, -0.070)^T$ | $\vec{\tilde{e}}_{12} = (1, 0, 0)^T$ |
| 13 | 30.85 | 0.15140 | 0.13700 | 0.02830 | $\vec{\tilde{r}}_{13,13} = (0, 0, 0.34)^T$;　　$\vec{\tilde{r}}_{13,14} = (0, 0.2, -0.6)^T$; <br> $\vec{\tilde{r}}_{13,16} = (0, -0.2, -0.06)^T$ | $\vec{\tilde{e}}_{13} = (0, 1, 0)^T$ |
| 14 | 2.07 | 0.00200 | 0.00200 | 0.00022 | $\vec{\tilde{r}}_{14,14} = (0, 0, -0.154)^T$;　　$\vec{\tilde{r}}_{14,15} = (0, 0, 0.154)$ | $\vec{\tilde{e}}_{14} = (1, 0, 0)^T$ |
| 15 | 1.14 | 0.00250 | 0.00425 | 0.00014 | $\vec{\tilde{r}}_{15,15} = (0, 0, -0.132)^T$ | $\vec{\tilde{e}}_{15} = (1, 0, 0)^T$ |
| 16 | 2.07 | 0.00200 | 0.00200 | 0.00022 | $\vec{\tilde{r}}_{16,16} = (0, 0, -0.154)^T$;　　$\vec{\tilde{r}}_{16,17} = (0, 0, 0.154)^T$ | $\vec{\tilde{e}}_{16} = (-1, 0, 0)^T$ |
| 17 | 1.14 | 0.00250 | 0.00425 | 0.00014 | $\vec{\tilde{r}}_{17,17} = (0, 0, -0.132)^T$ | $\vec{\tilde{e}}_{17} = (-1, 0, 0)^T$ |

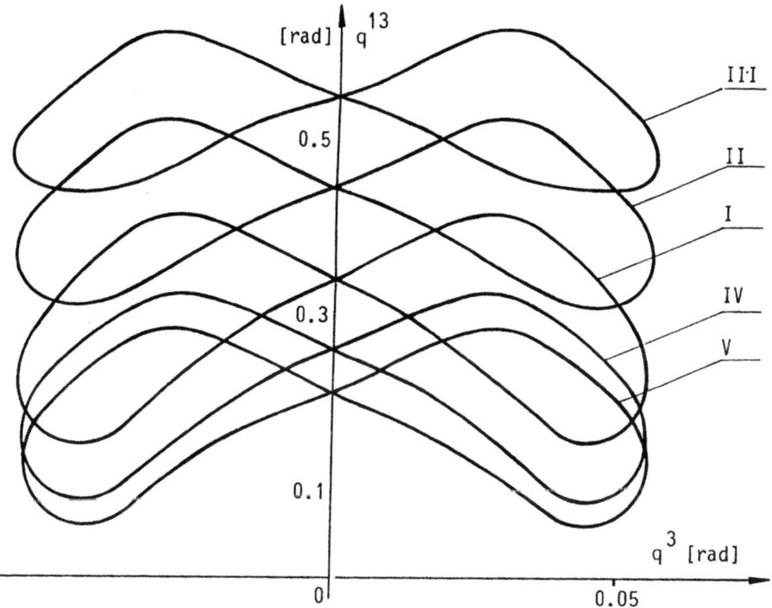



Fig. 2.33. Compensating movements for the single-support gait upon level ground for T=1.5, S=0.6 and different ZMP laws

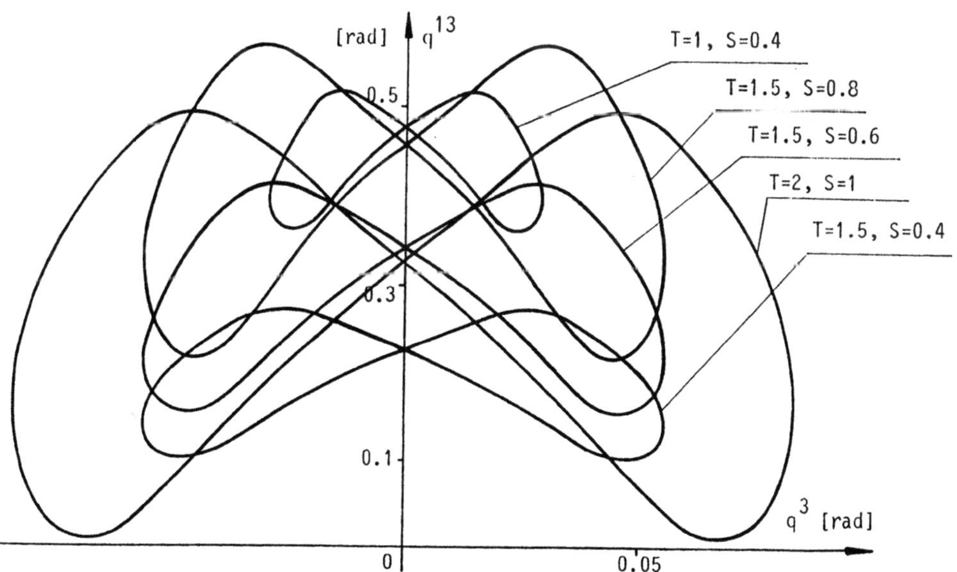



Fig. 2.34. Compensating movements for the single-support gait upon level ground for ZMP law I for different values of parameters T and S

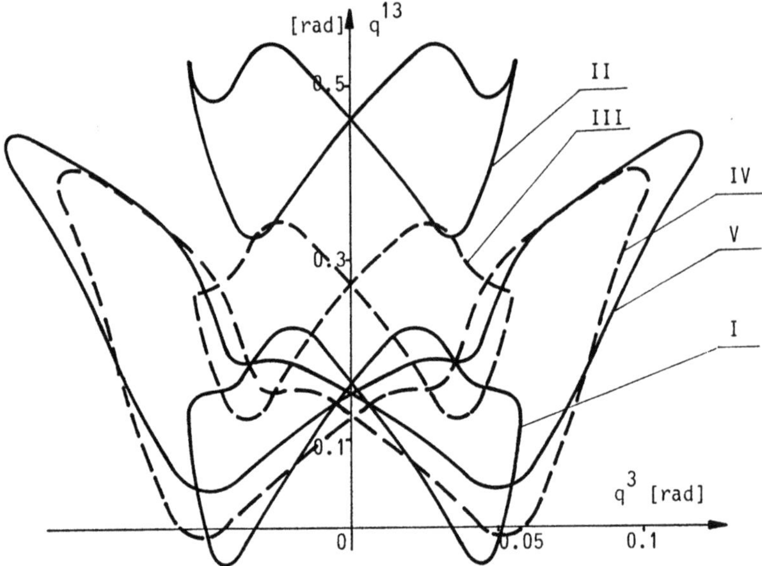

Fig. 2.35. Compensating movements for the single-support gait up the
stairs for T=1.5, S=0.6 and different ZMP laws

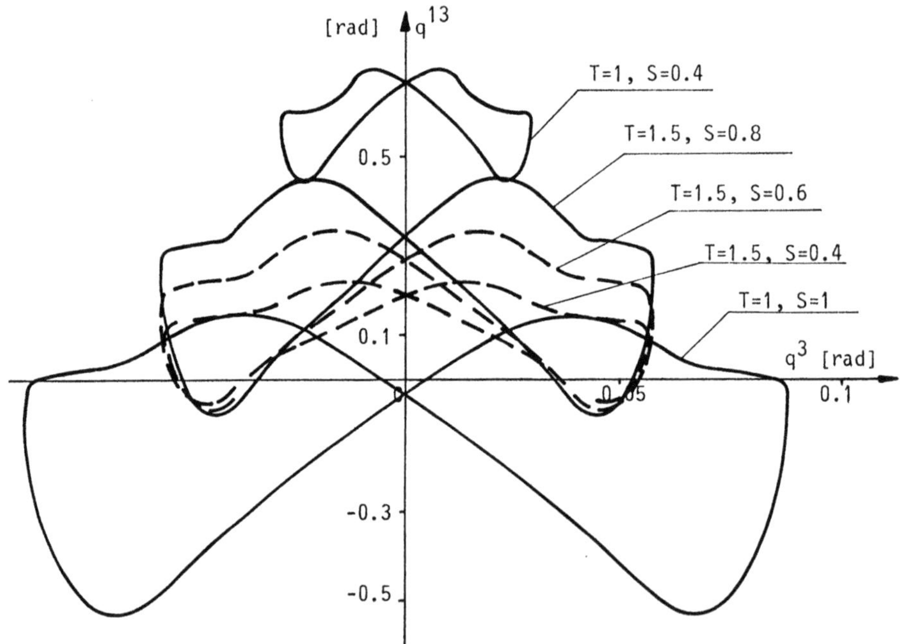

Fig. 2.36. Compensating movements for the single-support gait up the
stairs for ZMP law I and different values of parameters T
and S

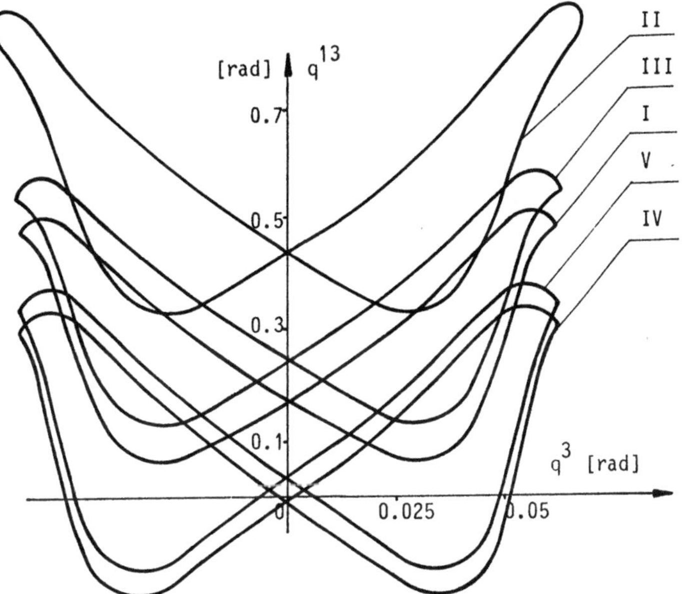



Fig. 2.37. Compensating movements for the single-support gait down the stairs for T=1.5, S=0.6 and different ZMP laws

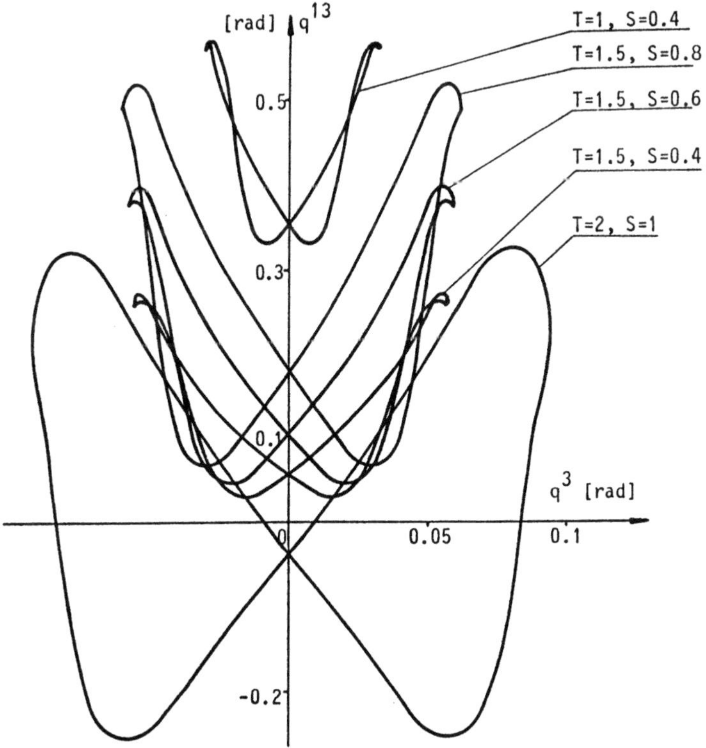



Fig. 2.38. Compensating movements for the single-support gait down the stairs for ZMP law I and different values of parameters T and S

$$MS = \begin{bmatrix} 1 & 2 & 3 & 4 & 5 & 6 & 7 & 8 & 9 & 10 & 11 & 12 \\ 1 & 2 & 3 & 4 & 5 & 6 & 7 & 13 & 14 & 15 & 16 & 0 \\ 1 & 2 & 3 & 4 & 5 & 6 & 7 & 13 & 17 & 18 & 19 & 0 \end{bmatrix} \qquad (2.8.5)$$

The kinematic and dynamic parameters for this mechanism are given in Table 2.3. It is assumed that the joints 15 and 18 are unpowered and the arms move freely (due to the coupling from the powered joints). For unpowered joints, the only moments considered are those arising from the forces of viscous friction

$$P_{15} = k\dot{q}^{15}, \qquad P_{18} = k\dot{q}^{18}, \qquad (k=-0.2)$$

where k is the viscous friction coefficient. In this case, the compensating dynamics comprises the trunk movements both in the sagittal and frontal plane and the arms movements in the sagittal plane only (Fig. 2.40).

The input data for the prescribed part of synergy for the motion of legs in single-support gait phase and the law of ZMP displacement are the same as in Example 2. The information on mechanism structure is obtained from matrix (2.8.5). Other input data are: $N=3$, $I_1=1$, $I_2=7$, $I_3=8$; $K_1=12$, $K_2=11$, $K_3=11$; $NT=19$; all the links that are body-type; $\xi_7^2=\xi_{13}^2=1$; the laws of motion of kinematic pairs referring to the compensating dynamics ($\xi_3^3 = \xi_{13}^3 = \xi_{15}^3 = \xi_{18}^3 = 1$), and the prescribed dynamics for other kinematic pairs; the moments of unpowered joints due only to viscous friction ($\xi_{15}^4 = \xi_{18}^4 = 1$); $\vec{r}_{o1} = (0, 0, -0.0001)^T$; $\vec{e}_1 = (1, 0, 0)^T$; the initial values for the positions ($q^3(0)$, $q^{13}(0)$, $q^{15}(0)$ and $q^{18}(0)$) and velocities ($\dot{q}^3(0)$, $\dot{q}^{13}(0)$, $\dot{q}^{15}(0)$ and $\dot{q}^{18}(0)$) of the compensating part of the system; the transformation matrix for repeatibility conditions at the half-step

$$W = \begin{bmatrix} -1 & 0 & 0 & 0 & 0 & 0 & 0 & 0 \\ 0 & 1 & 0 & 0 & 0 & 0 & 0 & 0 \\ 0 & 0 & -1 & 0 & 0 & 0 & 0 & 0 \\ 0 & 0 & 0 & -1 & 0 & 0 & 0 & 0 \\ 0 & 0 & 0 & 0 & -1 & 0 & 0 & 0 \\ 0 & 0 & 0 & 0 & 0 & 1 & 0 & 0 \\ 0 & 0 & 0 & 0 & 0 & 0 & -1 & 0 \\ 0 & 0 & 0 & 0 & 0 & 0 & 0 & -1 \end{bmatrix} \qquad (2.8.6)$$

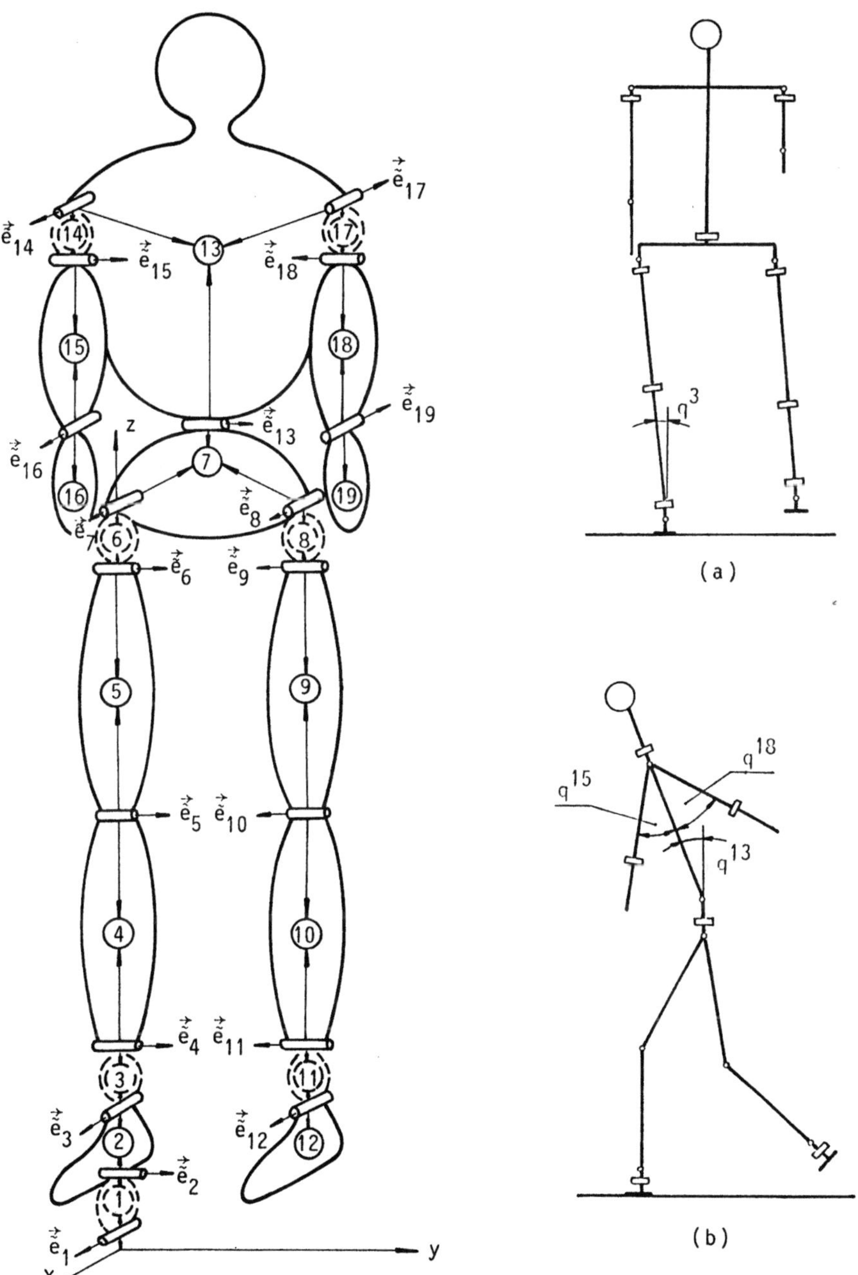



Fig. 2.39. Mechanical scheme of the antropomorphic mechanism with free arms

Fig. 2.40. Compensating d.o.f. in the frontal (a) and sagittal (b) plane

Table 2.3. Kinematic and dynamic parameters of the mechanism

| Link | Mass [kg] | Moment of inertia $[\mathrm{kgm}^2]$ | | | Distance of the axes centres of joints from the link centre [m] | Joint unit axes |
|---|---|---|---|---|---|---|
| | | $J_x$ | $J_y$ | $J_z$ | | |
| 1 | 2 | 3 | 4 | 5 | 6 | 7 |
| 1 | 0.0 | 0.0 | 0.0 | 0.0 | $\vec{\tilde{r}}_{1,1} = (0, 0, 0.0001)^T$;　$\vec{\tilde{r}}_{1,2} = (0, 0, -0.0001)^T$ | $\vec{\tilde{e}}_1 = (1, 0, 0)^T$ |
| 2 | 1.53 | 0.00006 | 0.00055 | 0.00045 | $\vec{\tilde{r}}_{2,2} = (0, 0, 0.030)^T$;　$\vec{\tilde{r}}_{2,3} = (0, 0, -0.070)^T$ | $\vec{\tilde{e}}_2 = (0, 1, 0)^T$ |
| 3 | 0.0 | 0.0 | 0.0 | 0.0 | $\vec{\tilde{r}}_{3,3} = (0, 0, 0.0001)^T$;　$\vec{\tilde{r}}_{3,4} = (0, 0, -0.0001)^T$ | $\vec{\tilde{e}}_3 = (1, 0, 0)^T$ |
| 4 | 3.21 | 0.00393 | 0.00393 | 0.00038 | $\vec{\tilde{r}}_{4,4} = (0, 0, 0.210)^T$;　$\vec{\tilde{r}}_{4,5} = (0, 0, -0.210)^T$ | $\vec{\tilde{e}}_4 = (0, 1, 0)^T$ |
| 5 | 8.41 | 0.01120 | 0.01200 | 0.00300 | $\vec{\tilde{r}}_{5,5} = (0, 0, 0.220)^T$;　$\vec{\tilde{r}}_{5,6} = (0, 0, -0.220)^T$ | $\vec{\tilde{e}}_5 = (0, 1, 0)^T$ |
| 6 | 0.0 | 0.0 | 0.0 | 0.0 | $\vec{\tilde{r}}_{6,6} = (0, 0, 0.0001)^T$;　$\vec{\tilde{r}}_{6,7} = (0, 0, -0.0001)^T$ | $\vec{\tilde{e}}_6 = (0, 1, 0)^T$ |
| 7 | 6.96 | 0.00700 | 0.00565 | 0.00627 | $\vec{\tilde{r}}_{7,7} = (0, 0.135, 0.1)^T$;　$\vec{\tilde{r}}_{7,8} = (0, -0.135, 0.1)^T$<br>$\vec{\tilde{r}}_{7,13} = (0, 0, -0.05)^T$ | $\vec{\tilde{e}}_7 = (1, 0, 0)^T$ |
| 8 | 0.0 | 0.0 | 0.0 | 0.0 | $\vec{\tilde{r}}_{8,8} = (0, 0, -0.0001)^T$;　$\vec{\tilde{r}}_{8,9} = (0, 0, 0.0001)^T$ | $\vec{\tilde{e}}_8 = (1, 0, 0)^T$ |
| 9 | 8.41 | 0.01120 | 0.01200 | 0.00300 | $\vec{\tilde{r}}_{9,9} = (0, 0, -0.220)^T$;　$\vec{\tilde{r}}_{9,10} = (0, 0, 0.220)^T$ | $\vec{\tilde{e}}_9 = (0, -1, 0)^T$ |
| 10 | 3.21 | 0.00393 | 0.00393 | 0.00038 | $\vec{\tilde{r}}_{10,10} = (0, 0, -0.210)^T$;　$\vec{\tilde{r}}_{10,11} = (0, 0, 0.210)^T$ | $\vec{\tilde{e}}_{10} = (0, -1, 0)^T$ |
| 11 | 0.0 | 0.0 | 0.0 | 0.0 | $\vec{\tilde{r}}_{11,11} = (0, 0, -0.0001)^T$;　$\vec{\tilde{r}}_{11,12} = (0, 0, 0.0001)^T$ | $\vec{\tilde{e}}_{11} = (0, -1, 0)^T$ |

Table 2.3. Cont.

| 1 | 2 | 3 | 4 | 5 | 6 | 7 |
|---|---|---|---|---|---|---|
| 12 | 1.53 | 0.00006 | 0.00055 | 0.00045 | $\vec{\tilde{r}}_{12,12} = (0, 0, -0.070)^T$ | $\vec{\tilde{e}}_{12} = (1, 0, 0)^T$ |
| 13 | 30.85 | 0.1514 | 0.13700 | 0.02830 | $\vec{\tilde{r}}_{13,13} = (0, 0, 0.34)^T$;  $\vec{\tilde{r}}_{13,14} = (0, 0.2, -0.06)^T$  $\vec{\tilde{r}}_{13,17} = (0, -0.2, -0.06)^T$ | $\vec{\tilde{e}}_{13} = (0, 1, 0)^T$ |
| 14 | 0.C | 0.0 | 0.0 | 0.0 | $\vec{\tilde{r}}_{14,14} = (0, 0, -0.0001)^T$;  $\vec{\tilde{r}}_{14,15} = (0, 0, 0.0001)^T$ | $\vec{\tilde{e}}_{14} = (1, 0, 0)^T$ |
| 15 | 2.07 | 0.00200 | 0.00200 | 0.00022 | $\vec{\tilde{r}}_{15,15} = (0, 0, -0.154)^T$;  $\vec{\tilde{r}}_{15,16} = (0, 0, 0.154)^T$ | $\vec{\tilde{e}}_{15} = (0, 1, 0)^T$ |
| 16 | 1.14 | 0.00250 | 0.00425 | 0.00014 | $\vec{\tilde{r}}_{16,16} = (0, 0, -0.132)^T$ | $\vec{\tilde{e}}_{16} = (1, 0, 0)^T$ |
| 17 | 0.0 | 0.0 | 0.0 | 0.0 | $\vec{\tilde{r}}_{17,17} = (0, 0, -0.0001)^T$;  $\vec{\tilde{r}}_{17,18} = (0, 0, 0.0001)^T$ | $\vec{\tilde{e}}_{17} = (-1, 0, 0)^T$ |
| 18 | 2.07 | 0.00200 | 0.00200 | 0.00022 | $\vec{\tilde{r}}_{18,18} = (0, 0, -0.154)^T$;  $\vec{\tilde{r}}_{18,19} = (0, 0, 154)^T$ | $\vec{\tilde{e}}_{18} = (0, -1, 0)^T$ |
| 19 | 1.14 | 0.00250 | 0.00425 | 0.00014 | $\vec{\tilde{r}}_{19,19} = (0, 0, -0.132)^T$ | $\vec{\tilde{e}}_{19} = (-1, 0, 0)^T$ |

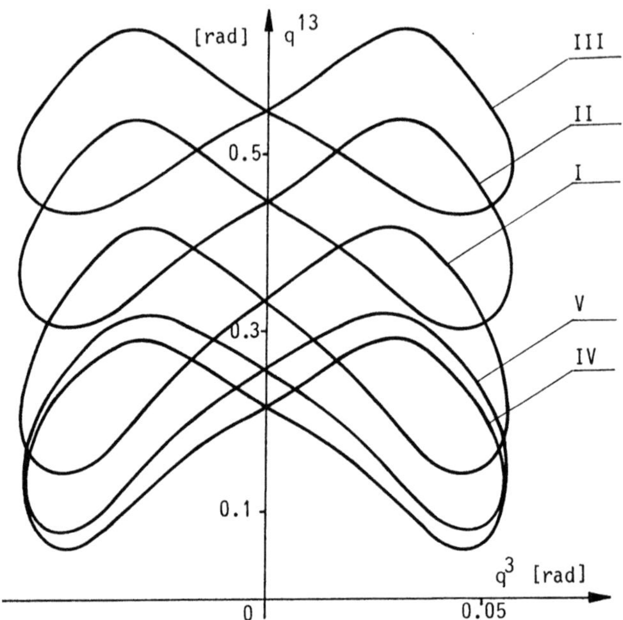

Fig. 2.41. Compensating movements for the single-support gait
upon level ground with free arms for T=1.5, S=0.6,
and different ZMP laws

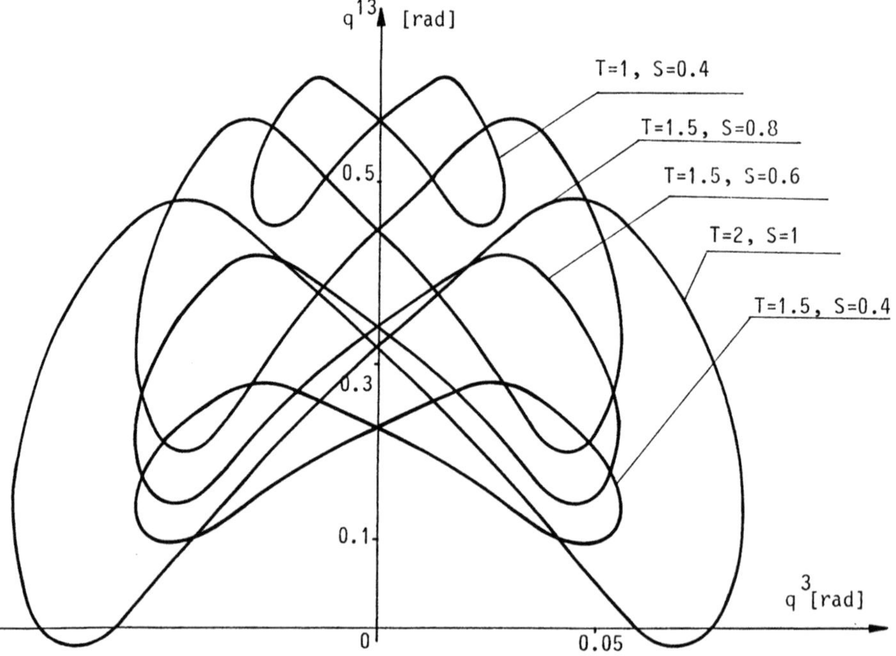

Fig. 2.42. Compensating movements for the single-support gait
upon level ground with free arms for ZMP law I and
different values of parameters T and S

On the basis of input data, the first two differential equations are obtained as in Examples 1 and 2. The other two second-order differential equations for $P_{15}$ and $P_{18}$ are extracted from the system (2.3.16). The system of four differential equations thus obtained is solved by iterative procedure for the prescribed boundary conditions.

Some of the results obtained for the anthropomorphic structure with free arms are shown in Figs. 2.41 - 2.43.

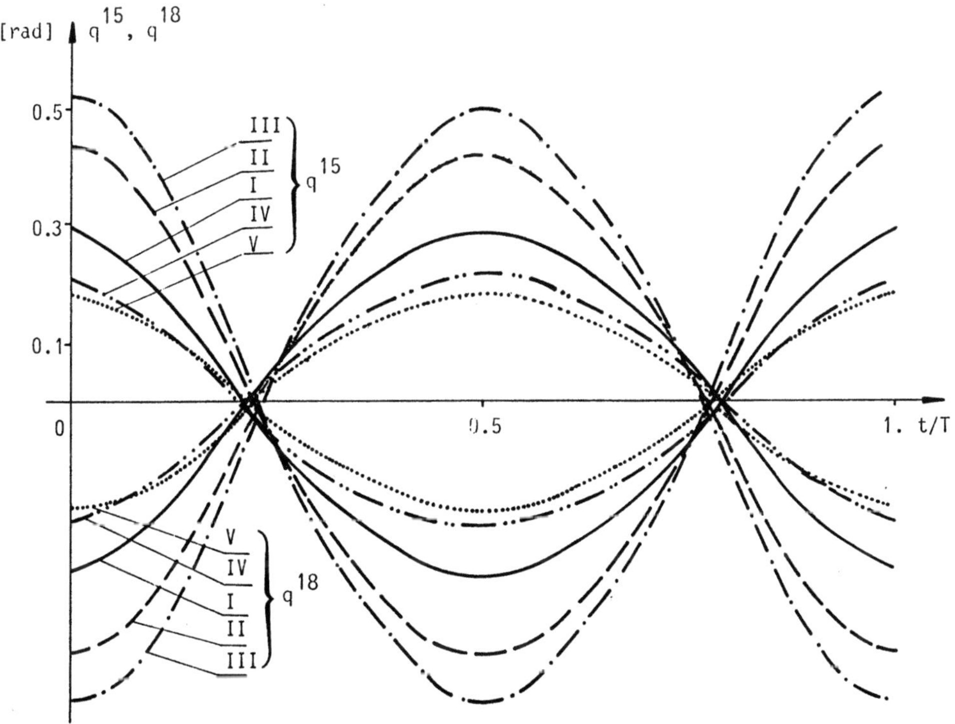


Fig. 2.43. Compensating movements of arms in the sagittal plane
for the single-support gait upon level ground for
T=1.5, S=0.6 and different ZMP laws

Example 4. We consider the type of gait in which single - and double--support phase appear alternatively. The mechanism structure is presented in Fig. 2.44. It consists of 20 links and 20 revolute joints of the 5-th class. The topological structure of the mechanism can be described by the matrix

$$MS = \begin{bmatrix} 1 & 2 & 3 & 4 & 5 & 6 & 7 & 8 & 9 & 10 & 11 & 12 & 13 & 14 \\ 1 & 2 & 3 & 4 & 5 & 6 & 7 & 8 & 15 & 16 & 17 & 18 & 0 & 0 \\ 1 & 2 & 3 & 4 & 5 & 6 & 7 & 8 & 15 & 16 & 19 & 20 & 0 & 0 \end{bmatrix} \qquad (2.8.7)$$

Numerical values for the mechanical part of the system are given in Table 2.4.

The kinematic chain number 1 represents legs. This chain is open during the single-support gait phase, while in the double-support phase it becomes a closed kinematic chain. The dynamics prescribed to the first chain is based on the results of biometric investigations of double-support phase gait.

The set of adopted trajectories for the ZMP displacement law is presented in Fig. 2.23. During the gait, the arms are fixed and the legs are parallel. This has been achieved by prescribing constant values to the positions: $q^1 = q^7 = q^{10} = q^{14} = 0$; $q^8 = q^9 = -0.933248$ [rad]; $q^{17} = q^{19} = 1.862253$ [rad]; $q^{18} = q^{20} = 1.5707$ [rad].

The compensating part of dynamics is the movement of the upper part of the body in the sagittal and frontal plane (as in Example 1). This part of the system is $q^{15}$ and $q^{16}$ (Fig. 2.45).

Apart from the mechanism structure (2.8.7) and the prescribed part of dynamics, the input data are: $N=3$; $I_1=1$, $I_2=8$, $I_3=10$; $K_1=14$, $K_2=K_3=12$; $NT=20$; all the links that are body-type; the branching links are $\xi_8^2 = 1$, $\xi_{16}^2 = 1$; the laws of motion for the kinematic pairs referring to the compensating dynamics $\xi_{15}^3 = 1$ and $\xi_{16}^3 = 1$, and the prescribed dynamics for other kinematic pairs; the prescribed moments (equal to zero) for the first chain ($\xi_2^5 = \xi_3^5 = \xi_7^5 = \xi_8^5 = \xi_9^5 = \xi_{10}^5 = 1$); $\vec{r}_{o1} = (0,\ 0,\ -0.0001)^T$; $\vec{e}_1 = (1,\ 0,\ 0)^T$; the initial values for positions ($q^{15}(0)$ and $q^{16}(0)$) and velocities ($\dot{q}^{15}(0)$ and $\dot{q}^{16}(0)$) of the open part of the system; the transformation matrix for the repeatibility coditions at half-step (2.7.2).

As in Example 1, a system of equations of the type (2.4.5) is formed for the single-support phase, while for the double-support gait phase the relations (2.5.23) and (2.5.25) hold. In the latter case, the calculation of the right-hand sides of differential equations is executed in two steps. First, 6 equations for the prescribed moments of the first chain are separated from the set of equations (2.5.23). Then, from these equations is calculated the vector $\sigma$ (ground reaction). The

Table 2.4. Kinematic and dynamic parameters of the mechanism

| Link | Mass [kg] | Moment of inertia [kgm$^2$] | | | Distance of the axes centres of joints from the link centre [m] | Joint unit axes |
| | | $J_x$ | $J_y$ | $J_z$ | | |
| 1 | 2 | 3 | 4 | 5 | 6 | 7 |
| 1 | 0.0 | 0.0 | 0.0 | 0.0 | $\vec{\tilde{r}}_{1,1} = (0, 0, 0.0001)^T$; $\quad \vec{\tilde{r}}_{1,2} = (0, 0, -0.0001)^T$ | $\vec{\tilde{e}}_1 = (1, 0, 0)^T$ |
| 2 | 1.53 | 0.00006 | 0.00055 | 0.00045 | $\vec{\tilde{r}}_{2,2} = (0, 0, 0.030)^T$; $\quad \vec{\tilde{r}}_{2,3} = (0, 0, -0.070)^T$ | $\vec{\tilde{e}}_2 = (0, 1, 0)^T$ |
| 3 | 0.0 | 0.0 | 0.0 | 0.0 | $\vec{\tilde{r}}_{3,3} = (0, 0, 0.0001)^T$; $\quad \vec{\tilde{r}}_{3,4} = (0, 0, -0.0001)^T$ | $\vec{\tilde{e}}_3 = (1, 0, 0)^T$ |
| 4 | 3.21 | 0.00393 | 0.00393 | 0.00038 | $\vec{\tilde{r}}_{4,4} = (0, 0, 0.210)^T$; $\quad \vec{\tilde{r}}_{4,5} = (0, 0, -0.210)^T$ | $\vec{\tilde{e}}_4 = (0, 1, 0)^T$ |
| 5 | 8.41 | 0.01120 | 0.01200 | 0.00300 | $\vec{\tilde{r}}_{5,5} = (0, 0, 0.220)^T$; $\quad \vec{\tilde{r}}_{5,6} = (0, 0, -0.220)^T$ | $\vec{\tilde{e}}_5 = (0, 1, 0)^T$ |
| 6 | 0.0 | 0.0 | 0.0 | 0.0 | $\vec{\tilde{r}}_{6,6} = (0, 0, 0.0001)^T$; $\quad \vec{\tilde{r}}_{6,7} = (0, 0, -0.0001)^T$ | $\vec{\tilde{e}}_6 = (0, 1, 0)^T$ |
| 7 | 0.0 | 0.0 | 0.0 | 0.0 | $\vec{\tilde{r}}_{7,7} = (0, 0, 0.0001)^T$; $\quad \vec{\tilde{r}}_{7,8} = (0, 0, -0.0001)^T$ | $\vec{\tilde{e}}_7 = (1, 0, 0)^T$ |
| 8 | 6.96 | 0.00700 | 0.00565 | 0.00625 | $\vec{\tilde{r}}_{8,8} = (0, 0.135, 0.1)^T$; $\quad \vec{\tilde{r}}_{8,9} = (0, -0.135, 0.1)^T$ <br> $\vec{\tilde{r}}_{8,15} = (0, 0, -0.05)^T$ | $\vec{\tilde{e}}_8 = (1, 0, 0)^T$ |
| 9 | 0.0 | 0.0 | 0.0 | 0.0 | $\vec{\tilde{r}}_{9,9} = (0, 0, -0.0001)^T$; $\quad \vec{\tilde{r}}_{9,10} = (0, 0, 0.0001)^T$ | $\vec{\tilde{e}}_9 = (1, 0, 0)^T$ |
| 10 | 0.0 | 0.0 | 0.0 | 0.0 | $\vec{\tilde{r}}_{10,10} = (0, 0, -0.0001)^T$; $\quad \vec{\tilde{r}}_{10,11} = (0, 0, 0.0001)^T$ | $\vec{\tilde{e}}_{10} = (0, 0, 1)^T$ |
| 11 | 8.41 | 0.01120 | 0.01200 | 0.00300 | $\vec{\tilde{r}}_{11,11} = (0, 0, -0.220)^T$; $\quad \vec{\tilde{r}}_{11,12} = (0, 0, 0.220)^T$ | $\vec{\tilde{e}}_{11} = (0, -1, 0)^T$ |

Table 2.4. Cont.



| 1 | 2 | 3 | 4 | 5 | 6 | 7 |
|---|---|---|---|---|---|---|
| 12 | 3.21 | 0.00393 | 0.00393 | 0.00038 | $\vec{\tilde{r}}_{12,12} = (0, 0, -0.210)^T$;   $\vec{\tilde{r}}_{12,13} = (0, 0, 0.210)^T$ | $\vec{\tilde{e}}_{12} = (0, -1, 0)^T$ |
| 13 | 0.0 | 0.0 | 0.0 | 0.0 | $\vec{\tilde{r}}_{13,13} = (0, 0, -0.0001)^T$; $\vec{\tilde{r}}_{13,14} = (0, 0, 0.0001)^T$ | $\vec{\tilde{e}}_{13} = (0, -1, 0)^T$ |
| 14 | 1.53 | 0.00006 | 0.00055 | 0.00045 | $\vec{\tilde{r}}_{14,14} = (0, 0, -0.070)^T$ | $\vec{\tilde{e}}_{14} = (1, 0, 0)^T$ |
| 15 | 0.0 | 0.0 | 0.0 | 0.0 | $\vec{\tilde{r}}_{15,16} = (0, 0, 0.0001)^T$; $\vec{\tilde{r}}_{15,\,16} = (0, 0, -0.0001)^T$ | $\vec{\tilde{e}}_{15} = (0, 1, 0)^T$ |
| 16 | 30.85 | 0.15140 | 0.13700 | 0.02830 | $\vec{\tilde{r}}_{16,16} = (0, 0, 0.34)^T$;   $\vec{\tilde{r}}_{16,17} = (0, 0.2, -0.06)^T$; <br> $\vec{\tilde{r}}_{16,19} = (0, -0.2, -0.06)^T$ | $\vec{\tilde{e}}_{16} = (1, 0, 0)^T$ |
| 17 | 2.07 | 0.00200 | 0.00200 | 0.00022 | $\vec{\tilde{r}}_{17,17} = (0, 0, -0.154)^T$; $\vec{\tilde{r}}_{17,18} = (0, 0, 0.154)^T$ | $\vec{\tilde{e}}_{17} = (1, 0, 0)^T$ |
| 18 | 1.14 | 0.00250 | 0.00425 | 0.00014 | $\vec{\tilde{r}}_{18,18} = (0, 0, -0.132)^T$ | $\vec{\tilde{e}}_{18} = (1, 0, 0)^T$ |
| 19 | 2.07 | 0.00200 | 0.00200 | 0.00022 | $\vec{\tilde{r}}_{19,19} = (0, 0, -0.154)^T$; $\vec{\tilde{r}}_{19,20} = (0, 0, 0.154)^T$ | $\vec{\tilde{e}}_{19} = (-1, 0, 0)^T$ |
| 20 | 1.14 | 0.00250 | 0.00425 | 0.00014 | $\vec{\tilde{r}}_{20,20} = (0, 0, -0.132)^T$ | $\vec{\tilde{e}}_{20} = (-1, 0, 0)^T$ |

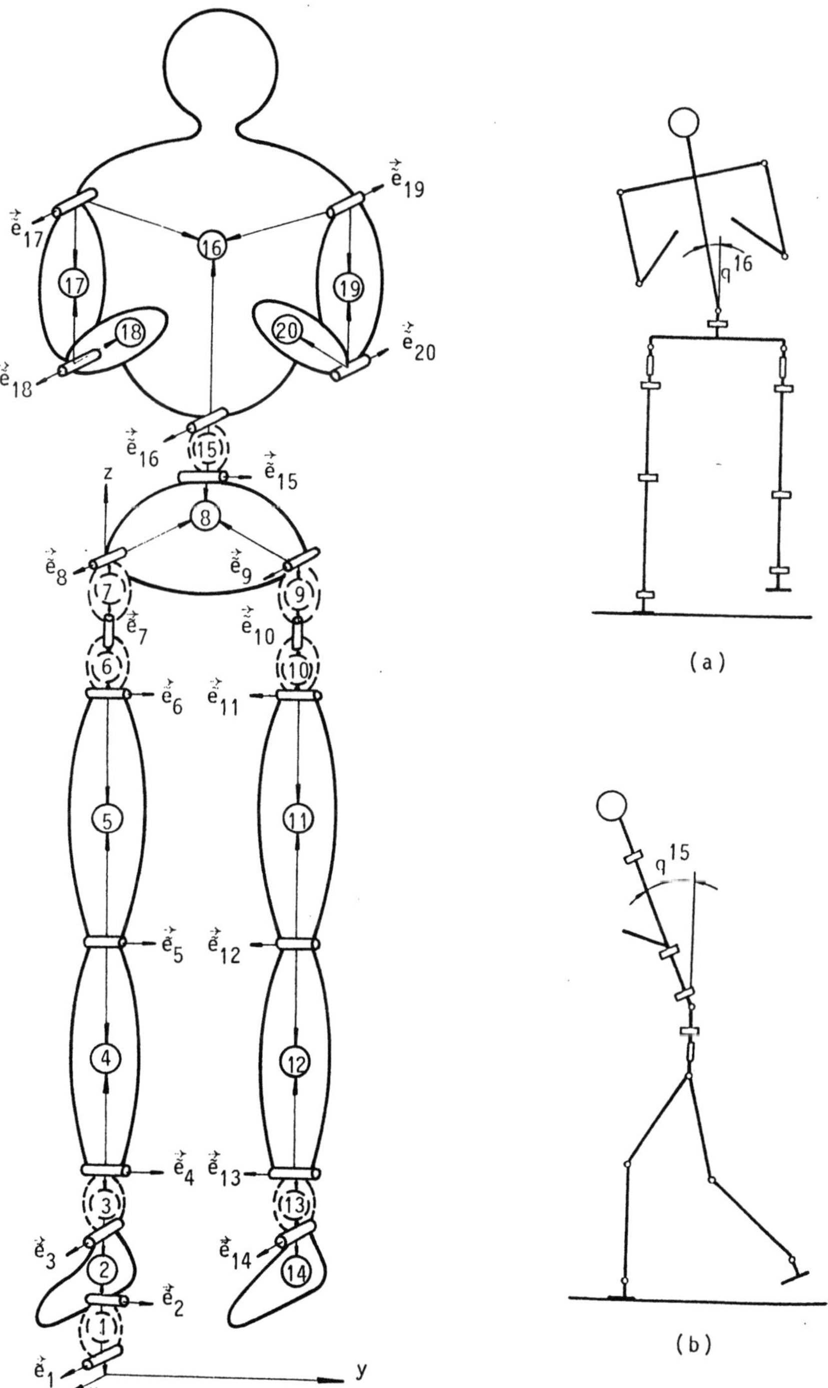

Fig. 2.44. Mechanical scheme of the antropomorphic mechanism with fixed arms

Fig. 2.45. Compensating d.o.f. in the frontal (a) and sagittal (b) plane

second step is to calculate the right-hand sides of the differential
equations describing the system's equilibrium in both the sagittal and
frontal plane.

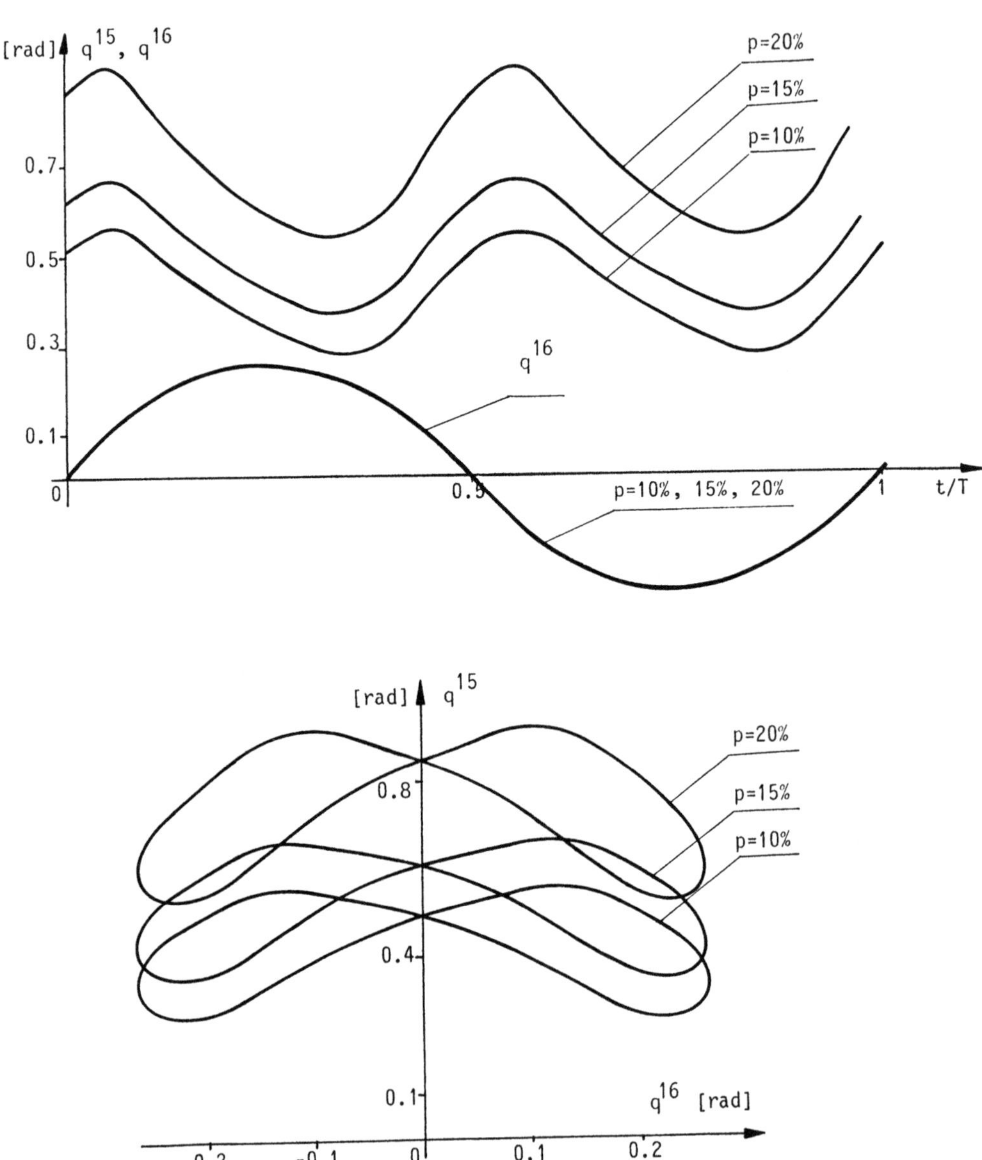



Fig. 2.46. Compensating movements for the double-support gait
upon level ground for ZMP law I and different duration of double-support phase

This procedure enables numerical integration of the system in both phases of the gait. As in the previous examples, an iterative procedure is to be employed for the synthesis of nominal dynamics.

Compensating movements for different duration periods p, and ZMP law (Fig. 2.23) of the double-support phase are presented in Fig. 2.46-2.47.

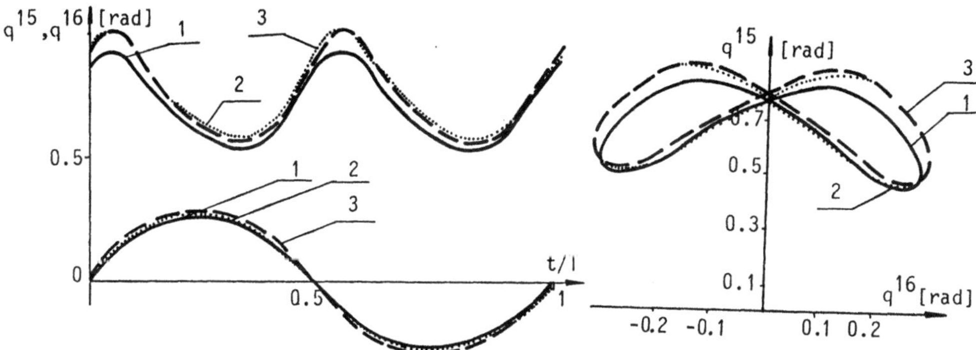

Fig. 2.47. Compensating movements for the double-support gait
upon level ground for different ZMP laws and duration of double-support phase p=20%

Example 5. This example is related to the modelling of a two-link foot Fig. 2.19. The mechanism consists of 19 links and 19 revolute joints (Fig. 2.48). Topological structure of the mechanism is given by the matrix:

$$MS = \begin{bmatrix} 1 & 2 & 3 & 4 & 5 & 6 & 7 & 8 & 9 & 10 & 11 & 12 & 13 & 14 \\ 1 & 2 & 3 & 4 & 5 & 6 & 7 & 8 & 15 & 16 & 17 & 0 & 0 & 0 \\ 1 & 2 & 3 & 4 & 5 & 6 & 7 & 8 & 15 & 18 & 19 & 0 & 0 & 0 \end{bmatrix} \qquad (2.8.8)$$

Table 2.5. contains the kinematic and dynamic parameters of the mechanism. Fig. 2.49. shows the $q^4$ and $q^{15}$ d.o.f. for the compensation in the frontal (2.49a) and sagittal (2.49b) plane, respectively. The right foot is modelled by links 2 and 3, the left one by links 13 and 14. The arms are fixed, i.e. $q^{16}=q^{18}=1.862253$ [rad] and $q^{17}=q^{19}=1.5707$ [rad]. The conditions $q^8: = q^8-q^4$ and $q^9: = q^9+q^4$ ensure the pelvis (link 8) is in the horizontal position and the legs are parallel in the frontal plane.

Other input data are: N=3; $I_1=1$, $I_2=8$, $I_3=9$; $K_1=14$, $K_2=11$, $K_3=11$, NT=19; $\xi_i^1=0$, i=1,...,19; $\xi_8^2=1$, $\xi_{15}^2=1$; value $\xi_i^3$ is 1 for i∈{4, 13} and 0 for other values of index i; $\vec{r}_{o1}=(0, 0, -0.0001)^T$; $\vec{e}_1=(1, 0, 0)^T$; the initial values of the open part of the system ($q^4(0)$, $q^{15}(0)$, $\dot{q}^4(0)$ and $\dot{q}^{15}(0)$); the transformation matrix W for repeatibility conditions is the same as in Example 2, i.e. (2.8.4).

Fig. 2.50. show the compensating movements for this example.

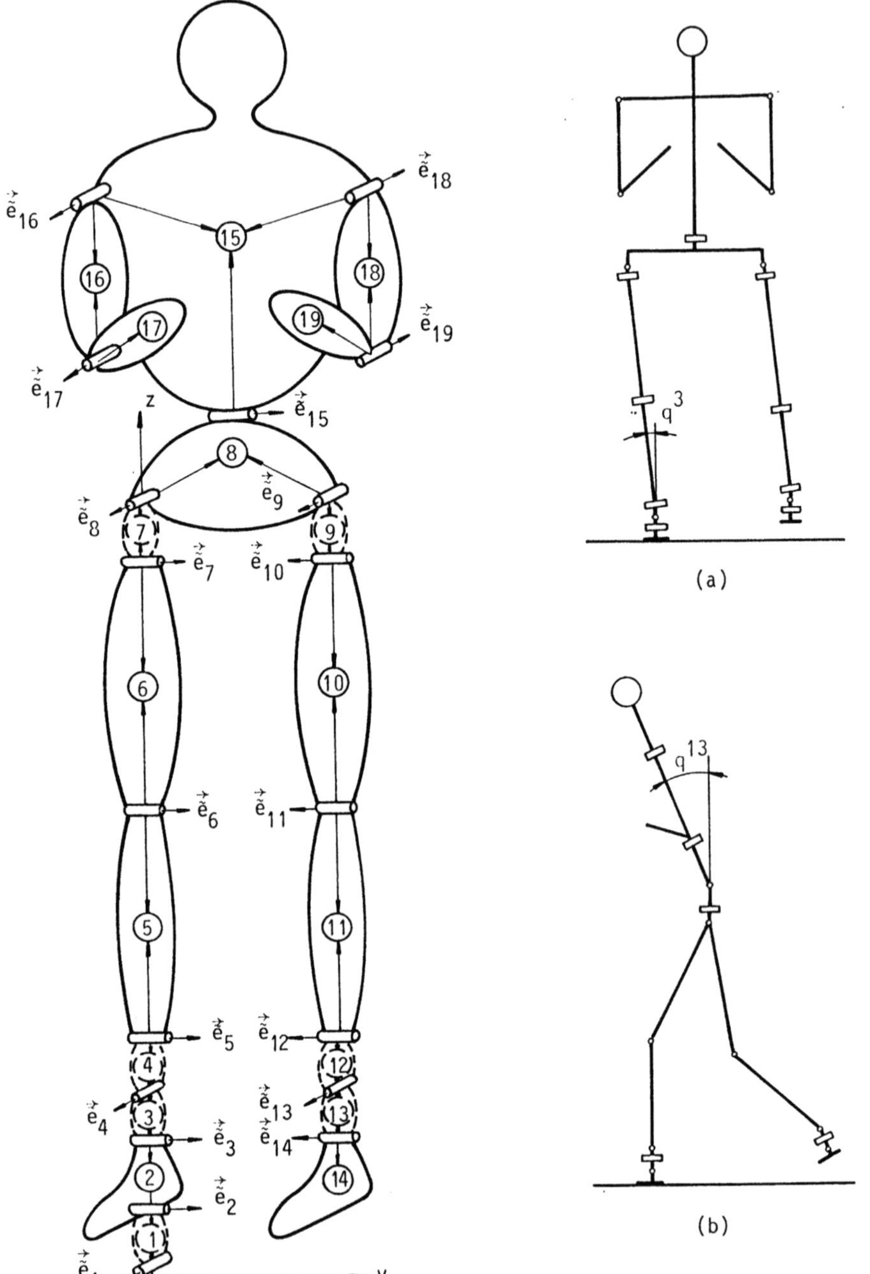



Fig. 2.48. Mechanical scheme of the anthropomorphic mechanism with fixed arms

Fig. 2.49. Compensating d.o.f. in the frontal (a) and sagittal (b) plane

Table 2.5. Kinematic and dynamic parameters of the mechanism

| Link | Mass [kg] | Moment of inertia $[kgm^2]$ | | | Distance of the axes centres of joints from the link centre [m] | Joint unit axes |
|---|---|---|---|---|---|---|
| | | $J_x$ | $J_y$ | $J_z$ | | |
| 1 | 2 | 3 | 4 | 5 | 6 | 7 |
| 1 | 0.0 | 0.0 | 0.0 | 0.0 | $\vec{\tilde{r}}_{1,1} = (0,\ 0,\ 0.0001)^T;\quad \vec{\tilde{r}}_{1,2} = (0,\ 0,\ -0.0001)^T$ | $\vec{\tilde{e}}_1 = (1,\ 0,\ 0)^T$ |
| 2 | 1.53 | 0.00006 | 0.00055 | 0.00045 | $\vec{\tilde{r}}_{2,2} = (0,\ 0,\ 0.030)^T;\quad \vec{\tilde{r}}_{2,3} = (0,\ 0,\ -0.070)^T$ | $\vec{\tilde{e}}_2 = (0,\ 1,\ 0)^T$ |
| 3 | 0.0 | 0.0 | 0.0 | 0.0 | $\vec{\tilde{r}}_{3,3} = (0,\ 0,\ 0.0001)^T;\quad \vec{\tilde{r}}_{3,4} = (0,\ 0,\ -0.0001)^T$ | $\vec{\tilde{e}}_3 = (0,\ 1,\ 0)^T$ |
| 4 | 0.0 | 0.0 | 0.0 | 0.0 | $\vec{\tilde{r}}_{4,4} = (0,\ 0,\ 0.0001)^T;\quad \vec{\tilde{r}}_{4,5} = (0,\ 0,\ -0.0001)^T$ | $\vec{\tilde{e}}_4 = (1,\ 0,\ 0)^T$ |
| 5 | 3.21 | 0.00393 | 0.00393 | 0.00038 | $\vec{\tilde{r}}_{5,5} = (0,\ 0,\ 0.210)^T;\quad \vec{\tilde{r}}_{5,6} = (0,\ 0,\ -0.210)^T$ | $\vec{\tilde{e}}_5 = (0,\ 1,\ 0)^T$ |
| 6 | 8.41 | 0.01120 | 0.01200 | 0.00300 | $\vec{\tilde{r}}_{6,6} = (0,\ 0,\ 0.220)^T;\quad \vec{\tilde{r}}_{6,7} = (0,\ 0,\ -0.220)^T$ | $\vec{\tilde{e}}_6 = (0,\ 1,\ 0)^T$ |
| 7 | 0.0 | 0.0 | 0.0 | 0.0 | $\vec{\tilde{r}}_{7,7} = (0,\ 0,\ 0.0001)^T;\quad \vec{\tilde{r}}_{7,8} = (0,\ 0,\ -0.0001)^T$ | $\vec{\tilde{e}}_7 = (0,\ 1,\ 0)^T$ |
| 8 | 6.96 | 0.00700 | 0.00565 | 0.00627 | $\vec{\tilde{r}}_{8,8} = (0,\ 0.135,\ 0.1)^T;\quad \vec{\tilde{r}}_{8,9} = (0,\ -0.135,\ 0.1)^T;$ $\vec{\tilde{r}}_{8,15} = (0,\ 0,\ -0.05)^T$ | $\vec{\tilde{e}}_8 = (1,\ 0,\ 0)^T$ |
| 9 | 0.0 | 0.0 | 0.0 | 0.0 | $\vec{\tilde{r}}_{9,9} = (0,\ 0,\ -0.0001)^T;\quad \vec{\tilde{r}}_{9,10} = (0,\ 0,\ 0.0001)^T$ | $\vec{\tilde{e}}_9 = (1,\ 0,\ 0)^T$ |
| 10 | 8.41 | 0.01120 | 0.01200 | 0.00300 | $\vec{\tilde{r}}_{10,10} = (0,\ 0,\ -0.220)^T;\quad \vec{\tilde{r}}_{10,11} = (0,\ 0,\ 0.220)^T$ | $\vec{\tilde{e}}_{10} = (0,\ -1,\ 0)^T$ |
| 11 | 3.21 | 0.00393 | 0.00393 | 0.00038 | $\vec{\tilde{r}}_{11,11} = (0,\ 0,\ -0.210)^T;\quad \vec{\tilde{r}}_{11,12} = (0,\ 0,\ 0.210)^T$ | $\vec{\tilde{e}}_{11} = (0,\ -1,\ 0)^T$ |

Table 2.5. Cont.

| 1 | 2 | 3 | 4 | 5 | 6 | 7 |
|---|---|---|---|---|---|---|
| 12 | 0.0 | 0.0 | 0.0 | 0.0 | $\vec{\tilde{r}}_{12,12} = (0,\,0,\,-0.0001)^{T}$; $\vec{\tilde{r}}_{12,13} = (0,\,0,\,0.0001)^{T}$ | $\vec{\tilde{e}}_{12} = (0,\,-1,\,0)^{T}$ |
| 13 | 0.0 | 0.0 | 0.0 | 0.0 | $\vec{\tilde{r}}_{13,13} = (0,\,0,\,-0.0001)^{T}$; $\vec{\tilde{r}}_{13,14} = (0,\,0,\,0.0001)^{T}$ | $\vec{\tilde{e}}_{13} = (1,\,0,\,0)^{T}$ |
| 14 | 1.53 | 0.00006 | 0.00055 | 0.00045 | $\vec{\tilde{r}}_{14,14} = (0,\,0,\,-0.070)^{T}$ | $\vec{\tilde{e}}_{14} = (0,\,-1,\,0)^{T}$ |
| 15 | 30.85 | 0.15140 | 0.13700 | 0.02830 | $\vec{\tilde{r}}_{15,15} = (0,\,0,\,0.34)^{T}$;  $\vec{\tilde{r}}_{15,16} = (0,\,0.2,\,-0.06)^{T}$ <br> $\vec{\tilde{r}}_{15,18} = (0,\,-0.2,\,-0.06)^{T}$ | $\vec{\tilde{e}}_{15} = (0,\,1,\,0)^{T}$ |
| 16 | 2.07 | 0.00200 | 0.00200 | 0.00022 | $\vec{\tilde{r}}_{16,16} = (0,\,0,\,-0.154)^{T}$; $\vec{\tilde{r}}_{16,17} = (0,\,0,\,0.154)^{T}$ | $\vec{\tilde{e}}_{16} = (1,\,0,\,0)^{T}$ |
| 17 | 1.14 | 0.00250 | 0.00425 | 0.00014 | $\vec{\tilde{r}}_{17,17} = (0,\,0,\,-0.132)^{T}$ | $\vec{\tilde{e}}_{17} = (1,\,0,\,0)^{T}$ |
| 18 | 2.07 | 0.00200 | 0.00200 | 0.00022 | $\vec{\tilde{r}}_{18,18} = (0,\,0,\,-0.154)^{T}$; $\vec{\tilde{r}}_{18,19} = (0,\,0,\,0.154)^{T}$ | $\vec{\tilde{e}}_{18} = (-1,\,0,\,0)^{T}$ |
| 19 | 1.14 | 0.00250 | 0.00425 | 0.00014 | $\vec{\tilde{r}}_{19,19} = (0,\,0,\,-0.132)^{T}$ | $\vec{\tilde{e}}_{19} = (-1,\,0,\,0)^{T}$ |

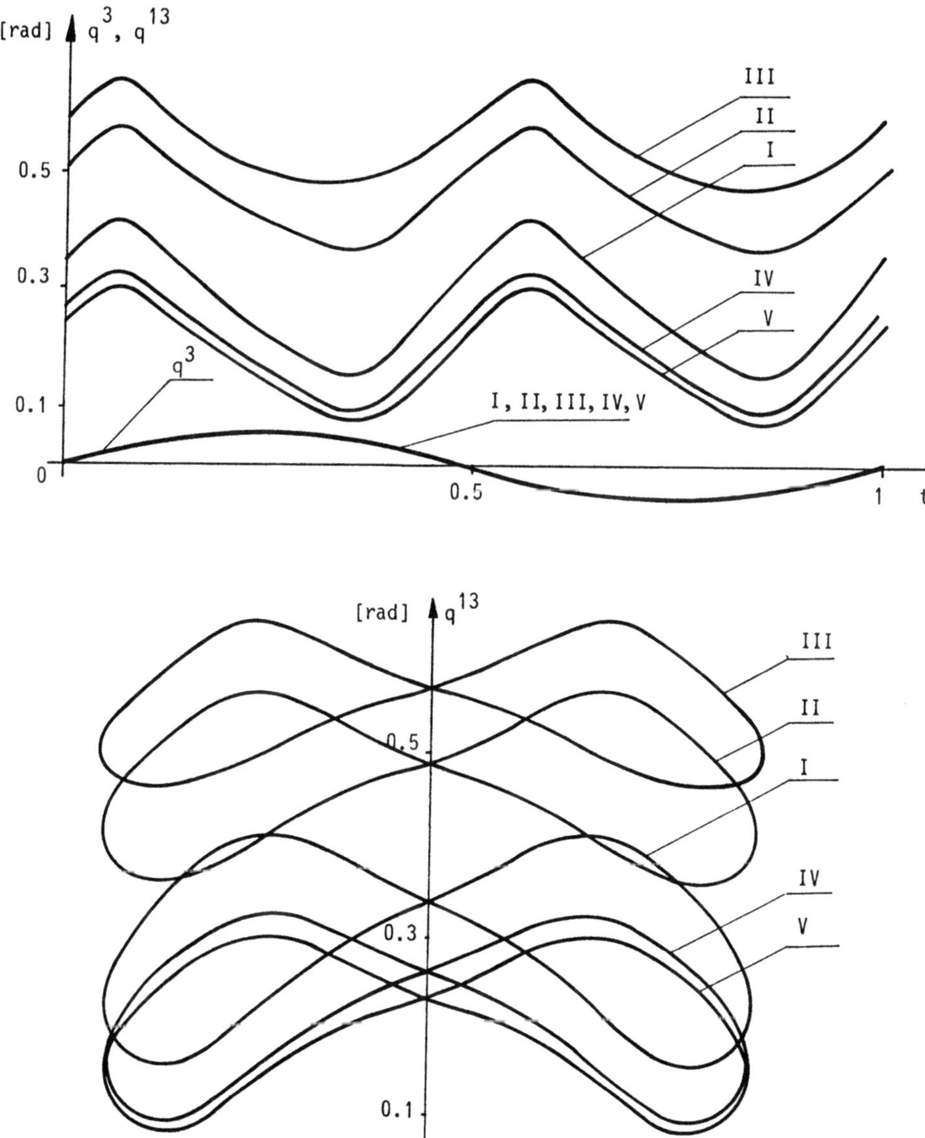



Fig. 2.50. Compensating movements for the single-support gait upon level ground with two-link foot for T=1.5, S=0.6 and different ZMP laws

# References

[1] Vukobratović M., Potkonjak V., _Dynamics of Manipulation Robots: Theory and Application_, Springer-Verlag, 1982.

[2] Vukobratović M., Stokić D., _Control of Manipulation Robots_, Springer-Verlag, 1982.

[3] Vukobratović M., Kirćanski N., _Real-Time Dynamics of Manipulation Robots_, Springer-Verlag, 1985.

[4] Vukobratović M., Juričić D., "Contribution to the Synthesis of Biped Gait", IEEE Trans. on Bio-Medical Engineering, Vol. 16, No 1, 1969.

[5] Juričić D., Vukobratović M., "Mathematical Modelling of a Bipedal Walking System", ASME Publ. 72-WA/BHF-13.

[6] Vukobratović M., Stepanenko Yu., "Mathematical Models of General Anthropomorphic Systems", Mathematical Biosciences, Vol. 17, pp. 215-258, 1973.

[7] Stepanenko Yu., _Dynamics of Spatial Mechanisms_, Research monograph, (in Russian), Belgrade Mathematical Institute, Belgrade, 1974.

[8] Vukobratović M., _Legged Locomotion Robots and Anthropomorphic Mechanisms_, research monograph, "Mihailo Pupin" Institute, Beograd, 1975 (in English), published in Russian, "Mir", Moscow, 1976, in Japanese, Nikkan Shimbun Ltd, Tokyo, 1975, also published in Chinese, Peking, 1983.

[9] Vukobratović M., Hristić D., Ćirić V., Zečević M., "Analysis of Energy Demand Distribution Within Anthropomorphic Systems", Trans. of the ASME, Series G, Journal of Dynamic Systems, Measurement and Control, Vol. 95, No 4, 1973.

[10] Vukobratović M., "How to Control Artificial Anthropomorphic Systems", IEEE Trans. on Systems, Man and Cybernetics, Vol. 3, No 5, 1973.

[11] Vukobratović M., "Notion of Algorithmic Control for One Class of Large-Scale Systems" (in Russian), Automatica and Remote Control, ANUSSR, No 7, 1975, Moscow.

[12] Vukobratović M., Stokić D., "One Engineering Concept of Dynamic Control of Manipulators", Trans. of the ASME, Journal of Dynamic Systems, Measurement and Control, Special Issue, Vol. 102, June, 1981.

# Appendix – Programme LOCDYN

The programme LOCDYN performs computation of nominal dynamics for locomotion mechanisms producing artificial anthropomorphic walk according to the algorithmic structures described in detail in Chapter 2.

The programme is writen in FORTRAN.

In the main programme an iterative procedure for nominal dynamics synthesis, which is desribed in detail in Section 2.6., is given. The global algorithmic structure of the main programme is of the following form :

1. Reading input data describing the mechanism's configuration and gait parameters.
2. Input initial vector state.
3. Integration of the system,
4. If repeatability conditions are satisfied, go to step 7., else go to step 5.
5. Determination of initial conditions increments.
6. Forming new initial conditions and go to step 3.
7. Computation and storing of the positions and velocities for compensating DOF-s.

A list of all subroutines together with their functions is given in Table A.1. and the hierarchical structure of the programme in Table A.2.

TABLE A.1.

---

LIST OF ALL SUBROUTINES WITH THEIR FUNCTIONS

---

Programme LOCDYN includes following subroutines :

| NAME | FUNCTION |
| --- | --- |
| INPUT | setting input data describing mechanical structure of the system and integration parameters |
| GMPRD | matrix multiplication |
| MINV | matrix inversion |
| NIODES | integration of the system of the N first-order differential equations |
| OUT | control of the integration process |
| OUT1 | printing of computed positions and velocities for compensating DOF |
| PERF | computation of performance index |
| SYSTEM | forming right-hand sides of the system of differential equations |
| MODEL | forms the mathematical model of the mechanism |
| SAI | interpolation of the prescribed dynamic |
| SUPP | presription of the ZMP law |
| CROSS | cross product of two vectors, one vector is the i-th row of 25x3 matrix and the other is the j-th column of 3x3 matrix |
| CROSSO | cross product of two vectors. First vector is 3x1 and the second is the i-th row of the matrix 25x3 |

TABLE A.1. (continued)

| | |
|---|---|
| CROSS1 | cross product of two vectors |
| CROSS2 | double cross product A x ( A x B ) |
| DOT | scalar product of two vectors<br>one vector is the i-th row of 25x3 matrix,<br>the other is the j-th column of 3x3 matrix |
| DOT0 | scalar product of the vector (3 x 1)<br>and the j-th column of the matrix<br>(3x3) |
| DOT1 | scalar product of two vectors |
| MOM12 | calculation of normal and parallel<br>component of the vector wrt longitudinal<br>axis |
| MOM2 | calculation of the coefficients B(i,j)<br>j=1,...,i and BO(i) for link i |
| ASSEM | assembly of kinematic pair P(i,k) |
| BRZUB | calculation of coefficients accompanying<br>angular and linear accelerations<br>BETA(i,j), j=1,...,KZ-1 and ALFA(i,j),<br>j=1,...,KZ-1 and coefficients<br>BETAO(i), ALFAO(i) dependent on<br>the velocities in the '+' joints array |
| MATH | calculation of inertial matrix H |
| MOMO | calculation of transformation matrix<br>for moments of inertia wrt fixed<br>coord. frame and calcultion of<br>velocity dependent component LAMBDA<br>( for the body ) |
| MOM10 | calculation of coefficients B(i,j),<br>BO(i) for the body |
| MOM11 | calculation of equivalent acceleration |

138

TABLE A.1. (continued)

| SPECIF | generation of assembly vector if corresponding vectors are colinear |
| TOTR | calculation of the position vector $R(i,j)$ $j=1,\ldots,i-1$ in the '+' joints array |
| TRANSF | calculation of transformation matrix Q |
| VEKH | calculation of vector HO |

TABLE A.2.

Hierachical structure of the programme LOCDYN

| LEVEL: | I | II | III | IV | V | VI |
|---|---|---|---|---|---|---|
| | GMPRD | | | | | |
| | INPUT | | | | | |
| | MINV | | | | | |
| | NIODES | | | | | |
| | | SYSTEM | | | | |
| | | | GMPRD | | | |
| | | | MINV | | | |
| | | | MODEL | | | |
| | | | | ASSEM | | |
| | | | | BRZUB | | |
| | | | | | CROSS0 | |
| | | | | | CROSS1 | |
| | | | | | CROSS2 | |
| | | | | | | DOT1 |
| | | | | MATH | | |
| | | | | | CROSS0 | |
| | | | | | CROSS1 | |
| | | | | MOMO | | |
| | | | | | DOT | |
| | | | | | GMPRD | |

TABLE A.2. (continued)

| LEVEL: I | II | III | IV | V | VI |
|---|---|---|---|---|---|
| | | | MOM10 | | |
| | | | MOM11 | | |
| | | | | CROSS1 | |
| | | | | DOT1 | |
| | | | MOM2 | | |
| | | | | MOM12 | |
| | | | SPECIF | | |
| | | | TOTR | | |
| | | | TRANSF | | |
| | | | | CROSS | |
| | | | | DOT | |
| | | | | GMPRD | |
| | | | VEKH | | |
| | | | | CROSS1 | |
| | | | | DOT1 | |
| | | SAI | | | |
| | | SUPP | | | |
| | OUT | | | | |
| | OUT1 | | | | |
| PERF | | | | | |

## DESCRIPTION OF SSP SUBROUTINES AND FILES USED

Programme LOCDYN uses the following standard subroutines .

### Subroutine NIODES

This subroutine integrates system of N first-order differential equations by ADAMS-MOULTON integration procedure with RUNGE-KUTTA starter.

#### Usage

CALL NIODES (SYSTEM,OUT,IFL,IER)

Parameter description :

SYSTEM - name of the subroutine forming right sides of differential equations

OUT  - name of the subroutine controlling integration process

IFL  - number of basic integration step halvings ( output value )

IER  - error parameter obtained from subroutine OUT, IER=1 means that the integration process was broken before the end of integration interval was reached

Description of common zone /ACOM/ :

COMMON /ACOM/ N,TO,TB,HA,POC(16),EPS(16),YY(16)


List desription :

N     - number of differential equations

TO,TB - integration limits,

HA    - starting integration step

POC   - initial state vector

EPS - vector of maximal intagration errors

YY   -   final state vector


### Subroutine MINV

This subroutine determines inverse matrix for the given matrix A (NxN) by standard Gauss-Jordan method.

### Usage

    CALL MINV (A, N, D, L , M)


Parameter description

A     - input matrix. Resulting (inverse) matrix is placed in the same place.

N     - order of matrix A

D        - determinant value

L        - working vector (1xN)

M        - working vector (1xN).


## Subroutine GMPRD

This subroutine performs multiplication of two matrices dimensioned NxL and LxM , respectively.


## Usage

CALL GMPRD (A, B, R, N, M, L)


## Parameter description

A        - input matrix dimension NxL

B        - input matrix dimension LxM

R        - output matrix dimension NxM ( product A.B )

N        - number of rows in matrix A

M        - number of columns in matrix B

L        - number of columns in matrix A and number of

         rows in matrix B


The programme LOCDYN uses files INDAT.DAT , PREDYN.DAT and OUTHEAD.DAT, as well as output file OUTFIL.DAT.

Input file INDAT.DAT contains data describing system configuration and data desribing integration parameters in the followng order :

1. Coordinate frame orientation ( DESNI ).

2. Number of kinematic chains consisting the mechanism

   ( NL ).

3. For every kinematic chain :

   3.1. ordinal number of the basic link ( IP )

   3.2. number of links within the chain ( IK )

4. System's mechanical structure matrix M. The element $M(i,j)$
   is the j-th joint in the i-th kinematic chain.

5. Total number of DOF consisting the mechanism ( NUZ )

6. For every link :

   6. 1. ordinal number ( I )

   6. 2. link type indicator

        ( KSI1=1-cane, KSI1=0-body),

   6. 3. branching indicator

        ( KSI2=1-branching link )

   6. 4. compensating dynamic indicator

        ( KSI3=1-compensating dynamic )

   6. 5. viscous friction indicator

        ( KSI4=1-unpowred DOF with moments only

          due to viscous friction )

   6. 6. indicator of prescribed moments within closed

        knematic   chain

        ( KSI5=1-prescribed moments within closed chain )

   6. 7. specificity indicator

        ( KSI6=i - ordinal number of joint with
        specificity, KSI6=0-specificity does not exist )

   6. 8. link mass ( MM )

   6. 9. moments of inertia for cane wrt orthogonal and
        longitudinal axis, for body wrt X,Y and Z axes
        ( JN,JS and JX,JY,JZ respectively )

   6.10. viscous friction coefficient for the link with
        moments due to viscous friction ( PRİG )

   6.11. the projection of rotational axis unit vector
        in the local coordinate frame ( EU )

   6.12. the distance vector from the i-th joint centre
        to the mass centre of link $\underline{i}$ ( $ROU_{ii}$ )

   6.13. the distance vector from thr centre of joint
        $\underline{k}$ to the mass centre of the link $\underline{i}$ at the

       kinematic pair $P_{ik}$ ( $ROU_{i,i+1}$ )

  6.14. the distance vector from the centre of joint $\underline{1}$ to the mass centre of the link $\underline{i}$ of the kinematic pair $P_{il}$ for branching link ( ROUC )

7. the "home" position of the first link in the absolute coordinate frame ( RI )

8. unit vector of the first joint axis in the absolute coordinate frame ( E )

9. For evrery joint with specificities, projections of the substitution vectors for corresponding kinematic pair ( RAS )

10. Integration parameters:

  10.1. maximal value of performance index for which initial conditions corrections are performed by local criteria ( COMP )

  10.2. maximal error for repeatability conditions ( EPSS1 )

  10.3. noramlization coefficient ( GM )

  10.4. initial time ( TO )

  10.5. terminal time ( TB )

  10.6. repeatability conditions matrix for compensating DOF ( W )

  10.7. maximal integration error ( EPS )

  10.8. initial vector state ( ZS )


The file PREDYN.DAT contains following data :


1. Number of discretization points ( NDIS )

2. Integration step ( DELT )

3. Step duration ( PER )

4. Step length parameter ( S )

5. For every d.o.f with prescribed dynamic : ordinal number (I), position (SA(K,I,1)), velocity ( SA(K,I,2),) and acceleration (SA(K,I,3))   where K=1,NDIS

The file OUTHEAD.DAT contains the header of output file.

The output file OUTFIL.DAT contains computed values of positions and velocities for compensating degrees of fredom.

In this Appendix, the listing of the programme LOCDYN, as well as the listings of the input and output files for  Example  1.  from Chapter 2., are given.

For examples 2. and 3. it is necessary to add statements which realize following relations :   $q^7 := q^7 - q^3$ , $\dot{q}^7 := \dot{q}^7 - \dot{q}^3$ ,

$q^8 := q^8 + q^3$ , $\dot{q}^8 := \dot{q}^8 + \dot{q}^3$ .

In order to realize Example 4., corresponding subroutines for double-support phase should be added to the existing programme.

The difference between the existing programme and programme which computes nominal dynamic for Example 5. is only in the subroutine SUPP which should be corrected in such a way to enable double-link feet modelling.

Listing of the programme LOCDYN and accompanying subroutines

```
C  *****************************************************************
C  *  *********************************************************** *
C  *  *                                                        * *
C  *  *     MAIN PROGRAM FOR FUNCTIONAL MOVEMENTS SYNTHESIS     * *
C  *  *     OF MECHANISM PRODUCING ARTIFICIAL ANTHROPOMORPHIC   * *
C  *  *                        WALK                             * *
C  *  *********************************************************** *
C  *****************************************************************
C
C-----------------------------------------------------------------
C
C      USED FILES
C
C      INPUT FILES :
C
C          PREDYN        - FILE CONTAINING PRESCRIBED DYNAMICS AND
C                          GAIT PARAMETERS
C          INDAT         - FILE CONTAINIG MECHANICAL CONFIGURATION
C                          OF THE SYSTEM AND INTEGRATION PARAMETERS
C          OUTHEAD       - FILE CONTAINING OUTPUT TITLE
C
C      OUTPUT FILES :
C
C          OUTFIL        - FILE CONTAINING COMPUTED DYNAMICS
C-----------------------------------------------------------------
C
C      CALLED SUBROUTINES
C
C              INPUT     reading mechanical cofiguration parameters,
C                        step parameters, integration parameters and
C                        prescrybed dynamics
C              NIODES    integration of the system
C              PERF      performance index computation
C              MINV      matrix inversion
C              GMPRD     matrix multiplication
C-----------------------------------------------------------------
C
       EXTERNAL SYSTEM,OUT,OUT1
C
       DIMENSION ZS(16),XS(16),B(16),W(16),A(256),LP(16),MP(16),G(16)
       DIMENSION JOINCM(25)
C
       COMMON /ACOM/ N,T0,TB,HA,ZN(16),EPS(16),DEP(16)
       COMMON /ZSIN/ SA(41,25,3),DELT,PER,S
       COMMON /SYS/ DESNI,NL,IP(3),NPL,JM
       COMMON /MIN/ MM,JS,JN,JX,JY,JZ
       COMMON /RAC/ IR,IW,IT,INDEX,IOUT
       COMMON /TIPL/ KSI1(25),KSI2(25),KSI3(25),RAS(5,2,3),KSI6(25)
      1,KSI4(25),KSI5(25)
       COMMON /UG/ SI(25),SIDOT(25),DSIDOT(25)
       COMMON /STRUKT/ JL,NUZ,NPZ(3,20),EU(25,3),ROU(49,3),IK(3)
       COMMON /ERIC/ EPOC(3),R0POC(3),R0UC(3,2,3)
       COMMON /VM/ VM(3),OMEGM(3)
       COMMON /PRIGUS/ PRIG(25)
       COMMON /HAHA/ H(25,25),H0(25),INTDM1,INTDM2
C
       REAL MM(25),JS(25),JN(25),JX(25),JY(25),JZ(25)
       INTEGER DESNI
       CHARACTER*80 TITLE(10)
       CHARACTER*10 POSIT,VEL,POSITJ,VELJ
       CHARACTER*8 DOFC
```

```fortran
C-----------------------------------------------------------------
C       FIXED INTEGRATION PARAMETERS
C-----------------------------------------------------------------
        DATA JB, JNN, STEP, DELTA, JNMX / 0, 0, 1., 0.005, 10/
        DATA POSIT,VEL,POSITJ,VELJ /'  POSITION','  VELOCITY',
       +                           '  (RAD) ',' (RAD/SEC)'/
        DATA DOFC /'   DOF  '/
C
100     FORMAT(I5)
110     FORMAT(8F10.0)
120     FORMAT('   TOO FAR FROM SOLUTION.CHOOSE ANOTHER VALUES',
       +          /,' DE1=',F15.6)
130     FORMAT(10X,'ITERATION',I6)
140     FORMAT(5X,' INCREMENT'/(5X,E13.3))
150     FORMAT(5X,'INCREMENT NORMALIZATION'/(5X,E13.3))
151     FORMAT(5X,'NEW INITIAL VALUES'/(5X,E15.5))
152     FORMAT(5X,'TERMINAL STATE VECTOR'/(5X,E15.5))
153     FORMAT(5X,'NUMBER OF HALVINGS',I5)
154     FORMAT(5X,'PERFORMANCE INDEX',F15.5)
160     FORMAT(10X,' NO CONVRGENCE, STEP TOO SMALL')
170     FORMAT(10X,'SLOW CONVERGENCE')
180     FORMAT(10X,'SOLUTION NOT OBTAINED WITH',I4,' ITERATIONS')
5100    FORMAT(' INITIAL VALUES',5(1X,F10.4))
13456   FORMAT(A80)
19001   FORMAT('1',A80,/)
19002   FORMAT(1X,'TIME ',6X,8(A10,3X))
19003   FORMAT(12X,8(A8,I2,3X))
19004   FORMAT(1X,'(SEC)',6X,8(A10,3X))
19005   FORMAT(1X,' ZMP LAW = ',I2,', STEP LENGTH = ',F4.2,
       +            ', HALF STEP DURATION = ',F4.2,///)
19006   FORMAT(' ',80('-')/)
19009   FORMAT(' ',A80)
C
C-----------------------------------------------------------------
C    DE1 - MAXIMAL INITIAL PERFORMANCE INDEX
C-----------------------------------------------------------------
C
        DE1 = 10.
C
C-----------------------------------------------------------------
C     OPEN  FILES
C-----------------------------------------------------------------
        IW = 9
        IR = 7
        INDEX = 8
        IOUT = 11
        JEDN = 10
        OPEN (unit=INDEX,FILE='PREDYN',STATUS='OLD')
        OPEN (unit=IR,FILE='INDAT',status='old')
        OPEN (unit=IW,FILE='OUTFIL',status='new')
        OPEN (unit=JEDN,FILE='OUTHEAD',status='old')
C
C-----------------------------------------------------------------
C     READ OUTPUT TITLE
C-----------------------------------------------------------------
        READ(JEDN,100) LINES
        DO 39777 KLINES = 1,LINES
39777   READ(JEDN,13456) TITLE(KLINES)
C
C-----------------------------------------------------------------
C     INPUT DATA DESCRIBING MECHANICAL CONFIGURATION,
C     GAIT PARAMETERS AND PRESCRIBED DYNAMICS
C-----------------------------------------------------------------
        CALL INPUT
C-----------------------------------------------------------------
```

```fortran
C        HA   -   INTEGRATION STEP
C---------------------------------------------------------------
         HA = DELT
C
C---------------------------------------------------------------
C        COMP   - IF DE>COMP, CORRECTION OF INITIAL CONDIDITONS IS
C                   PERFORMED BY GLOBAL CRITERIA, ELSE BY LOCAL CRITERIA
C        EPSS1 - MAXIMAL ERROR FOR REPEATABILITY CONDITIONS
C        GM     - NORMALIZATION COEFFICIENT
C---------------------------------------------------------------
         READ(IR,110) COMP
         READ(IR,110) EPSS1
         READ(IR,110) GM
C
C---------------------------------------------------------------
C        NPL - NUMBER OF D.O.F FOR COMPENSATION DYNAMIC
C        JM  - ZMP LAW
C---------------------------------------------------------------
         READ(IR,100) NPL,JM
         N=2*NPL
C
C---------------------------------------------------------------
C        INTEGRATION PARAMETERS
C        T0       - INITIAL TIME
C        TB       - TERMINAL TIME
C        W        - REPEATABILITY CONDITIONS
C        EPS      - MAXIMAL INTEGRATION ERROR (OUTPUT CRITERIA)
C---------------------------------------------------------------
         READ(IR,110) T0,TB
         READ(IR,110) (W(J),J=1,N)
         READ(IR,110) (EPS(J),J=1,N)
C
C---------------------------------------------------------------
C        PRINT OUT HEADER
C---------------------------------------------------------------
         WRITE(IW,19001) TITLE(1)
         IF (LINES-1) 18501,18501,18500
18500 DO 39778 KLINES = 2,LINES
39778 WRITE(IW,19009) TITLE(KLINES)
         NCOMPD = 0
18501 DO 39779 JOINTS = 1,NUZ
         IF (KSI3(JOINTS).NE.1) GO TO 39779
         NCOMPD = NCOMPD + 1
         JOINCM (NCOMPD) = JOINTS
39779  CONTINUE
C
         WRITE(IW,19005) JM,S,PER
         WRITE(IW,19002) ((POSIT), KJOIN=1,NCOMPD),
     +                   ((VEL), KJOIN=1,NCOMPD)
         WRITE(IW,19003) ((DOFC,JOINCM(KJOIN)), KJOIN = 1,NCOMPD),
     +                   ((DOFC,JOINCM(KJOIN)), KJOIN = 1,NCOMPD)
         WRITE(IW,19004) ((POSITJ), KJOIN=1,NCOMPD),
     +                   ((VELJ), KJOIN=1,NCOMPD)
         WRITE(IW,19006)
C---------------------------------------------------------------
C        INITIAL STATE VECTOR
C---------------------------------------------------------------
500      READ(IR,110) (ZN(J),J=1,N)
         DO 510 J=1,N
510      ZS(J)=ZN(J)
C
C---------------------------------------------------------------
C        INTEGRATION IN ORDER TO DETERMINE PERFORMANCE INDEX
C---------------------------------------------------------------
         CALL NIODES(SYSTEM,OUT,IFL,IER)
         IF(IER.EQ.1) GO TO 500
```

```
C
C-------------------------------------------------------------------
C       PERFORMANCE INDEX DETERMINATION
C-------------------------------------------------------------------
        CALL PERF(ZN,DEP,W,N,DE)
        IF(DE.LE.DE1) GO TO 520
        WRITE(IW,120) DE1
        GO TO 500
520     PR=DE
        IF(DE.LE.EPSS1) GO TO 710
C
C-------------------------------------------------------------------
C       FORMING THE SENSITIVITY MATRIX
C-------------------------------------------------------------------
        DO 530 J=1,N
        XS(J)=DEP(J)
530     B(J)=XS(J)-W(J)*ZS(J)
        K=0
        DO 560 JJ=1,N
        DO 540 J=1,N
540     ZN(J)=ZS(J)
        ZN(JJ)=ZN(JJ)+DELTA
C
C-------------------------------------------------------------------
C       INTEGRATION OF THE SYSTEM
C-------------------------------------------------------------------
        CALL NIODES(SYSTEM,OUT,IFL,IER)
        DO 550 J=1,N
        K=K+1
550     A(K)=(DEP(J)-XS(J))/DELTA
560     CONTINUE
        DO 570 J=1,N
        K=(J-1)*N+J
570     A(K)=A(K)-W(J)
        IF(DE.GT.COMP) GO TO 580
C
C-------------------------------------------------------------------
C       COMPUTE INCREMENT BY LOCAL CRITERIA
C-------------------------------------------------------------------
        CALL MINV(A,N,DP,LP,MP)
        CALL GMPRD(A,B,G,N,N,1)
        GO TO 610
C
C-------------------------------------------------------------------
C       COMPUTE INCREMENT BY GLOBAL CRITERIA
C-------------------------------------------------------------------
580     DO 600 J=1,N
        G(J)=0.
        DO 590 K=1,N
        KJ=(J-1)*N+K
590     G(J)=G(J)+A(KJ)*B(K)
600     G(J)=G(J)/DE
610     JB=JB+1
        IF(JB.GT.99) GO TO 710
C
C-------------------------------------------------------------------
C       NORMALIZATION
C-------------------------------------------------------------------
        GMX=ABS(G(1))
        DO 620 J=2,N
        IF(ABS(G(J)).GT.GMX) GMX=ABS(G(J))
620     CONTINUE
        SUM=GM*GMX
        IF(SUM.LE.1.) GO TO 640
        DO 630 J=1,N
630     G(J)=G(J)/SUM
```

```fortran
640     JNN=JNN+1
        DO 650 J=1,N
650     ZN(J)=ZS(J)-STEP*G(J)
C
C------------------------------------------------------------------------
C       INTEGRATION OF THE SYSTEM
C------------------------------------------------------------------------
        CALL NIODES(SYSTEM,OUT,IFL,IER)
        CALL PERF(ZN,DEP,W,N,DE)
        IF(DE.GT.PR) GO TO 670
        IF((PR-DE).LE.0.001.AND.DE.GT.1.) GO TO 500
        IF(DE.LE.EPSS1) GO TO 710
        JNN=0
        STEP=1.
        DO 660 J=1,N
660     ZS(J)=ZN(J)
        GO TO 520
670     STEP=0.5*STEP
        IF(JNN-JNMX) 640,640,680
680     WRITE(IW,160)
        GO TO 500
710     CONTINUE
C
C------------------------------------------------------------------------
C       COMPUTE AND PRRINT OUT NOMINAL DYNAMIC
C------------------------------------------------------------------------
        CALL NIODES(SYSTEM,OUT1,IFL,IER)
        CALL OUT1(TB,DEP,IER)
C
C------------------------------------------------------------------------
C       CLOSE FILES
C------------------------------------------------------------------------
        CLOSE(IR)
        CLOSE(INDEX)
        CLOSE(IW)
        CLOSE(JEDN)
C
C------------------------------------------------------------------------
C       FINISHED COMPUTATION
C------------------------------------------------------------------------
        STOP 'NOMINAL DYNAMIC COMPUTED'
        END
C
C**********************************************************************
C       SUBROUTINE : PERF                                            *
C------------------------------------------------------------------- *
C       FUNCTION:  CALCULATES PERFORMANCE INDEX                      *
C------------------------------------------------------------------- *
C       INPUT VARIABLES:                                            *
C               ZN          INITIAL STATE VECTOR                     *
C               W           DIAGONAL ELEMENTS OF REPEATABILITY MATRIX *
C               DEP         TERMINAL STATE VECTOR                     *
C------------------------------------------------------------------- *
C       OUTPUT VARIABLES:                                           *
C               DE          PERFORMANCE INDEX                        *
C------------------------------------------------------------------- *
C       SUBROUTINES CALLED : NONE                                   *
C**********************************************************************
        SUBROUTINE PERF(ZN,DEP,W,N,DE)
        COMMON /RAC/ IR,IW,IT,INDEX,IOUT
        DIMENSION ZN(1),DEP(1),W(1)
        DE=0.
        DO 10 J=1,N
        DE=DE+(DEP(J)-W(J)*ZN(J))**2
10      CONTINUE
        DE=SQRT(DE)
```

```
      RETURN
      END
C
C****************************************************************
C     SUBROUTINE : SAI                                          *
C---------------------------------------------------------------*
C     FUNCTION :  LINEAR INTERPOLATION OF PRESCRIBED DYNAMICS    *
C---------------------------------------------------------------*
C     INPUT VARIABLES:                                          *
C         J          ORDINAL NUMBER OF DISCRETIZATION POINT     *
C         I          ORDINAL NUMBER OF D.O.F.                   *
C         K          POSITION     ( K = 1)                      *
C                    VELOCITY     ( K = 2)                      *
C                    ACCELERATION ( K = 3)                      *
C         XXT        CURRENT INTEGRATION TIME                   *
C---------------------------------------------------------------*
C     OUTPUT VARIABLES:                                         *
C         SAI        INTERPOLATED VALUE                         *
C---------------------------------------------------------------*
C     SUBROUTINES CALLED:  NONE                                 *
C****************************************************************
      FUNCTION SAI(J,I,K,XXT)
      COMMON /ZSIN/ SA(41,25,3),DELT,PER,S
      IF(J.EQ.41) GO TO 10
      SAI=(SA(J+1,I,K)-SA(J,I,K))*XXT/DELT+SA(J,I,K)
      RETURN
10    SAI=SA(J,I,K)
      RETURN
      END
C
C****************************************************************
C     SUBROUTINE : SYSTEM                                       *
C---------------------------------------------------------------*
C     FUNCTION:  FORMING RIGHT SIDES OF DIFFERENTIAL EQUATIONS   *
C                FOR UNPRESCRIBED DYNAMIC                        *
C---------------------------------------------------------------*
C     INPUT VARIABLES:                                          *
C         XX         TIME                                       *
C         PD         STATE VECTOR OF UNPRESCRIBED PART OF SYSTEM *
C---------------------------------------------------------------*
C     OUTPUT VARIABLES:                                         *
C         DV         VECTOR OF THE RIGHT SIDES OF DIFFERENTIAL   *
C                    EQUATIONS                                  *
C---------------------------------------------------------------*
C      SUBROUTINES CALLED:                                      *
C      MODEL, SUPP, MINV, GMPRD                                 *
C****************************************************************
      SUBROUTINE SYSTEM(XX,DV,PD)
      DIMENSION DV(1),PD(1),EH(3,25),EH0(3),HKA(25)
      COMMON /ZSIN/ SA(41,25,3),DELT,PER,S
      COMMON /SYS/ DESNI,NL,IP(3),NPL,JM
      COMMON /MIN/ MM,JS,JN,JX,JY,JZ
      COMMON /RAC/ IR,IW,IT,INDEX,IOUT
      COMMON /TIPL/ KSI1(25),KSI2(25),KSI3(25),RAS(5,2,3),KSI6(25)
     1,KSI4(25),KSI5(25)
      COMMON /UG/ SI(25),SIDOT(25),DSIDOT(25)
      COMMON /STRUKT/ JL,NUZ,NPZ(3,20),EU(25,3),R0U(49,3),IK(3)
      COMMON /ERIC/ EPOC(3),R0POC(3),R0UC(3,2,3)
      COMMON /VM/ VM(3),OMEGM(3)
      COMMON /PRIGUS/ PRIG(25)
      COMMON /HAHA/ H,H0,INTDM1,INTDM2
      REAL MM(25),JS(25),JN(25),JX(25),JY(25),JZ(25)
      INTEGER DESNI
      DIMENSION LV(25),MV(25),R(25,3)
      DIMENSION H(25,25),H0(25),HK(25),A(150)
C-------------------------------------------------------------
```

```
C       INTERPOLATION OF PRESCRIBED DYNAMICS FOR XX
C--------------------------------------------------------------------
        XXJ=XX/DELT
        J=INT(XXJ)+1
        XXTEK=XX-DELT*J
        K=0
        DO 110 I=1,NUZ
        IF(KSI3(I)) 100,105,100
100     K=K+1
        SI(I)=DV(K)
        SIDOT(I)=DV(NPL+K)
        PD(K)=DV(NPL+K)
        GO TO 110
105     SI(I)=SAI(J,I,1,XXTEK)
        SIDOT(I)=SAI(J,I,2,XXTEK)
        DSIDOT(I)=SAI(J,I,3,XXTEK)
110     CONTINUE
C
C--------------------------------------------------------------------
C       ZMP LAW
C--------------------------------------------------------------------
        CALL SUPP(XX,JM)
C
C--------------------------------------------------------------------
C       FORMING THE MATHEMATICAL MODEL
C--------------------------------------------------------------------
        CALL MODEL(H,H0,EH0,EH,XX)
C
C--------------------------------------------------------------------
C       EXTRACTION OF UNPRESCRIBED PART OF DYNAMICS RELATED TO ZMP
C--------------------------------------------------------------------
        DO 760 ILI=1,NPL
760     HK(ILI)=0.
        K=0
        DO 145 I=1,NUZ
        IF(KSI4(I)) 120,145,120
120     K=K+1
        L=0
        HK(K+2)=H0(I)
        DO 140 J=1,NUZ
        IF(KSI3(J)) 125,130,125
125     L=L+1
        IA=K+NPL*(L-1)+2
        A(IA)=H(I,J)
        GO TO 140
130     HK(K+2)=HK(K+2)+H(I,J)*DSIDOT(J)
140     CONTINUE
        HK(K+2)=-HK(K+2)+PRIG(I)*SIDOT(I)
145     CONTINUE
        K=0
        DO 180 J=1,NUZ
        IF(KSI3(J)) 185,190,185
185     K=K+1
        IA1=(K-1)*NPL+1
        A(IA1)=EH(1,J)
        A(IA1+1)=EH(2,J)
        GO TO 180
190     HK(1)=HK(1)+EH(1,J)*DSIDOT(J)
        HK(2)=HK(2)+EH(2,J)*DSIDOT(J)
180     CONTINUE
        HK(1)=-HK(1)-EH0(1)
        HK(2)=-HK(2)-EH0(2)
181     CONTINUE
        CALL MINV(A,NPL,DETT,LV,MV)
        CALL GMPRD(A,HK,HKA,NPL,NPL,1)
        DO 155 I=1,NPL
```

```fortran
155     PD(NPL+I)=HKA(I)
        RETURN
        END
C
C***********************************************************************
C       SUBROUTINE : INPUT                                            *
C---------------------------------------------------------------------*
C       FUNCTION :  INPUT DATA DESCRIBING MECHANICAL CONFIGURATION    *
C                   AND PRESCRIBED DYNAMICS OF THE MECHANISM          *
C---------------------------------------------------------------------*
C       INPUT VARIABLES:                                              *
C            IR        LRN OF THE FILE INDAT                          *
C            INDEX     LRN OF THE FILE PREDYN                         *
C---------------------------------------------------------------------*
C       OUTPUT VARIABLES:                                             *
C           ALL VARIABLES DESCRIBING MECHANICAL CONFIGURATION AND     *
C           VECTOR SA(M,N,L) CONTAINING PRESCRIBED SYNERGY            *
C---------------------------------------------------------------------*
C       SUBROUTINES CALLED:   NONE                                    *
C***********************************************************************
        SUBROUTINE INPUT
        REAL MM(25),JS(25),JN(25),JX(25),JY(25),JZ(25)
        INTEGER DESNI
        COMMON /SYS/ DESNI,NL,IP(3),NPL,JM
        COMMON /RAC/ IR,IW,IT,INDEX,IOUT
        COMMON /MIN/ MM,JS,JN,JX,JY,JZ
        COMMON /TIPL/ KSI1(25),KSI2(25),KSI3(25),RAS(5,2,3),KSI6(25)
       1,KSI4(25),KSI5(25)
        COMMON /STRUKT/ JL,NUZ,NPZ(3,20),EU(25,3),ROU(49,3),IK(3)
        COMMON /UG/ SI(25),SIDOT(25),DSIDOT(25)
        COMMON /ERIC/ E(3),RI(3),ROUC(3,2,3)
        COMMON /VM/ VM(3),OMEGM(3)
        COMMON /PRIGUS/ PRIG(25)
        COMMON /ZSIN/ SA(41,25,3),DELT,PER,S
        CHARACTER*5 ORCSYS(2)
        DATA ORCSYS /'RIGHT','LEFT'/
C
100     FORMAT(16I5)
110     FORMAT(8F10.0)
4386    FORMAT(I2)
84301   FORMAT(I5)
84302   FORMAT(3F10.6)
84303   FORMAT(3E20.6)
C--------------------------------------------------------------------
C       INPUT VARIABLES EQUAL ZERO
C--------------------------------------------------------------------
        DO 231 I=1,25
        SI(I)=0.
        SIDOT(I)=0.
        PRIG(I)=0.
        DO 231 J=1,3
231     EU(I,J)=0.
        DO 232 I=1,49
        DO 232 J=1,3
232     ROU(I,J)=0.
        DO 233 I=1,3
        DO 233 J=1,2
        DO 233 K=1,3
233     ROUC(I,J,K)=0.
C
C--------------------------------------------------------------------
C       COORDINATE FRAME ORIENTATION
C--------------------------------------------------------------------
        READ(IR,100) DESNI
        IOR = DESNI
        IF (IOR.NE.1) IOR = 2
```

```fortran
C
C------------------------------------------------------------------------
C       NUMBER OF '+' JOINTS ARRAYS
C------------------------------------------------------------------------
        READ(IR,100) NL
C
C------------------------------------------------------------------------
C       ORDINAL NUMBER OF THE BASIC SEGMENT WITHIN THE CHAIN
C------------------------------------------------------------------------
        READ(IR,100) (IP(I),I=1,NL)
C
C------------------------------------------------------------------------
C       NUMBER OF JOINTS IN THE I-TH ARRAY OF '+' JOINTS
C------------------------------------------------------------------------
        READ(IR,100) (IK(I),I=1,NL)
C
C------------------------------------------------------------------------
C       MATRIX DESCRIBING CONFIGURATION OF THE MECHANISM
C------------------------------------------------------------------------
        DO 200 I=1,NL
        NZ=IK(I)
        READ(IR,100) (NPZ(I,K),K=1,NZ)
200     CONTINUE
        READ(IR,100) NUZ
C
C------------------------------------------------------------------------
C           DATA DESCRIBING PARTICULAR SEGMENT
C           I        - ORDINAL NUMBER OF SEGMENT
C           KSI1     - TYPE OF SEGMENT:
C                      KSI1 = 1 - CANE
C                      KSI1 = 0 - BODY
C           KSI2     - BRANCHING SEGMENT( KSI2 = 1)
C           KSI3     - COMPENSATION DYNAMIC( KSI3 = 1 )
C           KSI4     - UNPOWERED JOINTS WITH THE MOMENTS
C                      DUE ONLY TO VISCOUS FRICTION ( KSI4 = 1 )
C           KSI5     - PRESCRIBED TORQUES IN THE CLOSED CHAIN
C                      ( KSI5 = 1 )
C           KSI6     - SPECIFICITY (KSI6 = ORDINAL NUMBER DUE SPECIFICITY)
C           MM       - SEGMENT MASS
C           JN       - INERTIAL MOMENT WRT ORTHOGONAL AXIS
C           JS       - INERTIAL MOMENT WRT LONGITUDINAL AXIS
C           JX       - INERTIAL MOMENT FOR BODY WRT X AXIS
C           JY       - INERTIAL MOMENT FOR BODY WRT Y AXIS
C           JZ       - INERTIAL MOMENT FOR BODY WRT Z AXIS
C           PRIG     - COEFFICIENT OF VISCOSITY FOR KSI4=1
C------------------------------------------------------------------------
        DO 230 II=1,NUZ
        READ(IR,100) I,KSI1(I),KSI2(I),KSI3(I),KSI4(I),KSI5(I),KSI6(I)
        IF(KSI1(I)) 210,220,210
210     READ(IR,110) MM(I),JN(I),JS(I)
        IF(KSI4(I).EQ.1) READ(IR,110) PRIG(I)
115     FORMAT(' ',1X,7(I2,1X),7(F10.6,1X))
        GO TO 230
220     READ(IR,110) MM(I),JX(I),JY(I),JZ(I)
        IF(KSI4(I).EQ.1) READ(IR,110) PRIG(I)
230     CONTINUE
C
        IP1=IP(1)
        IK1=IK(1)
        DO 250 L=1,NL
        DO 240 J=IP1,IK1
        I=NPZ(L,J)
C
C------------------------------------------------------------------------
C           PROJECTIONS OF UNIT VECTOR AXES OF JOINTS
C           IN LOCAL COORDINATE FRAME
```

```
C---------------------------------------------------------------
      READ(IR,110) (EU(I,K),K=1,3)
C
C---------------------------------------------------------------
C         PROJECTIONS OF VECTORS FROM THE JOINT CENTRE
C         TO THE CENTRE OF MASS IN THE KINEMATIC PAIR R (I,I)
C---------------------------------------------------------------
      READ(IR,110) (R0U(I,K),K=1,3)
      IF(J.EQ.IK1) GO TO 240
      IN=I+NUZ
C
C---------------------------------------------------------------
C         PROJECTIONS OF VECTORS FROM THE JOINT CENTRE
C         TO THE CENTRE OF MASS IN THE KINEMATIC PAIR R (I,I+1)
C---------------------------------------------------------------
      READ(IR,110) (R0U(IN,K),K=1,3)
      IF(L.EQ.NL) GO TO 240
      IP2=IP(L+1)
      IF(I.EQ.NPZ(L+1,IP2)) GO TO 241
      GO TO 240
241   CONTINUE
      DO 235 K1=1,2
C
C---------------------------------------------------------------
C         DISTANCE FROM CENTER OF MASS TO JOINT,IF THE LINK
C         BELONGS TO MORE THAN ONE JOINT
C---------------------------------------------------------------
235   READ(IR,110) (R0UC(L,K1,K),K=1,3)
240   CONTINUE
      IF(L.EQ.NL) GO TO 250
      IP1=IP(L+1)+1
      IK1=IK(L+1)
250   CONTINUE
C
C---------------------------------------------------------------
C         UNIT VECTOR OF THE FIRST JOINT
C---------------------------------------------------------------
      READ(IR,110) (E(K),K=1,3)
C
C---------------------------------------------------------------
C         VECTOR R (0,1)
C---------------------------------------------------------------
      READ(IR,110) (RI(K),K=1,3)
      DO 2666 I=1,NUZ
      IF(KSI6(I).EQ.0) GO TO 2666
      J=KSI6(I)
C
C---------------------------------------------------------------
C         SPECIFICITIES
C---------------------------------------------------------------
      READ(IR,110) (RAS(J,1,K),K=1,3)
      READ(IR,110) (RAS(J,2,K),K=1,3)
2666  CONTINUE
C
C---------------------------------------------------------------
C      INPUT PRESCRIBED DYNAMICS
C---------------------------------------------------------------
      READ(INDEX,84301) NDIS
      READ(INDEX,84302) DELT,PER,S
260   CONTINUE
      READ(INDEX,84301) I
      IF (I.EQ.0) GO TO 261
      DO 265 K = 1,NDIS
265   READ(INDEX,84303) (SA(K,I,J),J=1,3)
      GO TO 260
261   CONTINUE
```

```fortran
      RETURN
      END
C
C*****************************************************************
C     SUBROUTINE : ASSEM                                         *
C---------------------------------------------------------------*
C     FUNCTION :  ASSEMBLY KINEMATIC PAIR P(I,K)                 *
C---------------------------------------------------------------*
C     INPUT VARIABES:                                            *
C        ROA        VECTOR R(I,K) IN THE ABSOLUTE COOR. FRAME    *
C        ROU        VECTOR R(I,I) IN THE ABSOLUTE COOR. FRAME    *
C        E1         VECTOR E(I) IN THE ABSOLUTE COOR. FRAME      *
C        E3         VECTOR E(I) IN THE LOCAL COOR. FRAME         *
C---------------------------------------------------------------*
C     OUTPUT VARIABLES:                                          *
C        A4         TRANSFORMATION MATRIX A(I) BEFORE ROTATION   *
C---------------------------------------------------------------*
C     SUBROUTINES CALLED:                                        *
C     CROSS0, CROSS2, GMPRD                                      *
C*****************************************************************
      SUBROUTINE ASSEM(ROA,ROU,E1,E3,A4)
      COMMON /RAC/ IR,IW,IT,INDEX,IOUT
      DIMENSION ROA(3),ROU(3),E1(3),E3(3),A4(3,3),A3(3,3),A2(3,3)
      DIMENSION TEMP0(3),TEMP1(3),TEMP2(3),TEMP3(3)
      CALL CROSS2(E1,ROA,TEMP1)
      CALL CROSS2(E3,ROU,TEMP3)
      T1=0.
      T3=0.
      DO 15 K=1,3
      T1=T1+TEMP1(K)*TEMP1(K)
      T3=T3+TEMP3(K)*TEMP3(K)
15    CONTINUE
      T1=SQRT(T1)
      T3=SQRT(T3)
       T1=1./T1
       T3=1./T3
      CALL CROSS1(TEMP1,E1,TEMP0)
      CALL CROSS1(TEMP3,E3,TEMP2)
      DO 20 K=1,3
      A2(K,1)=TEMP1(K)*T1
      A2(K,2)=E1(K)
      A2(K,3)=TEMP0(K)*T1
      A3(1,K)=-TEMP3(K)*T3
      A3(2,K)=E3(K)
      A3(3,K)=-TEMP2(K)*T3
20    CONTINUE
      CALL GMPRD(A2,A3,A4,3,3,3)
      RETURN
      END
C
C*****************************************************************
C     SUBROUTINE : MOM10                                         *
C---------------------------------------------------------------*
C     FUNCTION:   CALCULATION OF COEFFICIENT B(I,J),B0(I)        *
C                 FOR THE BODY                                   *
C---------------------------------------------------------------*
C     INPUT VARIABLES:                                           *
C            KZ         INDEX OF THE LINK BEING PROCESSED        *
C            T          MOMENT OF INERTIA WRT FIXED COORD. FRAME *
C            ALP        COEFFICIENTS ALFA(I,J) FOR THE BODY      *
C            ALP0       COEFFICIENTS ALFA0(I) FOR THE BODY       *
C            LAMD       COMPONENET DEPENDING ON VELOCITY FOR THE BODY *
C---------------------------------------------------------------*
C     OUTPUT VARIABLES:                                          *
C            BII        COEFFICIENTS BETA(I,J) FOR THE BODY      *
```

```
C                 BI0      COEFFICIENTS BETA0(I) FOR THE BODY              *
C-----------------------------------------------------------------------*
C      SUBROUTINE CALLED: NONE                                           *
C***********************************************************************
       SUBROUTINE MOM10(BII,BIO,T,ALP,ALPO,LAMD,KZ)
       REAL LAMD(3)
       DIMENSION BII(25,3),BIO(3),T(3,3)
       DIMENSION ALP(25,3),ALPO(3)
       DIMENSION TEMP3(3),TEMP1(3)
       COMMON /STRUKT/ JL,NUZ,NPZ(3,20),EU(25,3),ROU(49,3),IK(3)
       DO 10 J=1,KZ
       DO 10 K=1,3
       BII(J,K)=0.
       DO 10 L=1,3
       BII(J,K)=BII(J,K)-T(K,L)*ALP(J,L)
10     CONTINUE
       DO 20 IP=1,3
       BIO(IP)=LAMD(IP)
       DO 20 K=1,3
       BIO(IP)=BIO(IP)-T(IP,K)*ALPO(K)
20     CONTINUE
       RETURN
       END
C
C***********************************************************************
C      SUBROUTINE : MOM0                                                 *
C-----------------------------------------------------------------------*
C      FUNCTION :   CALCULATION OF TRANSFORMATION MATRIX FOR MOMENTS     *
C                   OF INERTIA WRT FIXED COORD. FRAME AND CALCULATION    *
C                   OF VELOCITY DEPENDENT COMPONENT LAMDA                *
C                   (FOR THE BODY)                                       *
C-----------------------------------------------------------------------*
C      INPUT VARIABLES:                                                  *
C             KZ         INDEX OF THE LINK BEING PROCESSED               *
C             OMEGI      ANGULAR VELOCITY OMEGA-KZ                       *
C             Q          TRANSFORMATION MATRIX Q-KZ                      *
C-----------------------------------------------------------------------*
C      OUTPUT VARIABLES:                                                 *
C             T        - MOMENT OF INERTIA WRT FIXED COORD. FRAME        *
C             LAMD     - COMPONENT LAMDA                                 *
C-----------------------------------------------------------------------*
C      SUBROUTINES CALLED: DOT0                                          *
C***********************************************************************
       SUBROUTINE MOM0(OMEGI,Q,KZ,T,LAMD)
       REAL LAMD(3)
       REAL MM(25),JS(25),JN(25),JX(25),JY(25),JZ(25)
       DIMENSION TEMP1(3),TEMP9(3,3)
       DIMENSION OMEGI(3),Q(3,3),T(3,3)
       COMMON /MIN/ MM,JS,JN,JX,JY,JZ
       COMMON /STRUKT/ JL,NUZ,NPZ(3,20),EU(25,3),ROU(49,3),IK(3)
       I=NPZ(JL,KZ)
       OMEG1=DOT0(OMEGI,Q,1)
       OMEG2=DOT0(OMEGI,Q,2)
       OMEG3=DOT0(OMEGI,Q,3)
       TEMP1(1)=(JY(I)-JZ(I))*OMEG2*OMEG3
       TEMP1(2)=(JZ(I)-JX(I))*OMEG1*OMEG3
       TEMP1(3)=(JX(I)-JY(I))*OMEG1*OMEG2
       CALL GMPRD(Q,TEMP1,LAMD,3,3,1)
       DO 12 J=1,3
       TEMP9(J,1)=JX(I)*Q(J,1)
       TEMP9(J,2)=JY(I)*Q(J,2)
       TEMP9(J,3)=JZ(I)*Q(J,3)
12     CONTINUE
       DO 13 J=1,3
       DO 13 K=1,3
       T(J,K)=0.
```

```
      DO 13 L0=1,3
      T(J,K)=T(J,K)+TEMP9(J,L0)*Q(K,L0)
13    CONTINUE
      RETURN
      END
C
C*******************************************************************
C     SUBROUTINE : MATH                                            *
C-----------------------------------------------------------------*
C     FUNCTION :   CALCULATION OF INERTIAL MATRX H                 *
C-----------------------------------------------------------------*
C     INPUT VARIABLES:                                             *
C           KZ        INDEX OF THE LINK BEING PROCESSED            *
C           R         POSITION VECTORS R(I,J), I,J=1,...,KZ        *
C           E         JOINT AXES                                   *
C           AI        INERTIAL FORCES COEFFICIENTS A(I,J)          *
C           BII       INERTIAL MOMENTS COEFFICIENTS BETA(I,J)      *
C           XX        TIME                                         *
C-----------------------------------------------------------------*
C     OUTPUT VARIABLES:                                            *
C           H         INERTIAL MATRIX H                            *
C           EH        ELEMENTS CORRESPONDING TO ZMP                *
C-----------------------------------------------------------------*
C     SUBROUTINES CALLED:                                          *
C           CROSS1, CROSS0                                         *
C*******************************************************************
      SUBROUTINE MATH(R,AI,BII,E,H,KZ,EH,XX)
      REAL MM(25),JS(25),JN(25),JX(25),JY(25),JZ(25)
      DIMENSION R(25,3),AI(25,3),BII(25,3),E(25,3),H(25,25),TEMP0(3)
      COMMON /TIPL/ KSI1(25),KSI2(25),KSI3(25),RAS(5,2,3),KSI6(25)
     1,KSI4(25),KSI5(25)
      COMMON /MIN/ MM,JS,JN,JX,JY,JZ
      COMMON /STRUKT/ JL,NUZ,NPZ(3,20),EU(25,3),ROU(49,3),IK(3)
      COMMON /RAC/ IR,IW,IT,INDEX,IOUT
      COMMON /ERIC/ EPOC(3),ROPOC(3),ROUC(3,2,3)
      DIMENSION TEMP1(3),EH(3,25),RIR(3)
      J=NPZ(JL,KZ)
      DO 10 II=1,KZ
      I=NPZ(JL,II)
      DO 10 K=1,KZ
      KK=NPZ(JL,K)
      DO 100 K1=1,3
100   TEMP1(K1)=-MM(J)*AI(K,K1)
      CALL CROSS0(TEMP1,R,TEMP0,II)
      DO 101 K1=1,3
101   TEMP0(K1)=-TEMP0(K1)+BII(K,K1)
      DH=-(E(I,1)*TEMP0(1)+E(I,2)*TEMP0(2)+E(I,3)*TEMP0(3))
      H(I,KK)=H(I,KK)+DH
      IF(II.NE.1) GO TO 10
      DO 800 K12 = 1,3
800   RIR (K12) = R(1,K12)-ROPOC(K12)
      CALL CROSS1(RIR,TEMP1,TEMP0)
      EH(1,KK)=EH(1,KK)+TEMP0(1)+BII(K,1)
      EH(2,KK)=EH(2,KK)+TEMP0(2)+BII(K,2)
10    CONTINUE
      RETURN
      END
C
C*******************************************************************
C     SUBROUTINE : VEKH                                            *
C-----------------------------------------------------------------*
C     FUNCTION :   CALCULATION OF VECTOR H0                        *
C-----------------------------------------------------------------*
C     INPUT VARIABLES:                                             *
C           KZ        INDEX OF THE LINK BEING PROCESSED            *
C           R         POSITION VECTORS R(I,J)                      *
```

```
C            AIO       COEFFICIENTS A0(I)                                     *
C            BIO       COEFFICIENTS BETA0(I)                                  *
C            E         JOINT AXES                                            *
C            XX        CURRENT TIME                                          *
C-------------------------------------------------------------------------*
C     OUTPUT VARIABLES:                                                     *
C            H0        VECTOR H0                                            *
C            EH0       PART OF THE VECTOR H0 CORRESPONDING TO ZMP           *
C-------------------------------------------------------------------------*
C     SUBROUTINES CALLED: CROSS1                                           *
C************************************************************************
      SUBROUTINE VEKH(R,AIO,BIO,E,H0,KZ,EH0,XX)
      REAL MM(25),JS(25),JN(25),JX(25),JY(25),JZ(25)
      INTEGER DESNI
      DIMENSION R(25,3),AIO(3),BIO(3),E(25,3),H0(25)
     1,TEMP1(3),TEMP2(3),TEMP3(3),TEMP4(3),GM(3),TEMP5(3)
     2,EH0(3),RIR(3)
      COMMON /SYS/ DESNI,NL,IP(3),NPL,JM
      COMMON /MIN/ MM,JS,JN,JX,JY,JZ
      COMMON /TIPL/ KSI1(25),KSI2(25),KSI3(25),RAS(5,2,3),KSI6(25)
     1,KSI4(25),KSI5(25)
      COMMON /STRUKT/ JL,NUZ,NPZ(3,20),EU(25,3),ROU(49,3),IK(3)
      COMMON /RAC/ IR,IW,IT,INDEX,IOUT
      COMMON /ERIC/ EPOC(3),ROPOC(3),ROUC(3,2,3)
      J=NPZ(JL,KZ)
      TEMP4(1)=AIO(1)
      TEMP4(2)=AIO(2)
      TEMP4(3)=AIO(3)-MM(J)*9.81*DESNI
      DO 40 II=1,KZ
      I=NPZ(JL,II)
      DO 10 K=1,3
      TEMP1(K)=E(I,K)
 10   TEMP2(K)=R(II,K)
      CALL CROSS1(TEMP2,TEMP4,TEMP3)
      DO 20 K=1,3
 20   TEMP5(K)=TEMP3(K)+BIO(K)
      H0(I)=H0(I)-DOT1(TEMP1,TEMP5)
 40   CONTINUE
      DO 800 K12 = 1,3
 800  RIR(K12) = R(1,K12) - ROPOC(K12)
      CALL CROSS1(RIR,TEMP4,TEMP1)
      EH0(1)=EH0(1)+TEMP1(1)+BIO(1)
      EH0(2)=EH0(2)+TEMP1(2)+BIO(2)
      RETURN
      END
C
C************************************************************************
C     SUBROUTINE : MOM2                                                     *
C-------------------------------------------------------------------------*
C     FUNCTION :  CALCULATION OF COEFFICIENTS B(I,J), J=1,...,I AND         *
C                 B0(I) FOR SEGMENT I                                       *
C-------------------------------------------------------------------------*
C     INPUT VARIABLES:                                                      *
C            KZ        INDEX OF THE LINK BEING PROCESSED                    *
C            ALP       COEFFICIENTS ALFA(I,J)                              *
C            ALP0      COEFFICIENTS ALFA0(I)                               *
C            TAU       EQUIVALENT ACCELERATION                             *
C            S         LINK UNIT VECTOR WRT FIXED COORD. FRAME             *
C-------------------------------------------------------------------------*
C     OUTPUT VARIABLES:                                                     *
C            BII       COEFFICIENTS B(I,J)                                 *
C            B0(I)     COEFFICIENTS B0(I)                                  *
C-------------------------------------------------------------------------*
C     SUBROUTINES CALLED: MOM12                                            *
C************************************************************************
      SUBROUTINE MOM2(KZ,BII,BIO,ALP,ALP0,S,TAU)
```

```fortran
      REAL MM(25),JS(25),JN(25),JX(25),JY(25),JZ(25)
      DIMENSION BII(25,3),BIO(3),ALP(25,3),ALP0(3),S(3),TAU(3)
      DIMENSION EPS(3),EPSN(3),EPSS(3)
      COMMON /MIN/MM,JS,JN,JX,JY,JZ
      COMMON /STRUKT/ JL,NUZ,NPZ(3,20),EU(25,3),R0U(49,3),IK(3)
      I=NPZ(JL,KZ)
      DO 10 K=1,3
10    EPS(K)=ALP(J,K)
      CALL MOM12(EPS,EPSS,EPSN,S)
      DO 15 K=1,3
      BII(J,K)=-(JN(I)*EPSN(K)+JS(I)*EPSS(K))
15    CONTINUE
      DO 20 K=1,3
20    EPS(K)=ALP0(K)
      CALL MOM12(EPS,EPSS,EPSN,S)
      DO 25 K=1,3
      BIO(K)=-(JN(I)*(EPSN(K)+TAU(K))+JS(I)*EPSS(K))
25    CONTINUE
      RETURN
      END
C
C***************************************************************************
C     SUBROUTINE : MOM11                                                  *
C-------------------------------------------------------------------------*
C     FUNCION :   CALCULATION OF THE EQUIVALENT ACCELERATION              *
C-------------------------------------------------------------------------*
C     INPUT VARIABLES:                                                    *
C            I          INDEX OF THE LINK BEING PROCESSED                 *
C            RI         RII VECTOR FOR THE LINK                           *
C            OMEGI      ANGULAR VELOCITY WRT FIXED COORD. FRAME           *
C                       FOR THE LINK                                      *
C-------------------------------------------------------------------------*
C     OUTPUT VARIABLES:                                                   *
C            S          UNIT VECTOR                                       *
C            TAU        EQUIVALENT ACCELERATION                           *
C-------------------------------------------------------------------------*
C     SUBROUTINES CALLED: CROSS1                                          *
C-------------------------------------------------------------------------*
      SUBROUTINE MOM11(I,RI,OMEGI,S,TAU)
      DIMENSION RI(3),OMEGI(3),S(3),TEMP(3),TAU(3)
      RAS=0.
      DO 10 K=1,3
10    RAS=RAS+RI(K)*RI(K)
      RAS=1./SQRT(RAS)
      DO 15 K=1,3
15    S(K)=RAS*RI(K)
      CALL CROSS1(S,OMEGI,TEMP)
      TEMP1=DOT1(OMEGI,S)
      DO 20 K=1,3
20    TAU(K)=TEMP1*TEMP(K)
      RETURN
      END
C
C***************************************************************************
C     SUBROUTINE : TOTR                                                   *
C-------------------------------------------------------------------------*
C     FUNCTION :   CALCULATION OF POSITION VECTORS R(I,J),                *
C                  J=1,...,I-1 IN THE '+' JOINTS ARRAY                    *
C-------------------------------------------------------------------------*
C     INPUT VARIABLES:                                                    *
C            I          INDEX OF THE LINK BEING PROCESSED                 *
C            RI         VECTOR RII                                        *
C            R0         VECTOR RI-1,I                                     *
C-------------------------------------------------------------------------*
C     OUTPUT VARIABLES:                                                   *
C            R          J-TH ROW OF MATRIX R REPRESENTS POSITION          *
```

```fortran
C                         VECTOR FROM THE J-TH TO THE I-TH SEGMENT IN THE*
C                         '+' ARRAY JOINTS                               *
C------------------------------------------------------------------------*
C     CALLED SUBROUTINES: NONE                                           *
C************************************************************************
      SUBROUTINE TOTR(R,RI,R0,I)
      DIMENSION R(25,3),RI(3),R0(3)
      DO 15 J=1,I
      DO 15 K=1,3
      IF(J-I) 12,20,12
20    R(J,K)=RI(K)
      GO TO 15
12    R(J,K)=R(J,K)+RI(K)-R0(K)
15    CONTINUE
      RETURN
      END
C
C***********************************************************************
C     SUBROUTINE : MODEL                                                *
C----------------------------------------------------------------------*
C     FUNCTION :   COMPUTES MATHEMATICAL MODEL OF THE MECHANISM         *
C                  IN THE FORM                                          *
C                      P = H(Q)Q' + H0(Q,Q')                           *
C----------------------------------------------------------------------*
C     INPUT VARIABLES:                                                  *
C                XX            INTEGRATION TIME                         *
C----------------------------------------------------------------------*
C     OUTPUT:                                                           *
C                H             INERTIAL MATRIX H(Q)                     *
C                H0            VECTOR H(Q,Q')                           *
C                EH            THE PART OF MATRIX H CORRESPONDING TO ZMP *
C                EH0           THE PART OF VECTOR H0 CORRESPONDING TO ZMP *
C----------------------------------------------------------------------*
C     SUBROUTINES CALLED:                                               *
C       ASSEM, TRANSF, TOTR,BRZUB,MOM0,MOM10,                           *
C       MOM11, MOM2, MATH, VEKH                                         *
C************************************************************************
      SUBROUTINE MODEL(H,H0,EH0,EH,XX)
      REAL MM(25),JS(25),JN(25),JX(25),JY(25),JZ(25)
      REAL LAMD(3)
      INTEGER DESNI
      DIMENSION E(25,3),R(25,3),BII(25,3),ALP(25,3),BETI(25,3)
      DIMENSION ROA(3),ROU(3),E1(3),E3(3),RI(3),R00(3),OMEG0(3),OMEGI
     1(3),S(3),TAU(3),BI0(3),ALP0(3),BETI0(3)
      DIMENSION Q(3,3),T(3,3),QP(3,3)
      DIMENSION H(25,25),H0(25),RP(25,3),EH(3,25),EH0(3)
      DIMENSION R0(3)
      DIMENSION R0P(3),R00P(3)
      DIMENSION ALPP(25,3),BETIP(25,3),ALP0P(3),BETI0P(3),BI0P(3),
     1OMEG0P(3)
      DIMENSION BETIV(3)
      DIMENSION RNOVO(3)
      DIMENSION RPOM(3)
      COMMON /SYS/ DESNI,NL,IP(3),NPL,JM
      COMMON /MIN/ MM,JS,JN,JX,JY,JZ
      COMMON /RAC/ IR,IW,IT,INDEX,IOUT
      COMMON /TIPL/ KSI1(25),KSI2(25),KSI3(25),RAS(5,2,3),KSI6(25)
     1,KSI4(25),KSI5(25)
      COMMON /UG/ SI(25),SIDOT(25),DSIDOT(25)
      COMMON /STRUKT/ JL,NUZ,NPZ(3,20),EU(25,3),R0U(49,3),IK(3)
      COMMON /ERIC/ EPOC(3),R0POC(3),R0UC(3,2,3)
C
      DO 7000 J=1,3
      DO 7001 I=1,NUZ
7001  EH(J,I)=0.
7000  EH0(J)=0.
```

```
       DO 6000 I=1,NUZ
       H0(I)=0.
       DO 6000 J=1,NUZ
 6000  H(I,J)=0.
       DO 6001 K=1,3
       R0(K)=R0POC(K)
       E(1,K)=EPOC(K)
       OMEG0(K)=0.
       ALP0(K)=0.
 6001  BETI0(K)=0.
       DO 6005 I=1,3
       DO 6005 J=1,3
       Q(I,J)=0.
       IF(I.EQ.J) Q(I,J)=1.
 6005  CONTINUE
       IP1=IP(1)
       IK1=IK(1)
       DO 160 JL=1,NL
       IF(JL.GE.NL) GO TO 65
       IP2=IP(JL+1)
       IP2=NPZ(JL+1,IP2)
       DO 120 K=1,3
       IP2N=IP2+NUZ
 120   ROU IP2N,K)=ROUC(JL,1,K)
 65    CONTINUE
       DO 2200 LUP=1,IK1
       IR2=NPZ(JL,LUP)
       IR2N=IR2+NUZ
 2200  CONTINUE
C
       DO 150 I=IP1,IK1
       CALL SPEC⁻F(R0,ROA,ROU,E1,E3,Q,E,I)
       CALL ASSEM(ROA,ROU,E1,E3,Q)
       CALL TRANSF(E,Q,I,RI,R0,R00)
       CALL TOTR(R,RI,R00,I)
       CALL BRZUB(OMEGI,OMEG0,E,ALP,ALP0,BETI,BETI0,R00,RI,R0,I)
       JLI=NPZ(JL,I)
       IF(KSI1(JL ).EQ.0) GO TO 990
       CALL MOM11(I,RI,OMEGI,S,TAU)
       CALL MOM2(I,BII,BI0,ALP,ALP0,S,TAU)
       GO TO 991
 990   CONTINUE
       CALL MOM0(OMEGI,Q,I,T,LAMD)
       CALL MOM10(BII,BI0,T,ALP,ALP0,LAMD,I)
 991   CONTINUE
       DO 2001 K=1,3
 2001  BETIV(K)=-MM(JLI)*BETI0(K)
       CALL MATH(R,BETI,BII,E,H,I,EH,XX)
       CALL VEKH(R,BETIV,BI0,E,H0,I,EH0,XX)
       IF(NPZ(JL,I).EQ.IP2) GO TO 996
       GO TO 150
 996   IMP=I
       DO 9000 JW=1,IMP
       DO 9000 KW=1,3
       ALPP(JW,KW)=ALP(JW,KW)
       BETIP(JW,KW)=BETI(JW,KW)
 9000  RP(JW,KW)=R(JW,KW)
       DO 9001 JW=1,3
       BETIOP(JW)=BETI0(JW)
       ALP0P(JW)=ALP0(JW)
       BI0P(JW)=BI0(JW)
       OMEG0P(JW)=OMEG0(JW)
       R0P(JW)=R0(JW)
       R00P(JW)=R00(JW)
       DO 9001 KW=1,3
 9001  QP(JW,KW)=Q(JW,KW)
```

```
150     CONTINUE
        DO 180 I=1,IMP
        DO 180 K=1,3
        BETI(I,K)=BETIP(I,K)
        ALP(I,K)=ALPP(I,K)
180     R(I,K)=RP(I,K)
        DO 190 K=1,3
190     ROU(IP2N,K)=ROUC(JL,2,K)
        DO 200 I=1,3
        BETIO(I)=BETIOP(I)
        ALPO(I)=ALPOP(I)
        BIO(I)=BIOP(I)
        OMEGO(I)=OMEGOP(I)
        ROO(I)=ROOP(I)
        DO 200 J=1,3
        Q(I,J)=QP(I,J)
200     CONTINUE
        IF(JL.EQ.NL) GO TO 160
        IP2=IP(JL+1)
        JP2=NPZ(JL+1,IP2+1)
        DO 201 I=1,3
        RO(I)=0.
        E(JP2,I)=0.
        DO 201 J=1,3
        RO(I)=RO(I)+Q(I,J)*ROU(IP2N,J)
201     E(JP2,I)=E(JP2,I)+Q(I,J)*EU(JP2,J)
        IP1=IP(JL+1)+1
        IK1=IK(JL+1)
160     CONTINUE
        RETURN
        END
C*****************************************************************************
C       SUBROUTINE : BRZUB                                                  *
C---------------------------------------------------------------------------*
C       FUNCTION :   CALCULATION OF COEFFICIENTS ACCOMPANYING               *
C                    ANGULAR AND LINEAR ACCELARATIONS BETA(I,J),            *
C                    J=1,...,KZ-1 AND ALFA(I,J),J=1,...,KZ-1 AND            *
C                    COEFFICIENTS BETA0(I), ALFA0(I) DEPENDENT ON           *
C                    VELOCITIES IN THE + JOINTS ARRAY                       *
C---------------------------------------------------------------------------*
C       INPUT VARIABLES:                                                    *
C                KZ          INDEX OF THE LINK BEING PROCESSED              *
C                OMEGO       ANGULAR VELOCITY OMEGA-K                       *
C                E           JOINT AXES                                     *
C                ROO         VECTOR R-K,I                                   *
C                RI          VECTOR R-I,I                                   *
C                RO          VECTOR R-L,I                                   *
C---------------------------------------------------------------------------*
C       OUTPUT VARIABLES:                                                   *
C                OMEGI       ANGULAR VELOCITY OMEGA(I)                      *
C                ALP         COEFFICIENTS ALFA(I,J), J=1,...,KZ             *
C                ALPO        COEFFICIENTS ALFA0(I)                          *
C                BETI        COEFFICIENTS BETA(I,J), J=1,...,KZ             *
C                BETIO       COEFFICIENTS BETA0(I)                          *
C                WHERE       I=NPZ(.,KZ), K=NPZ(.,KZ-1), L=NPZ(.,KZ+1)      *
C---------------------------------------------------------------------------*
C       SUBROUTINES CALLED:                                                 *
C          CROSS0, CROSS1, CROSS2                                           *
C*****************************************************************************
        SUBROUTINE BRZUB(OMEGI,OMEGO,E,ALP,ALPO,BETI,BETIO,ROO,RI,RO,KZ)
        COMMON /TIPL/ KSI1(25),KSI2(25),KSI3(25),RAS(5,2,3),KSI6(25)
       1,KSI4(25),KSI5(25)
        COMMON /UG/ SI(25),SIDOT(25),DSIDOT(25)
        COMMON /STRUKT/ JL,NUZ,NPZ(3,20),EU(25,3),ROU(49,3),IK(3)
        COMMON /RAC/ IR,IW,IT,INDEX,IOUT
        DIMENSION OMEGO(3),RI(3),ROO(3),RO(3),TEMP6(3),
```

```
     1OMEGI(3),ALP(25,3),BETI(25,3),BETI0(3),ALP0(3),
     2TEMP7(3),TEMP8(3),TEMP9(3),TEMP10(3)
      DIMENSION TEMP0(3),TEMP2(3),E(25,3),TEMP1(3)
      I=NPZ(JL,KZ)
      DO 30 K=1,3
      TEMP6(K)=RI(K)-R00(K)
 30   CONTINUE
      CALL CROSS0(OMEG0,E,TEMP0,I)
      CALL CROSS0(RI,E,TEMP2,I)
      DO 40 K=1,3
      TEMP0(K)=SIDOT(I)*TEMP0(K)
      OMEGI(K)=OMEG0(K)+SIDOT(I)*E(I,K)
      ALP(KZ,K)=E(I,K)
      BETI(KZ,K)=-TEMP2(K)
 40   CONTINUE
      IF(KZ.EQ.1) GO TO 70
      I2=KZ-1
      DO 60 J=1,I2
      CALL CROSS0(TEMP6,ALP,TEMP1,J)
      DO 50 K=1,3
      BETI(J,K)=BETI(J,K)-TEMP1(K)
 50   CONTINUE
 60   CONTINUE
 70   CONTINUE
      DO 611 K=1,3
      TEMP1(K)=RI(K)
      TEMP2(K)=R00(K)
 611  CONTINUE
      CALL CROSS1(ALP0,TEMP6,TEMP7)
      CALL CROSS1(TEMP0,TEMP1,TEMP8)
      CALL CROSS2(OMEG0,TEMP2,TEMP9)
      CALL CROSS2(OMEGI,TEMP1,TEMP10)
      DO 61 K=1,3
      OMEG0(K)=OMEGI(K)
      BETI0(K)=BETI0(K)+TEMP7(K)+TEMP8(K)
     1-TEMP9(K)+TEMP10(K)
 61   ALP0(K)=ALP0(K)+TEMP0(K)
 100  CONTINUE
      RETURN
      END
C*****************************************************************************
C     SUBROUTINE : TRANSF                                                   *
C---------------------------------------------------------------------------*
C     FUNCTION :   CALCULATION OF TRANSFORMATION MATRIX Q0I (AI)            *
C---------------------------------------------------------------------------*
C     INPUT VARIABLES:                                                      *
C             KZ          INDEX OF THE LINK CURRENTLY PROCESSED             *
C             E           JOINT AXES WRT ABSOLUTE COORD. FRAME              *
C             Q           TRANSFORMATION MATRIX Q PRIOR ROT.                *
C---------------------------------------------------------------------------*
C     OUTPUT VARIABLES:                                                     *
C             Q           TRANSFORMATION MATRIX AFTER ROT.                  *
C             RI          VECTOR R-I,I                                      *
C             R0          VECTOR R-L,I                                      *
C             R00         VECTOR R-K,I                                      *
C             WHERE I=NPZ(.,KZ), K=NPZ(.,KZ-1), L=NPZ(.,KZ+1)              *
C             NOTE : ALL VECTORS (OUTPUT) ARE ROTATED                       *
C---------------------------------------------------------------------------*
C     SUBROUTINES CALLED:                                                   *
C        CROSS, GMPRD                                                       *
C*****************************************************************************
      SUBROUTINE TRANSF(E,Q,KZ,RI,R0,R00)
      INTEGER DESNI
      DIMENSION E(25,3),Q(3,3),RI(3),R0(3),R00(3)
      DIMENSION TEMP0(3),TEMP1(3),TEMP2(3),TEMP3(3)
      COMMON /STRUKT/ JL,NUZ,NPZ(3,20),EU(25,3),R0U(49,3),IK(3)
```

164

```fortran
      COMMON /UG/ SI(25),SIDOT(25),DSIDOT(25)
      COMMON/TIPL/ KSI1(25),KSI2(25),KSI3(25),RAS(5,2,3),KSI6(25)
     1,KSI4(25),KSI5(25)
      COMMON /SYS/ DESNI,NL,IP(3),NPL,JM
      I=NPZ(JL,KZ)
      CS=COS(SI(I))
      SN=SIN(SI(I))
      DO 40 K=1,3
      TEMP0(K)=EU(I,K)
      TEMP2(K)=R0U(I,K)
      IF(KZ.EQ.IK(JL)) GO TO 40
      I1=NPZ(JL,KZ+1)
      TEMP1(K)=EU(I1,K)
      IN=I+NUZ
      TEMP3(K)=R0U(IN,K)
 40   CONTINUE
      DO 50 K=1,3
      R00(K)=R0(K)
 50   CONTINUE
      DO 60 J=1,3
      TEMP=DOT(E,Q,I,J)
      CALL CROSS(E,Q,TEMP0,I,J)
      DO 60 K=1,3
      Q(K,J)=Q(K,J)*CS+(1.-CS)*TEMP*E(I,K)+SN*TEMP0(K)*DESNI
 60   CONTINUE
 61   CONTINUE
      CALL GMPRD(Q,TEMP2,RI,3,3,1)
      IF(KZ.EQ.IK(JL)) GO TO 200
      CALL GMPRD(Q,TEMP3,R0,3,3,1)
      CALL GMPRD(Q,TEMP1,TEMP3,3,3,1)
 200  CONTINUE
      DO 70 K=1,3
      IF(KZ.EQ.IK(JL)) GO TO 70
      I1=NPZ(JL,KZ+1)
      E(I1,K)=TEMP3(K)
 70   CONTINUE
      RETURN
      END
C
C****************************************************************************
C     SUBROUTINE : MOM12                                                   *
C----------------------------------------------------------------------------*
C     FUNCTION :   CALCULATE NORMAL AND PARALLEL COMPONENT OF VECTOR       *
C                  EPS WRT LONGITUDINAL AXE S                              *
C----------------------------------------------------------------------------*
C     INPUT VARIABLES:                                                     *
C                EPS       INPUT VECTOR (3X1)                              *
C                S         UNIT VECTOR OF AXE                              *
C----------------------------------------------------------------------------*
C     OUTPUT VARIABLES:                                                    *
C                EPSS      PARALLEL COMPONENT                              *
C                EPSN      NORMAL COMPONENT                                *
C----------------------------------------------------------------------------*
C     SUBROUTINES CALLED: NONE                                            *
C****************************************************************************
      SUBROUTINE MOM12(EPS,EPSS,EPSN,S)
      DIMENSION EPS(3),EPSS(3),EPSN(3),S(3)
      CALL CROSS2(S,EPS,EPSN)
      TEMP1=DOT1(EPS,S)
      DO 10 K=1,3
      EPSN(K)=-EPSN(K)
 10   EPSS(K)=TEMP1*S(K)
      RETURN
      END
C
C****************************************************************************
```

```fortran
C        SUBROUTINE : OUT                                                  *
C----------------------------------------------------------------------*
C        FUNCTION : DETERMINE THE ERROR CODE IER ACCORDING TO            *
C                   SQUARE CRITERIA                                       *
C----------------------------------------------------------------------*
C        INPUT VARIABLES:                                                 *
C                 T         INTEGRATION TIME                              *
C                 DEP       CURRENT VALUE OF SOLUTION                     *
C----------------------------------------------------------------------*
C        OUTPUT VARIABLES:                                                *
C                 IER       ERROR CODE                                    *
C----------------------------------------------------------------------*
C        CALLED SUBROUTINES: NONE                                         *
C**********************************************************************
       SUBROUTINE OUT(T,DEP,IER)
       COMMON /ACOM/ N,DUM(51)
       DIMENSION DEP(16)
       AA=DEP(1)**2+DEP(2)**2
       IF(AA.GT.10.) IER=1
       RETURN
       END
C
C**********************************************************************
C        SUBROUTINE : OUT1                                                *
C----------------------------------------------------------------------*
C        FUNCTION :  OUTPUT COPMPUTED POSITIONS AND VELOCITIES            *
C----------------------------------------------------------------------*
C        INPUT VARIABLES:                                                 *
C                 T         INTEGRATION TIME                              *
C                 DEP       COMPUTED SOLUTION                             *
C                 IER       ERROR CODE (NOT CONSIDERED)                   *
C----------------------------------------------------------------------*
C        OUTPUT VARIABLES: NONE                                           *
C----------------------------------------------------------------------*
C        SUBROUTINES CALLED: NONE                                         *
C**********************************************************************
       SUBROUTINE OUT1 (T,DEP,IER)
       COMMON /ACOM/ N,DUM(51)
       COMMON /RAC/ IR,IW,IT,INDEX,IOUT
       DIMENSION DEP(16)
C
       WRITE(IW,409) T,(DEP(I),I=1,N)
409    FORMAT(' ',1X,F6.4,3X,F10.4,3X,F10.4,3X,F10.4,3X,F10.4)
       RETURN
       END
C
C**********************************************************************
C        SUBROUTINE : SUPP                                                *
C----------------------------------------------------------------------*
C        FUNCTION :  PRESCRIBES THE ZMP LAW                               *
C----------------------------------------------------------------------*
C        INPUT VARIABLES:                                                 *
C                 XX        INTEGRATION TIME                              *
C                 JM        THE FEET POSITION INDICATOR                   *
C----------------------------------------------------------------------*
C        OUTPUT VARIABLES:                                                *
C                 R0U       VECTOR R11                                    *
C                 R0POC     VECTOR R01                                    *
C----------------------------------------------------------------------*
C        CALLED SUBROUTINES: NONE                                         *
C**********************************************************************
       SUBROUTINE SUPP(XX,JM)
       COMMON /UG/ SI(25),SIDOT(25),DSIDOT(25)
       COMMON /STRUKT/ JL,NUZ,NPZ(3,20),EU(25,3),R0U(49,3),IK(3)
       COMMON /ERIC/ EPOC(3),R0POC(3),R0UC(3,2,3)
```

```fortran
      DIMENSION TT1(5),TT2(5),BB1(5),BB2(5)
      DATA TT1/0.0,0.000,0.000, 0.20, 0.200/
      DATA TT2/1.0,0.300,0.500, 1.00, 0.60/
      DATA BB2/0.0,0.000,0.000, 0.02, 0.020/
      DATA BB1/0.0,0.035,0.035, 0.00, 0.035/
      T1=TT1(JM)
      T2=TT2(JM)
      B1=BB1(JM)
      B2=BB2(JM)
      IF(XX.LE.T1) GO TO 5011
      IF(XX.LE.T2) GO TO 5012
      GO TO 5013
C-----------------------------------------------------------------------
C     FIRST PHASE ( ZMP UNDER HILL)
C-----------------------------------------------------------------------
5011  CONTINUE
      ROU(2,1) = B2
      ROU(2,2) = 0.
      ROU(2,3) = 0.03
      SI (2) = SI(2) + ATAN(B2/0.03)
C
      RETURN
C-----------------------------------------------------------------------
C     SECOND PHASE ( ZMP UNDER FEET )
C-----------------------------------------------------------------------
5012  CONTINUE
      ROU(2,1) = 0.
      ROU(2,2) = 0.
      ROU(2,3) = 0.03
      SI (2) = 0.
      SIDOT(2) = 0.
      DSIDOT(2) =0.
C
      RETURN
C-----------------------------------------------------------------------
C     THIRD PHASE ( ZMP UNDER TIPTOES)
C-----------------------------------------------------------------------
5013  CONTINUE
      ROU(2,1)=-B1
      ROU(2,2)=0.
      ROU(2,3)=0.03
      SI(2)=-(-SI(2)+3.141592/2.-ATAN(0.03/B1))
      RETURN
      END
C
C*********************************************************************
C     SUBROUTINE : SPECIF                                           *
C-------------------------------------------------------------------*
C     FUNCTION :   GENERATES ASSEMBLY VECTOR IF                     *
C                  CORRESPONDING VECTORS ARE COLINEAR               *
C-------------------------------------------------------------------*
C     INPUT VARIABLES:                                              *
C            RO         VECTOR R(I-1,I) IN ABS. COORD. FRAME        *
C            Q          TRANSFORMATION MATRIX PRIOR ROT.            *
C            E          JOINT AXES IN ABS. COORD. FRAME             *
C            KZ         INDEX OF THE LINK CURRENTLY PROC.           *
C-------------------------------------------------------------------*
C     OUTPUT VARIABLES:                                             *
C            ROA        VECTOR R(I,K) WRT. ABS. COORD. FRAME        *
C            ROU        VECTOR R(I,I) WRT. ABS. COORD. FRAME        *
C            E1         I-TH JOINT AXE WRT. ABS. COORD. FRAME       *
C            E3         I-TH JOINT AXE WRT. LOC. COORD. FRAME       *
C-------------------------------------------------------------------*
C     SUBROUTINES CALLED: NONE                                      *
C*********************************************************************
      SUBROUTINE SPECIF(R0,ROA,ROU,E1,E3,Q,E,KZ)
```

```
      DIMENSION R0(3),E1(3),E3(3),Q(3,3),E(25,3),TEMP0(3),
     1ROA(3),ROU(3)
      COMMON /STRUKT/ JL,NUZ,NPZ(3,20),EU(25,3),R0U(49,3),IK(3)
      COMMON /TIPL/ KSI1(25),KSI2(25),KSI3(25),RAS(5,2,3),KSI6(25)
     1,KSI4(25),KSI5(25)
      COMMON /RAC/ IR,IW,IT,INDEX
      I=NPZ(JL,KZ)
      DO 10 K=1,3
      E1(K)=E(I,K)
   10 E3(K)=EU(I,K)
      IF(KSI6(I).NE.0) GO TO 30
      DO 20 K=1,3
      ROA(K)=R0(K)
      ROU(K)=R0U(I,K)
   20 CONTINUE
      RETURN
   30 J=KSI6(I)
      DO 40 K=1,3
      ROU(K)=RAS(J,1,K)
   40 TEMP0(K)=RAS(J,2,K)
      DO 50 K=1,3
      ROA(K)=0.
      DO 50 L=1,3
   50 ROA(K)=ROA(K)+Q(K,L)*TEMP0(L)
      RETURN
      END
C***********************************************************************
C     SUBROUTINE : CROSS1                                              *
C---------------------------------------------------------------------*
C     FUNCTION :  CROSS PRODUCT OF TWO VECTORS                         *
C---------------------------------------------------------------------*
C     INPUT VARIABLES:                                                 *
C          E0        FIRST VECTOR DIMENSION 3x1                        *
C          R0        SECOND VECTOR DIMENSION 3x1                       *
C---------------------------------------------------------------------*
C     OUTPUT VARIABLES:                                                *
C          TEMP2     RESULTING VECTOR                                  *
C---------------------------------------------------------------------*
C     SUBROUTINES CALLED: NONE                                         *
C***********************************************************************
      SUBROUTINE CROSS1(E0,R0,TEMP2)
      DIMENSION E0(3),R0(3),TEMP2(3)
      TEMP2(1)=E0(2)*R0(3)-E0(3)*R0(2)
      TEMP2(2)=E0(3)*R0(1)-E0(1)*R0(3)
      TEMP2(3)=E0(1)*R0(2)-E0(2)*R0(1)
      RETURN
      END
C***********************************************************************
C     SUBROUTINE : CROSS2                                              *
C---------------------------------------------------------------------*
C     FUNCTION :  DOUBLE CROSS PRODUCT a x (a x b)                     *
C                 a AND b ARE 3x1 VECTORS                              *
C---------------------------------------------------------------------*
C     INPUT VARIABLES:                                                 *
C          OMEG0     VECTOR 3x1                                        *
C          R0        VECTOR 3x1                                        *
C---------------------------------------------------------------------*
C     OUTPUT VARIABLES:                                                *
C          TEMP4     RESULT VECTOR OF DOUBLE CROSS                     *
C                    PRODUCT                                           *
C                    OMEG0(x) X (OMEG0(x) X R0(x))                     *
C---------------------------------------------------------------------*
C     SUBROUTINES CALLED: DOT1                                         *
C***********************************************************************
      SUBROUTINE CROSS2(OMEG0,R0,TEMP4)
      DIMENSION OMEG0(3),R0(3),TEMP4(3)
```

```fortran
      TEMP=DOT1(OMEG0,R0)
      TEMP6=DOT1(OMEG0,OMEG0)
      DO 211 K=1,3
      TEMP4(K)= TEMP*OMEG0(K)-TEMP6*R0(K)
211   CONTINUE
      RETURN
      END
C***********************************************************************
C       SUBROUTINE : DOT                                               *
C----------------------------------------------------------------------*
C       FUNCTION : SCALAR PRODUCT OF TWO VECTORS                       *
C                  ONE VECTOR IS THE I-TH ROW OF 25x3 MATRIX, THE      *
C                  OTHER IS THE J-TH COLUMN OF 3x3 MATRIX              *
C----------------------------------------------------------------------*
C       INPUT VARIABLES:                                               *
C               A         MATRIX 25x3                                  *
C               B         MATRIX 3x3                                   *
C               I         INDEX OF THE I-TH ROW                        *
C               J         INDEX OF THE J-TH COLUMN                     *
C----------------------------------------------------------------------*
C       OUTPUT VARIABLES:                                              *
C               DOT       SCALAR PRODUCT A(i,x).B(x,j)                 *
C----------------------------------------------------------------------*
C       SUBROUTINES CALLED: NONE                                       *
C***********************************************************************
      FUNCTION DOT(A,B,I,J)
      DIMENSION A(25,3),B(3,3)
      DOT=A(I,1)*B(1,J)+A(I,2)*B(2,J)+A(I,3)*B(3,J)
      RETURN
      END
C***********************************************************************
C       SUBROUTINE : CROSS                                             *
C----------------------------------------------------------------------*
C       FUNCTION :  CROSS PRODUCT OF TWO VECTORS                       *
C                   (VECTOR PRODUCT). ONE VECTOR IS THE I-TH           *
C                   ROW OF 25x3 MATRIX AND THE OTHER IS THE J-TH       *
C                   COLUMN OF 3x3 MATRIX.                              *
C----------------------------------------------------------------------*
C       INPUT VARIABLES:                                               *
C               E         MATRIX 25x3                                  *
C               Q         MATRIX 3x3                                   *
C               I         INDEX OF THE I-TH ROW OF MATRIX E            *
C               J         INDEX OF THE J-TH COLUMN OF MATRIX Q         *
C----------------------------------------------------------------------*
C       OUTPUT VARIABLES:                                              *
C               TEMP0     CROSS PRODUCT E(i,x) X Q(x,j)                *
C----------------------------------------------------------------------*
C       SUBROUTTINES CALLED: NONE                                      *
C***********************************************************************
      SUBROUTINE CROSS(E,Q,TEMP0,I,J)
      DIMENSION E(25,3),Q(3,3),TEMP0(3)
      TEMP0(1)=E(I,2)*Q(3,J)-E(I,3)*Q(2,J)
      TEMP0(2)=E(I,3)*Q(1,J)-E(I,1)*Q(3,J)
      TEMP0(3)=E(I,1)*Q(2,J)-E(I,2)*Q(1,J)
      RETURN
      END
C***********************************************************************
C       SUBROUTINE : CROSS0                                            *
C----------------------------------------------------------------------*
C       FUNCTION :  CROSS PRODUCT OF TWO VECTORS                       *
C                   FIRST VECTOR IS 3x1 AND THE SECOND IS             *
C                   THE I-TH ROW OF THE MATRIX 25x3                    *
C----------------------------------------------------------------------*
C       INPUT VARIABLES:                                               *
C               OMEG0     VECTOR 3x1                                   *
C               E         MATRIX 25x3                                  *
```

```fortran
C                I        INDEX OF I-TH ROW OF THE MATRIX E            *
C--------------------------------------------------------------------*
C     OUTPUT VARIABLES:                                               *
C                TEMP1    CROSS PRODUCT OMEG0(x) X E(i,x)              *
C--------------------------------------------------------------------*
C     SUBROUTINES CALLED: NONE                                        *
C********************************************************************************
      SUBROUTINE CROSS0(OMEG0,E,TEMP1,I)
      DIMENSION OMEG0(3),E(25,3)
      DIMENSION TEMP1(3)
      TEMP1(1)=OMEG0(2)*E(I,3)-OMEG0(3)*E(I,2)
      TEMP1(2)=OMEG0(3)*E(I,1)-OMEG0(1)*E(I,3)
      TEMP1(3)=OMEG0(1)*E(I,2)-OMEG0(2)*E(I,1)
      RETURN
      END
C********************************************************************************
C     SUBROUTINE : DOT1                                               *
C--------------------------------------------------------------------*
C     FUNCTION :   SCALAR PRODUCT OF TWO VECTORS                      *
C--------------------------------------------------------------------*
C     INPUT VARIABLES:                                                *
C                OMEG0    VECTOR (3x1)                                 *
C                R0       VECTOR (3x1)                                 *
C--------------------------------------------------------------------*
C     OUTPUT VARIABLES:                                               *
C                DOT1     DOT PRODUCT OMEG0 . R0                       *
C--------------------------------------------------------------------*
C     SUBROUTINES CALLED: NONE                                        *
C********************************************************************************
      FUNCTION DOT1(OMEG0,R0)
      DIMENSION OMEG0(3),R0(3)
      DOT1=OMEG0(1)*R0(1)+OMEG0(2)*R0(2)+OMEG0(3)*R0(3)
      RETURN
      END
C********************************************************************************
C     SUBROUTINE : DOT0                                               *
C--------------------------------------------------------------------*
C     FUNCTION :   SCALAR PRODUCT OF THE VECTOR (3x1)                 *
C                  AND THE J-TH COLUMN OF THE MATRIX (3x3)            *
C--------------------------------------------------------------------*
C     INPUT VARIABLES:                                                *
C                OMEGI    VECTOR (3x1)                                 *
C                Q        MATRIX (3x3)                                 *
C                J        INDEX OF THE J-TH COLUMN OF MATRIX Q         *
C--------------------------------------------------------------------*
C     OUTPUT VARIABLES:                                               *
C                DOT0     DOT PRODUCT OMEGI(x) . Q(x,j)               *
C--------------------------------------------------------------------*
C     SUBROUTINES CALLED: NONE                                        *
C********************************************************************************
      FUNCTION DOT0(OMEGI,Q,J)
      DIMENSION OMEGI(3),Q(3,3)
      DOT0=OMEGI(1)*Q(1,J)+OMEGI(2)*Q(2,J)+OMEGI(3)*Q(3,J)
      RETURN
      END
```

Listing of the file INDAT.DAT for Example 1.

```
      1
      3
      1      5      7
      8      9      9
      1      2      3      4      5      6      7      8
      1      2      3      4      5      9     10     11     12
      1      2      3      4      5      9     10     13     14
     14
      1      0      0      0      0      0      0
           0.            0.            0.            0.
      2      0      0      0      0      0      0
         .153        .00006        .00055        .00045
      3      0      0      0      0      0      0
         .321        .00393        .00393        .00038
      4      0      0      0      0      0      0
        .841         .0112         .012          .003
      5      0      0      0      0      0      0
         .696          .007        .00565        .00627
      6      0      0      0      0      0      0
         .841        .01120        .01200         .0030
      7      0      0      0      0      0      0
         .321        .00393        .00393        .00038
      8      0      0      0      0      0      0
         .153        .00006        .00055        .00045
      9      0      0      1      0      0      0
           0.            0.            0.
     10      0      0      1      0      0      0
        3.085      0.15140       0.13700       0.02830
     11      0      0      0      0      0      0
        0.207      0.00200       0.00200       0.00022
     12      0      0      0      0      0      0
        0.114      0.00250       0.00425       0.00014
     13      0      0      0      0      0      0
        0.207      0.00200       0.00200       0.00022
     14      0      0      0      0      0      0
        0.114      0.00250       0.00425       0.00014
           1.            0.            0.
           0.            0.         0.0001
           0.            0.        -0.0001
           0.            1.            0.
           0.            0.          .03
         .000            0.        -0.070
           0.            1.            0.
           0.            0.         0.210
           0.            0.         -.210
           0.            1.            0.
         0.00            0.         0.220
         0.00            0.        -0.220
           0.            1.            0.
         0.00        0.135         0.100
         0.00       -0.135         0.100
         0.00       -0.135         0.100
         .000            0.         -.050
           0.           -1.            0.
         0.00            0.        -0.220
         0.00            0.         0.220
           0.           -1.            0.
           0.            0.         -.210
           0.            0.         0.210
           0.           -1.            0.
           0.            0.         -.070
```

```
       0.           1.           0.
       0.           0.          .00001
       0.           0.         -.00001
       1.           0.           0.
     .000           0.          .340
     .00          .200         -0.060
     .000         .200         -0.060
     .000        -.200         -0.060
       1.           0.           0.
       0.           0.         -0.154
       0.           0.          0.154
       1.           0.           0.
       0.           0.          -.132
      -1.           0.           0.
       0.           0.         -0.154
       0.           0.          0.154
      -1.           0.           0.
      .0.           0.          -.132
       1.           0.           0.
       0.           0.          -.0001
     1.
   0.001
    40.
  2
  5
     0.          0.75
     1.          -1.           1.          -1.
.001,0.001,0.001,0.001
.2265,0.0118,0.8859,1.4549
```

Listing of the file PREDYN.DAT for Example 1.

```
  41
0.018750  0.750000  0.600000
   1
      0.000000E+00     0.000000E+00     0.000000E+00
      0.000000E+00     0.000000E+00     0.000000E+00
      0.000000E+00     0.000000E+00     0.000000E+00
      0.000000E+00     0.000000E+00     0.000000E+00
      0.000000E+00     0.000000E+00     0.000000E+00
      0.000000E+00     0.000000E+00     0.000000E+00
      0.000000E+00     0.000000E+00     0.000000E+00
      0.000000E+00     0.000000E+00     0.000000E+00
      0.000000E+00     0.000000E+00     0.000000E+00
      0.000000E+00     0.000000E+00     0.000000E+00
      0.000000E+00     0.000000E+00     0.000000E+00
      0.000000E+00     0.000000E+00     0.000000E+00
      0.000000E+00     0.000000E+00     0.000000E+00
      0.000000E+00     0.000000E+00     0.000000E+00
      0.000000E+00     0.000000E+00     0.000000E+00
      0.000000E+00     0.000000E+00     0.000000E+00
      0.000000E+00     0.000000E+00     0.000000E+00
      0.000000E+00     0.000000E+00     0.000000E+00
      0.000000E+00     0.000000E+00     0.000000E+00
      0.000000E+00     0.000000E+00     0.000000E+00
      0.000000E+00     0.000000E+00     0.000000E+00
      0.000000E+00     0.000000E+00     0.000000E+00
      0.000000E+00     0.000000E+00     0.000000E+00
      0.000000E+00     0.000000E+00     0.000000E+00
```

| | | |
|---|---|---|
| 0.000000E+00 | 0.000000E+00 | 0.000000E+00 |
| 0.000000E+00 | 0.000000E+00 | 0.000000E+00 |
| 0.000000E+00 | 0.000000E+00 | 0.000000E+00 |
| 0.000000E+00 | 0.000000E+00 | 0.000000E+00 |
| 0.000000E+00 | 0.000000E+00 | 0.000000E+00 |
| 0.000000E+00 | 0.000000E+00 | 0.000000E+00 |
| 0.000000E+00 | 0.000000E+00 | 0.000000E+00 |
| 0.000000E+00 | 0.000000E+00 | 0.000000E+00 |
| 0.000000E+00 | 0.000000E+00 | 0.000000E+00 |
| 0.000000E+00 | 0.000000E+00 | 0.000000E+00 |
| 0.000000E+00 | 0.000000E+00 | 0.000000E+00 |
| 0.000000E+00 | 0.000000E+00 | 0.000000E+00 |
| 0.000000E+00 | 0.000000E+00 | 0.000000E+00 |
| 0.000000E+00 | 0.000000E+00 | 0.000000E+00 |
| 0.000000E+00 | 0.000000E+00 | 0.000000E+00 |
| 0.000000E+00 | 0.000000E+00 | 0.000000E+00 |
| 0.000000E+00 | 0.000000E+00 | 0.000000E+00 |
| 0.000000E+00 | 0.000000E+00 | 0.000000E+00 |

2

| | | |
|---|---|---|
| -0.349200E-01 | 0.465600E+00 | -0.571733E-04 |
| -0.261600E-01 | 0.465600E+00 | -0.123733E+01 |
| -0.174600E-01 | 0.418400E+00 | -0.372267E+01 |
| -0.104400E-01 | 0.325600E+00 | -0.496000E+01 |
| -0.523200E-02 | 0.232800E+00 | -0.496000E+01 |
| -0.174600E-02 | 0.139200E+00 | -0.496000E+01 |
| 0.000000E+00 | 0.465600E-01 | -0.372267E+01 |
| 0.000000E+00 | 0.000000E+00 | -0.123733E+01 |
| 0.000000E+00 | 0.000000E+00 | 0.000000E+00 |
| 0.000000E+00 | 0.000000E+00 | 0.000000E+00 |
| 0.000000E+00 | 0.000000E+00 | 0.000000E+00 |
| 0.000000E+00 | 0.000000E+00 | 0.000000E+00 |
| 0.000000E+00 | 0.000000E+00 | 0.000000E+00 |
| 0.000000E+00 | 0.000000E+00 | 0.000000E+00 |
| 0.000000E+00 | 0.000000E+00 | 0.000000E+00 |
| 0.000000E+00 | 0.000000E+00 | 0.000000E+00 |
| 0.000000E+00 | 0.000000E+00 | 0.000000E+00 |
| 0.000000E+00 | 0.000000E+00 | 0.000000E+00 |
| 0.000000E+00 | 0.000000E+00 | 0.000000E+00 |
| 0.000000E+00 | 0.000000E+00 | 0.000000E+00 |
| 0.000000E+00 | 0.000000E+00 | 0.000000E+00 |
| 0.000000E+00 | 0.000000E+00 | 0.000000E+00 |
| 0.000000E+00 | 0.000000E+00 | 0.000000E+00 |
| 0.000000E+00 | 0.000000E+00 | 0.000000E+00 |
| 0.000000E+00 | 0.000000E+00 | 0.000000E+00 |
| 0.000000E+00 | 0.000000E+00 | 0.000000E+00 |
| 0.000000E+00 | 0.000000E+00 | 0.000000E+00 |
| 0.000000E+00 | 0.000000E+00 | 0.000000E+00 |
| 0.000000E+00 | 0.000000E+00 | 0.000000E+00 |
| 0.000000E+00 | 0.000000E+00 | 0.000000E+00 |
| 0.000000E+00 | 0.000000E+00 | 0.000000E+00 |
| 0.572400E-01 | 0.763200E+00 | -0.126933E-05 |
| 0.714000E-01 | 0.763200E+00 | 0.000000E+00 |
| 0.858000E-01 | 0.763200E+00 | 0.000000E+00 |
| 0.100200E+00 | 0.763200E+00 | 0.000000E+00 |
| 0.114600E+00 | 0.763200E+00 | -0.762667E-04 |
| 0.129000E+00 | 0.763200E+00 | -0.508800E+01 |
| 0.142800E+00 | 0.572000E+00 | -0.152533E+02 |
| 0.150000E+00 | 0.190400E+00 | -0.203733E+02 |
| 0.150000E+00 | -0.191200E+00 | -0.203733E+02 |

3

| | | |
|---|---|---|
| -0.412800E-01 | -0.848000E+00 | 0.571733E-04 |
| -0.572400E-01 | -0.848000E+00 | 0.338133E+01 |
| -0.732000E-01 | -0.722400E+00 | 0.101440E+02 |
| -0.840000E-01 | -0.468800E+00 | 0.135467E+02 |

| | | |
|---|---|---|
| -0.900000E-01 | -0.215200E+00 | 0.135467E+02 |
| -0.918000E-01 | 0.384000E-01 | 0.135467E+02 |
| -0.888000E-01 | 0.292000E+00 | 0.101440E+02 |
| -0.810000E-01 | 0.419200E+00 | 0.338133E+01 |
| -0.732000E-01 | 0.419200E+00 | -0.253867E-04 |
| -0.654000E-01 | 0.418400E+00 | -0.253867E-04 |
| -0.576000E-01 | 0.418400E+00 | 0.253867E-04 |
| -0.497400E-01 | 0.419200E+00 | 0.762667E-05 |
| -0.418800E-01 | 0.418400E+00 | -0.292267E-04 |
| -0.340200E-01 | 0.418400E+00 | 0.126933E-05 |
| -0.261600E-01 | 0.418400E+00 | 0.178133E-04 |
| -0.183000E-01 | 0.418400E+00 | 0.508800E-05 |
| -0.105000E-01 | 0.418400E+00 | -0.635733E-05 |
| -0.261600E-02 | 0.418400E+00 | -0.635733E-05 |
| 0.523200E-02 | 0.418400E+00 | 0.635733E-05 |
| 0.130800E-01 | 0.418400E+00 | 0.635733E-05 |
| 0.209400E-01 | 0.418400E+00 | -0.508800E-05 |
| 0.288000E-01 | 0.418400E+00 | -0.178133E-04 |
| 0.366600E-01 | 0.418400E+00 | -0.126933E-05 |
| 0.445200E-01 | 0.418400E+00 | 0.292267E-04 |
| 0.523200E-01 | 0.419200E+00 | -0.762667E-05 |
| 0.600000E-01 | 0.418400E+00 | -0.253867E-04 |
| 0.678000E-01 | 0.418400E+00 | 0.253867E-04 |
| 0.762000E-01 | 0.419200E+00 | 0.253867E-04 |
| 0.840000E-01 | 0.419200E+00 | -0.253867E-04 |
| 0.918000E-01 | 0.418400E+00 | -0.253867E-04 |
| 0.996000E-01 | 0.418400E+00 | 0.253867E-04 |
| 0.107400E+00 | 0.419200E+00 | -0.508800E-04 |
| 0.579600E-01 | -0.344800E+00 | -0.749867E-04 |
| 0.516000E-01 | -0.344800E+00 | 0.167467E+01 |
| 0.450000E-01 | -0.281600E+00 | 0.502400E+01 |
| 0.408000E-01 | -0.156000E+00 | 0.669867E+01 |
| 0.390000E-01 | -0.304000E-01 | 0.669867E+01 |
| 0.396000E-01 | 0.928000E-01 | 0.118400E+02 |
| 0.432000E-01 | 0.412000E+00 | 0.202667E+02 |
| 0.552000E-01 | 0.856000E+00 | 0.220800E+02 |
| 0.750000E-01 | 0.296000E+00 | 0.203733E+02 |

4

| | | |
|---|---|---|
| -0.504000E-01 | 0.677600E+00 | 0.000000E+00 |
| -0.378000E-01 | 0.677600E+00 | -0.192000E+01 |
| -0.252000E-01 | 0.605600E+00 | -0.574933E+01 |
| -0.156000E-01 | 0.461600E+00 | -0.766933E+01 |
| -0.840000E-02 | 0.317600E+00 | -0.766933E+01 |
| -0.360000E-02 | 0.174400E+00 | -0.766933E+01 |
| -0.180000E-02 | 0.304000E-01 | -0.574933E+01 |
| -0.240000E-02 | -0.424000E-01 | -0.192000E+01 |
| -0.300000E-02 | -0.424000E-01 | 0.253867E-04 |
| -0.420000E-02 | -0.416000E-01 | 0.253867E-04 |
| -0.480000E-02 | -0.416000E-01 | -0.253867E-04 |
| -0.546000E-02 | -0.424000E-01 | -0.762667E-05 |
| -0.630000E-02 | -0.416000E-01 | 0.305067E-04 |
| -0.708000E-02 | -0.416000E-01 | -0.101653E-04 |
| -0.786000E-02 | -0.416000E-01 | -0.152533E-04 |
| -0.864000E-02 | -0.416000E-01 | 0.760533E-05 |
| -0.942000E-02 | -0.416000E-01 | 0.253867E-05 |
| -0.102000E-01 | -0.416000E-01 | -0.126933E-05 |
| -0.109800E-01 | -0.416000E-01 | -0.120533E-04 |
| -0.117600E-01 | -0.416000E-01 | -0.508800E-05 |
| -0.125400E-01 | -0.416000E-01 | 0.126933E-04 |
| -0.133800E-01 | -0.416000E-01 | 0.242133E-04 |
| -0.141600E-01 | -0.416000E-01 | 0.126933E-05 |
| -0.149400E-01 | -0.416000E-01 | -0.317867E-04 |
| -0.156600E-01 | -0.424000E-01 | -0.890667E-05 |
| -0.162600E-01 | -0.416000E-01 | -0.134400E+01 |
| -0.170400E-01 | -0.920000E-01 | -0.402133E+01 |

| | | |
|---|---|---|
| -0.202200E-01 | -0.192800E+00 | -0.536533E+01 |
| -0.247200E-01 | -0.293600E+00 | -0.536533E+01 |
| -0.312000E-01 | -0.393600E+00 | -0.536533E+01 |
| -0.396000E-01 | -0.493600E+00 | -0.402133E+01 |
| -0.495600E-01 | -0.544800E+00 | -0.134400E+01 |
| -0.597000E-01 | -0.544000E+00 | 0.762667E-04 |
| -0.696000E-01 | -0.544000E+00 | -0.219733E+01 |
| -0.798000E-01 | -0.626400E+00 | -0.659200E+01 |
| -0.936000E-01 | -0.791200E+00 | -0.878933E+01 |
| -0.109800E+00 | -0.960000E+00 | -0.878933E+01 |
| -0.129000E+00 | -0.112000E+01 | -0.878933E+01 |
| -0.151800E+00 | -0.128800E+01 | -0.659200E+01 |
| -0.177000E+00 | -0.136800E+01 | -0.219733E+01 |
| -0.202800E+00 | -0.425600E+00 | 0.101760E-03 |

5

| | | |
|---|---|---|
| 0.126600E+00 | -0.293600E+00 | 0.000000E+00 |
| 0.121200E+00 | -0.293600E+00 | -0.222933E+00 |
| 0.115800E+00 | -0.301600E+00 | -0.668800E+00 |
| 0.109800E+00 | -0.318400E+00 | -0.891734E+00 |
| 0.103800E+00 | -0.335200E+00 | -0.891734E+00 |
| 0.972000E-01 | -0.352000E+00 | -0.891734E+00 |
| 0.906000E-01 | -0.368800E+00 | -0.668800E+00 |
| 0.834000E-01 | -0.376800E+00 | -0.222933E+00 |
| 0.762000E-01 | -0.376800E+00 | 0.000000E+00 |
| 0.696000E-01 | -0.376800E+00 | 0.000000E+00 |
| 0.624000E-01 | -0.376800E+00 | 0.000000E+00 |
| 0.552000E-01 | -0.376800E+00 | 0.000000E+00 |
| 0.481800E-01 | -0.376800E+00 | -0.126933E-05 |
| 0.411000E-01 | -0.376800E+00 | 0.889600E-05 |
| 0.340200E-01 | -0.376800E+00 | -0.253867E-05 |
| 0.269400E-01 | -0.376800E+00 | -0.126933E-04 |
| 0.199200E-01 | -0.376800E+00 | 0.381867E-05 |
| 0.128400E-01 | -0.376800E+00 | 0.762667E-05 |
| 0.575400E-02 | -0.376800E+00 | 0.573867E-05 |
| -0.131400E-02 | -0.376800E+00 | -0.126933E-05 |
| -0.840000E-02 | -0.376800E+00 | -0.762667E-05 |
| -0.154200E-01 | -0.376800E+00 | -0.635733E-05 |
| -0.225000E-01 | -0.376800E+00 | 0.000000E+00 |
| -0.295800E-01 | -0.376800E+00 | 0.253867E-05 |
| -0.366600E-01 | -0.376800E+00 | 0.165333E-04 |
| -0.437400E-01 | -0.376800E+00 | 0.134400E+01 |
| -0.507600E-01 | -0.326400E+00 | 0.402133E+01 |
| -0.559800E-01 | -0.226400E+00 | 0.536533E+01 |
| -0.592800E-01 | -0.125600E+00 | 0.536533E+01 |
| -0.606000E-01 | -0.251200E-01 | 0.536533E+01 |
| -0.600000E-01 | 0.753600E-01 | 0.402133E+01 |
| -0.578400E-01 | 0.125600E+00 | 0.134400E+01 |
| -0.555000E-01 | 0.125600E+00 | 0.000000E+00 |
| -0.531600E-01 | 0.125600E+00 | 0.521600E+00 |
| -0.507600E-01 | 0.144800E+00 | 0.156800E+01 |
| -0.477000E-01 | 0.184000E+00 | 0.209067E+01 |
| -0.438600E-01 | 0.223200E+00 | 0.209067E+01 |
| -0.393000E-01 | 0.262400E+00 | 0.209067E+01 |
| -0.340200E-01 | 0.301600E+00 | 0.156800E+01 |
| -0.280200E-01 | 0.320800E+00 | 0.521600E+00 |
| -0.220200E-01 | 0.320800E+00 | 0.000000E+00 |

6

| | | |
|---|---|---|
| -0.220200E-01 | 0.320800E+00 | 0.000000E+00 |
| -0.159600E-01 | 0.320800E+00 | 0.000000E+00 |
| -0.996000E-02 | 0.320800E+00 | -0.888533E-15 |
| -0.393000E-02 | 0.320800E+00 | -0.381867E-05 |
| 0.209400E-02 | 0.320800E+00 | -0.253867E-05 |
| 0.810000E-02 | 0.320800E+00 | -0.635733E-05 |
| 0.141600E-01 | 0.320800E+00 | -0.508800E-05 |
| 0.201600E-01 | 0.320800E+00 | -0.296533E-15 |

| | | |
|---|---|---|
| 0.261600E-01 | 0.320800E+00 | -0.229333E-04 |
| 0.322200E-01 | 0.320800E+00 | -0.178133E-04 |
| 0.382200E-01 | 0.320800E+00 | 0.343467E-04 |
| 0.442200E-01 | 0.320800E+00 | 0.292267E-04 |
| 0.502800E-01 | 0.320800E+00 | 0.000000E+00 |
| 0.562800E-01 | 0.320800E+00 | 0.000000E+00 |
| 0.624000E-01 | 0.320800E+00 | 0.000000E+00 |
| 0.684000E-01 | 0.320800E+00 | 0.000000E+00 |
| 0.744000E-01 | 0.320800E+00 | 0.000000E+00 |
| 0.804000E-01 | 0.320800E+00 | 0.000000E+00 |
| 0.864000E-01 | 0.320800E+00 | 0.000000E+00 |
| 0.924000E-01 | 0.320800E+00 | 0.000000E+00 |
| 0.984000E-01 | 0.320800E+00 | 0.000000E+00 |
| 0.104400E+00 | 0.320800E+00 | 0.000000E+00 |
| 0.110400E+00 | 0.320800E+00 | -0.762667E-04 |
| 0.116400E+00 | 0.320800E+00 | -0.236800E-13 |
| 0.122400E+00 | 0.320800E+00 | 0.126933E-03 |
| 0.128400E+00 | 0.320800E+00 | 0.178133E-03 |
| 0.134400E+00 | 0.320800E+00 | -0.762667E-04 |
| 0.140400E+00 | 0.320800E+00 | -0.101760E-03 |
| 0.146400E+00 | 0.320800E+00 | -0.178133E-13 |
| 0.152400E+00 | 0.320800E+00 | -0.164267E+01 |
| 0.158400E+00 | 0.260000E+00 | -0.491733E+01 |
| 0.162600E+00 | 0.136800E+00 | -0.654933E+01 |
| 0.163800E+00 | 0.139200E-01 | -0.654933E+01 |
| 0.162600E+00 | -0.108800E+00 | -0.654933E+01 |
| 0.159600E+00 | -0.232000E+00 | -0.491733E+01 |
| 0.154200E+00 | -0.293600E+00 | -0.164267E+01 |
| 0.148800E+00 | -0.293600E+00 | -0.762667E-04 |
| 0.143400E+00 | -0.293600E+00 | 0.000000E+00 |
| 0.137400E+00 | -0.293600E+00 | 0.762667E-04 |
| 0.132000E+00 | -0.293600E+00 | -0.762667E-04 |
| 0.126600E+00 | -0.293600E+00 | 0.000000E+00 |

7

| | | |
|---|---|---|
| -0.202800E+00 | -0.136800E+01 | 0.101760E-03 |
| -0.228600E+00 | -0.136800E+01 | 0.596267E+01 |
| -0.254400E+00 | -0.114400E+01 | 0.179200E+02 |
| -0.271200E+00 | -0.697600E+00 | 0.237867E+02 |
| -0.280200E+00 | -0.251200E+00 | 0.237867E+02 |
| -0.281400E+00 | 0.196000E+00 | 0.237867E+02 |
| -0.273600E+00 | 0.639200E+00 | 0.179200E+02 |
| -0.257400E+00 | 0.864000E+00 | 0.596267E+01 |
| -0.241200E+00 | 0.864000E+00 | -0.788267E-04 |
| -0.225000E+00 | 0.864000E+00 | -0.214400E+01 |
| -0.208800E+00 | 0.783200E+00 | -0.642133E+01 |
| -0.195600E+00 | 0.623200E+00 | -0.856533E+01 |
| -0.184800E+00 | 0.464800E+00 | -0.856533E+01 |
| -0.177600E+00 | 0.304000E+00 | -0.856533E+01 |
| -0.173400E+00 | 0.143200E+00 | -0.642133E+01 |
| -0.172200E+00 | 0.632000E-01 | -0.214400E+01 |
| -0.171000E+00 | 0.632000E-01 | 0.253867E-04 |
| -0.169800E+00 | 0.632000E-01 | -0.253867E-04 |
| -0.168600E+00 | 0.632000E-01 | -0.253867E-04 |
| -0.167400E+00 | 0.632000E-01 | 0.000000E+00 |
| -0.166200E+00 | 0.632000E-01 | 0.000000E+00 |
| -0.165000E+00 | 0.632000E-01 | 0.000000E+00 |
| -0.163800E+00 | 0.632000E-01 | 0.762667E-04 |
| -0.162600E+00 | 0.632000E-01 | 0.126933E-05 |
| -0.161400E+00 | 0.632000E-01 | -0.125867E-03 |
| -0.160200E+00 | 0.632000E-01 | -0.179200E-03 |
| -0.159000E+00 | 0.632000E-01 | 0.749867E-04 |
| -0.157800E+00 | 0.632000E-01 | 0.105600E-03 |
| -0.157200E+00 | 0.632000E-01 | 0.762667E-05 |
| -0.155400E+00 | 0.632000E-01 | 0.164267E+01 |
| -0.154200E+00 | 0.124000E+00 | 0.491733E+01 |

|  |  |  |
|---|---|---|
| -0.151200E+00 | 0.247200E+00 | 0.654933E+01 |
| -0.145800E+00 | 0.370400E+00 | 0.654933E+01 |
| -0.136800E+00 | 0.492800E+00 | 0.654933E+01 |
| -0.127200E+00 | 0.616000E+00 | 0.491733E+01 |
| -0.114600E+00 | 0.677600E+00 | 0.164267E+01 |
| -0.101400E+00 | 0.677600E+00 | 0.125867E-03 |
| -0.888000E-01 | 0.677600E+00 | -0.592000E-14 |
| -0.756000E-01 | 0.677600E+00 | -0.101653E-03 |
| -0.630000E-01 | 0.677600E+00 | 0.762667E-04 |
| -0.504000E-01 | 0.677600E+00 | 0.000000E+00 |

8

|  |  |  |
|---|---|---|
| 0.750000E-01 | 0.124000E+01 | 0.203733E+02 |
| 0.102000E+00 | 0.162400E+01 | 0.144000E+02 |
| 0.135600E+00 | 0.177600E+01 | -0.266667E+01 |
| 0.168000E+00 | 0.152000E+01 | -0.186667E+02 |
| 0.192600E+00 | 0.107200E+01 | -0.237867E+02 |
| 0.208800E+00 | 0.627200E+00 | -0.268800E+02 |
| 0.216600E+00 | 0.720000E-01 | -0.270933E+02 |
| 0.211200E+00 | -0.384000E+00 | -0.181333E+02 |
| 0.202200E+00 | -0.612000E+00 | -0.121600E+02 |
| 0.188400E+00 | -0.840000E+00 | -0.100160E+02 |
| 0.170400E+00 | -0.992000E+00 | -0.274133E+01 |
| 0.151200E+00 | -0.944000E+00 | 0.551467E+01 |
| 0.134400E+00 | -0.785600E+00 | 0.856533E+01 |
| 0.121200E+00 | -0.624800E+00 | 0.856533E+01 |
| 0.111000E+00 | -0.464000E+00 | 0.642133E+01 |
| 0.103800E+00 | -0.384000E+00 | 0.214400E+01 |
| 0.966000E-01 | -0.384000E+00 | -0.253867E-04 |
| 0.894000E-01 | -0.384000E+00 | 0.253867E-04 |
| 0.822000E-01 | -0.384000E+00 | 0.253867E-04 |
| 0.750000E-01 | -0.384000E+00 | 0.000000E+00 |
| 0.678000E-01 | -0.384000E+00 | 0.000000E+00 |
| 0.606000E-01 | -0.384000E+00 | 0.000000E+00 |
| 0.536400E-01 | -0.384000E+00 | 0.000000E+00 |
| 0.464400E-01 | -0.384000E+00 | -0.126933E-05 |
| 0.392400E-01 | -0.384000E+00 | -0.126933E-05 |
| 0.320400E-01 | -0.384000E+00 | 0.372267E+01 |
| 0.248400E-01 | -0.244800E+00 | 0.112000E+02 |
| 0.228600E-01 | 0.344000E-01 | 0.149333E+02 |
| 0.262200E-01 | 0.314400E+00 | 0.149333E+02 |
| 0.346800E-01 | 0.592000E+00 | 0.993067E+01 |
| 0.484200E-01 | 0.688000E+00 | -0.372267E+01 |
| 0.606000E-01 | 0.456000E+00 | -0.161067E+02 |
| 0.654000E-01 | 0.816000E-01 | -0.198400E+02 |
| 0.636000E-01 | -0.291200E+00 | -0.198400E+02 |
| 0.543000E-01 | -0.663200E+00 | -0.149333E+02 |
| 0.387000E-01 | -0.848000E+00 | -0.496000E+01 |
| 0.225000E-01 | -0.848000E+00 | -0.242133E-04 |
| 0.690000E-02 | -0.848000E+00 | 0.508800E-04 |
| -0.948000E-02 | -0.848000E+00 | 0.228267E-03 |
| -0.253800E-01 | -0.848000E+00 | 0.125867E-03 |
| -0.412800E-01 | -0.848000E+00 | 0.571733E-04 |

11

|  |  |  |
|---|---|---|
| 0.186000E+01 | 0.000000E+00 | 0.000000E+00 |
| 0.186000E+01 | 0.000000E+00 | 0.000000E+00 |
| 0.186000E+01 | 0.000000E+00 | 0.000000E+00 |
| 0.186000E+01 | 0.000000E+00 | 0.000000E+00 |
| 0.186000E+01 | 0.000000E+00 | 0.000000E+00 |
| 0.186000E+01 | 0.000000E+00 | 0.000000E+00 |
| 0.186000E+01 | 0.000000E+00 | 0.000000E+00 |
| 0.186000E+01 | 0.000000E+00 | 0.000000E+00 |
| 0.186000E+01 | 0.000000E+00 | 0.000000E+00 |
| 0.186000E+01 | 0.000000E+00 | 0.000000E+00 |
| 0.186000E+01 | 0.000000E+00 | 0.000000E+00 |
| 0.186000E+01 | 0.000000E+00 | 0.000000E+00 |

| | | |
|---|---|---|
| 0.186000E+01 | 0.000000E+00 | 0.000000E+00 |
| 0.186000E+01 | 0.000000E+00 | 0.000000E+00 |
| 0.186000E+01 | 0.000000E+00 | 0.000000E+00 |
| 0.186000E+01 | 0.000000E+00 | 0.000000E+00 |
| 0.186000E+01 | 0.000000E+00 | 0.000000E+00 |
| 0.186000E+01 | 0.000000E+00 | 0.000000E+00 |
| 0.186000E+01 | 0.000000E+00 | 0.000000E+00 |
| 0.186000E+01 | 0.000000E+00 | 0.000000E+00 |
| 0.186000E+01 | 0.000000E+00 | 0.000000E+00 |
| 0.186000E+01 | 0.000000E+00 | 0.000000E+00 |
| 0.186000E+01 | 0.000000E+00 | 0.000000E+00 |
| 0.186000E+01 | 0.000000E+00 | 0.000000E+00 |
| 0.186000E+01 | 0.000000E+00 | 0.000000E+00 |
| 0.186000E+01 | 0.000000E+00 | 0.000000E+00 |
| 0.186000E+01 | 0.000000E+00 | 0.000000E+00 |
| 0.186000E+01 | 0.000000E+00 | 0.000000E+00 |
| 0.186000E+01 | 0.000000E+00 | 0.000000E+00 |
| 0.186000E+01 | 0.000000E+00 | 0.000000E+00 |
| 0.186000E+01 | 0.000000E+00 | 0.000000E+00 |
| 0.186000E+01 | 0.000000E+00 | 0.000000E+00 |
| 0.186000E+01 | 0.000000E+00 | 0.000000E+00 |
| 0.186000E+01 | 0.000000E+00 | 0.000000E+00 |
| 0.186000E+01 | 0.000000E+00 | 0.000000E+00 |
| 0.186000E+01 | 0.000000E+00 | 0.000000E+00 |
| 0.186000E+01 | 0.000000E+00 | 0.000000E+00 |
| 0.186000E+01 | 0.000000E+00 | 0.000000E+00 |
| 0.186000E+01 | 0.000000E+00 | 0.000000E+00 |
| 0.186000E+01 | 0.000000E+00 | 0.000000E+00 |

12

| | | |
|---|---|---|
| 0.157000E+01 | 0.000000E+00 | 0.000000E+00 |
| 0.157000E+01 | 0.000000E+00 | 0.000000E+00 |
| 0.157000E+01 | 0.000000E+00 | 0.000000E+00 |
| 0.157000E+01 | 0.000000E+00 | 0.000000E+00 |
| 0.157000E+01 | 0.000000E+00 | 0.000000E+00 |
| 0.157000E+01 | 0.000000E+00 | 0.000000E+00 |
| 0.157000E+01 | 0.000000E+00 | 0.000000E+00 |
| 0.157000E+01 | 0.000000E+00 | 0.000000E+00 |
| 0.157000E+01 | 0.000000E+00 | 0.000000E+00 |
| 0.157000E+01 | 0.000000E+00 | 0.000000E+00 |
| 0.157000E+01 | 0.000000E+00 | 0.000000E+00 |
| 0.157000E+01 | 0.000000E+00 | 0.000000E+00 |
| 0.157000E+01 | 0.000000E+00 | 0.000000E+00 |
| 0.157000E+01 | 0.000000E+00 | 0.000000E+00 |
| 0.157000E+01 | 0.000000E+00 | 0.000000E+00 |
| 0.157000E+01 | 0.000000E+00 | 0.000000E+00 |
| 0.157000E+01 | 0.000000E+00 | 0.000000E+00 |
| 0.157000E+01 | 0.000000E+00 | 0.000000E+00 |
| 0.157000E+01 | 0.000000E+00 | 0.000000E+00 |
| 0.157000E+01 | 0.000000E+00 | 0.000000E+00 |
| 0.157000E+01 | 0.000000E+00 | 0.000000E+00 |
| 0.157000E+01 | 0.000000E+00 | 0.000000E+00 |
| 0.157000E+01 | 0.000000E+00 | 0.000000E+00 |
| 0.157000E+01 | 0.000000E+00 | 0.000000E+00 |
| 0.157000E+01 | 0.000000E+00 | 0.000000E+00 |
| 0.157000E+01 | 0.000000E+00 | 0.000000E+00 |
| 0.157000E+01 | 0.000000E+00 | 0.000000E+00 |
| 0.157000E+01 | 0.000000E+00 | 0.000000E+00 |
| 0.157000E+01 | 0.000000E+00 | 0.000000E+00 |
| 0.157000E+01 | 0.000000E+00 | 0.000000E+00 |
| 0.157000E+01 | 0.000000E+00 | 0.000000E+00 |
| 0.157000E+01 | 0.000000E+00 | 0.000000E+00 |
| 0.157000E+01 | 0.000000E+00 | 0.000000E+00 |
| 0.157000E+01 | 0.000000E+00 | 0.000000E+00 |
| 0.157000E+01 | 0.000000E+00 | 0.000000E+00 |
| 0.157000E+01 | 0.000000E+00 | 0.000000E+00 |

| | | |
|---|---|---|
| 0.157000E+01 | 0.000000E+00 | 0.000000E+00 |
| 0.157000E+01 | 0.000000E+00 | 0.000000E+00 |
| 0.157000E+01 | 0.000000E+00 | 0.000000E+00 |
| 0.157000E+01 | 0.000000E+00 | 0.000000E+00 |
| 0.157000E+01 | 0.000000E+00 | 0.000000E+00 |
| 0.157000E+01 | 0.000000E+00 | 0.000000E+00 |

13

| | | |
|---|---|---|
| 0.186000E+01 | 0.000000E+00 | 0.000000E+00 |
| 0.186000E+01 | 0.000000E+00 | 0.000000E+00 |
| 0.186000E+01 | 0.000000E+00 | 0.000000E+00 |
| 0.186000E+01 | 0.000000E+00 | 0.000000E+00 |
| 0.186000E+01 | 0.000000E+00 | 0.000000E+00 |
| 0.186000E+01 | 0.000000E+00 | 0.000000E+00 |
| 0.186000E+01 | 0.000000E+00 | 0.000000E+00 |
| 0.186000E+01 | 0.000000E+00 | 0.000000E+00 |
| 0.186000E+01 | 0.000000E+00 | 0.000000E+00 |
| 0.186000E+01 | 0.000000E+00 | 0.000000E+00 |
| 0.186000E+01 | 0.000000E+00 | 0.000000E+00 |
| 0.186000E+01 | 0.000000E+00 | 0.000000E+00 |
| 0.186000E+01 | 0.000000E+00 | 0.000000E+00 |
| 0.186000E+01 | 0.000000E+00 | 0.000000E+00 |
| 0.186000E+01 | 0.000000E+00 | 0.000000E+00 |
| 0.186000E+01 | 0.000000E+00 | 0.000000E+00 |
| 0.186000E+01 | 0.000000E+00 | 0.000000E+00 |
| 0.186000E+01 | 0.000000E+00 | 0.000000E+00 |
| 0.186000E+01 | 0.000000E+00 | 0.000000E+00 |
| 0.186000E+01 | 0.000000E+00 | 0.000000E+00 |
| 0.186000E+01 | 0.000000E+00 | 0.000000E+00 |
| 0.186000E+01 | 0.000000E+00 | 0.000000E+00 |
| 0.186000E+01 | 0.000000E+00 | 0.000000E+00 |
| 0.186000E+01 | 0.000000E+00 | 0.000000E+00 |
| 0.186000E+01 | 0.000000E+00 | 0.000000E+00 |
| 0.186000E+01 | 0.000000E+00 | 0.000000E+00 |
| 0.186000E+01 | 0.000000E+00 | 0.000000E+00 |
| 0.186000E+01 | 0.000000E+00 | 0.000000E+00 |
| 0.186000E+01 | 0.000000E+00 | 0.000000E+00 |
| 0.186000E+01 | 0.000000E+00 | 0.000000E+00 |
| 0.186000E+01 | 0.000000E+00 | 0.000000E+00 |
| 0.186000E+01 | 0.000000E+00 | 0.000000E+00 |
| 0.186000E+01 | 0.000000E+00 | 0.000000E+00 |
| 0.186000E+01 | 0.000000E+00 | 0.000000E+00 |
| 0.186000E+01 | 0.000000E+00 | 0.000000E+00 |
| 0.186000E+01 | 0.000000E+00 | 0.000000E+00 |
| 0.186000E+01 | 0.000000E+00 | 0.000000E+00 |

14

| | | |
|---|---|---|
| 0.157000E+01 | 0.000000E+00 | 0.000000E+00 |
| 0.157000E+01 | 0.000000E+00 | 0.000000E+00 |
| 0.157000E+01 | 0.000000E+00 | 0.000000E+00 |
| 0.157000E+01 | 0.000000E+00 | 0.000000E+00 |
| 0.157000E+01 | 0.000000E+00 | 0.000000E+00 |
| 0.157000E+01 | 0.000000E+00 | 0.000000E+00 |
| 0.157000E+01 | 0.000000E+00 | 0.000000E+00 |
| 0.157000E+01 | 0.000000E+00 | 0.000000E+00 |
| 0.157000E+01 | 0.000000E+00 | 0.000000E+00 |
| 0.157000E+01 | 0.000000E+00 | 0.000000E+00 |
| 0.157000E+01 | 0.000000E+00 | 0.000000E+00 |
| 0.157000E+01 | 0.000000E+00 | 0.000000E+00 |
| 0.157000E+01 | 0.000000E+00 | 0.000000E+00 |
| 0.157000E+01 | 0.000000E+00 | 0.000000E+00 |
| 0.157000E+01 | 0.000000E+00 | 0.000000E+00 |
| 0.157000E+01 | 0.000000E+00 | 0.000000E+00 |

```
        0.157000E+01          0.000000E+00          0.000000E+00
        0.157000E+01          0.000000E+00          0.000000E+00
        0.157000E+01          0.000000E+00          0.000000E+00
        0.157000E+01          0.000000E+00          0.000000E+00
        0.157000E+01          0.000000E+00          0.000000E+00
        0.157000E+01          0.000000E+00          0.000000E+00
        0.157000E+01          0.000000E+00          0.000000E+00
        0.157000E+01          0.000000E+00          0.000000E+00
        0.157000E+01          0.000000E+00          0.000000E+00
        0.157000E+01          0.000000E+00          0.000000E+00
        0.157000E+01          0.000000E+00          0.000000E+00
        0.157000E+01          0.000000E+00          0.000000E+00
        0.157000E+01          0.000000E+00          0.000000E+00
        0.157000E+01          0.000000E+00          0.000000E+00
        0.157000E+01          0.000000E+00          0.000000E+00
        0.157000E+01          0.000000E+00          0.000000E+00
        0.157000E+01          0.000000E+00          0.000000E+00
        0.157000E+01          0.000000E+00          0.000000E+00
        0.157000E+01          0.000000E+00          0.000000E+00
        0.157000E+01          0.000000E+00          0.000000E+00
        0.157000E+01          0.000000E+00          0.000000E+00
        0.157000E+01          0.000000E+00          0.000000E+00
        0.157000E+01          0.000000E+00          0.000000E+00
        0.157000E+01          0.000000E+00          0.000000E+00
        0.157000E+01          0.000000E+00          0.000000E+00
0
```

Listing of the file OUTFIL.DAT for Example 1.


COMPENSATING MOVEMENTS FOR THE SINGLE-SUPPORT PHASE

GAIT UPON LEVEL GROUND FOR
ZMP LAW = 5, STEP LENGTH = 0.60, HALF STEP DURATION = 0.75


| TIME<br>(SEC) | POSITION<br>DOF 9<br>(RAD) | POSITION<br>DOF 10<br>(RAD) | VELOCITY<br>DOF 9<br>(RAD/SEC) | VELOCITY<br>DOF 10<br>(RAD/SEC) |
|---|---|---|---|---|
| 0.0000 | 0.2275 | 0.0118 | 0.8831 | 1.4555 |
| 0.0188 | 0.2444 | 0.0383 | 0.9193 | 1.3701 |
| 0.0375 | 0.2618 | 0.0632 | 0.9001 | 1.2861 |
| 0.0563 | 0.2771 | 0.0865 | 0.7073 | 1.1998 |
| 0.0750 | 0.2876 | 0.1082 | 0.4059 | 1.1125 |
| 0.0938 | 0.2922 | 0.1282 | 0.0999 | 1.0271 |
| 0.1125 | 0.2908 | 0.1467 | -0.2868 | 0.9420 |
| 0.1313 | 0.2823 | 0.1636 | -0.5660 | 0.8614 |
| 0.1500 | 0.2708 | 0.1790 | -0.6160 | 0.7868 |
| 0.1688 | 0.2595 | 0.1931 | -0.6116 | 0.7143 |
| 0.1875 | 0.2479 | 0.2058 | -0.6251 | 0.6424 |
| 0.2062 | 0.2361 | 0.2172 | -0.6238 | 0.5713 |
| 0.2250 | 0.2244 | 0.2273 | -0.6361 | 0.5011 |
| 0.2437 | 0.2124 | 0.2360 | -0.6390 | 0.4317 |
| 0.2625 | 0.2004 | 0.2435 | -0.6406 | 0.3631 |

| | | | | |
|---|---|---|---|---|
| 0.2813 | 0.1884 | 0.2496 | -0.6383 | 0.2951 |
| 0.3000 | 0.1765 | 0.2545 | -0.6331 | 0.2278 |
| 0.3188 | 0.1647 | 0.2582 | -0.6252 | 0.1610 |
| 0.3375 | 0.1530 | 0.2606 | -0.6131 | 0.0944 |
| 0.3563 | 0.1417 | 0.2617 | -0.5973 | 0.0279 |
| 0.3750 | 0.1307 | 0.2616 | -0.5776 | -0.0385 |
| 0.3938 | 0.1201 | 0.2603 | -0.5540 | -0.1049 |
| 0.4125 | 0.1099 | 0.2577 | -0.5264 | -0.1714 |
| 0.4313 | 0.1003 | 0.2538 | -0.4949 | -0.2381 |
| 0.4500 | 0.0913 | 0.2488 | -0.4751 | -0.3051 |
| 0.4688 | 0.0823 | 0.2424 | -0.4912 | -0.3723 |
| 0.4875 | 0.0733 | 0.2348 | -0.4403 | -0.4399 |
| 0.5063 | 0.0667 | 0.2259 | -0.2419 | -0.5086 |
| 0.5250 | 0.0644 | 0.2157 | -0.0093 | -0.5786 |
| 0.5438 | 0.0664 | 0.2042 | 0.2251 | -0.6498 |
| 0.5625 | 0.0732 | 0.1913 | 0.5149 | -0.7218 |
| 0.5813 | 0.0854 | 0.1772 | 0.7634 | -0.7912 |
| 0.6000 | 0.1012 | 0.1617 | 0.8995 | -0.8573 |
| 0.6188 | 0.1188 | 0.1449 | 0.9738 | -0.9319 |
| 0.6375 | 0.1376 | 0.1268 | 1.0443 | -1.0041 |
| 0.6563 | 0.1570 | 0.1073 | 0.9752 | -1.0781 |
| 0.6750 | 0.1741 | 0.0863 | 0.8681 | -1.1548 |
| 0.6938 | 0.1892 | 0.0639 | 0.7273 | -1.2344 |
| 0.7125 | 0.2016 | 0.0401 | 0.5988 | -1.3125 |
| 0.7313 | 0.2130 | 0.0148 | 0.6823 | -1.3831 |
| 0.7500 | 0.2275 | -0.0118 | 0.8831 | -1.4555 |

# Chapter 3:
# Control and Stability

## 3.1. Introduction

In this chapter we shall present the theoretical foundations of control
synthesis for two-leg locomotion systems. Because of the presence of
unpowered degrees of freedom (d.o.f.), the most serious problem which
has to be solved is the overall system stability. This is the reason
why the control synthesis at two stages has been adopted. At the first
stage, the stage of nominal regimes, such control has to be synthesized to ensure the system's motion in the absence of any disturbance
along the exact nominal trajectories calculated in advance. It should
be derived in such a way to satisfy the conditions of both the desired
gait type and overall system equilibrium. At the second stage, the stage
of perturbed regimes, only deviation of the actual state vector from
its nominal value is considered, and additional control is applied to
force the system state to its nominal. However, the movement thus realized, can induce an additional inertial force which, on the other
hand, can produce rotation of the whole system around the foot edge.
The movement should not make the situation worse, by producing some
additional inertial forces. As the nominal system motion is synthesized under the condition of the overall system equilibrium, the best way
to realize the system's return from a disturbed to its nominal regime
is to prevent the excursion of the system state out of a certain finite region. Deviation of the zero-moment point from its nominal position is adopted to be the measure of the system's overall deviation from
nominal trajectories. The actual position of ZMP can be computed
from the detected values of vertical reaction forces acting from the
ground upon the supporting feet. By measuring these forces, an additional feedback can be introduced to enable the system to maintain itself
in the equilibrium state, i.e. to keep the ZMP position within a limited area. This task of maintaining the system's overall equilibrium can
be assigned to different joints, and the question, which of them is
more suitable for this purpose, will also be considered.

Another point, characteristic of biped systems, should also be kept in
mind. Namely, by the action of the joint actuators, only the relative
position of two adjacent links, or, the so-called internal synergy,

changed. However, the maintenance of the biped system stability requires to consider the external synergy (i.e. angles of the biped links with respect to Cartesian frame). Even if the system perfectly realizes nominal trajectories of the joints angles (i.e. internal synergy), due to action of perturbations upon the system, it may rotate around the foot edge. The internal synergy guarantees system's equilibrium only if the unpowered d.o.f. behaves in desired way. This confirms the importance of an effective indirect control of the unpowered d.o.f. Namely, since the d.o.f. formed in the contact of the foot and ground, cannot be directly controlled, it has to be controlled by appropriate movements of powered joints.

A second important problem whose solution will be presented in this chapter is the stability analysis for a biped system. The well-known aggregation-decomposition method of stability analysis via Lyapunov's vector functions in finite regions of state space will be employed. In this method, which is very convenient for computer use, it is assumed that each d.o.f. is powered by its "own" actuator. However, in the case of the d.o.f. that are formed in the contact of the foot and ground surface, such actuator cannot be applied, and the method is not directly applicable. This problem, however, can be solved by the appropriate modelling of the mechanism. With manipulation robots each subsystem model usually contains one powered d.o.f. In order to include unpowered d.o.f., the system should be modelled in such a way to unite the model of some powered and some unpowered d.o.f. into one subsystem model. The subsystem thus modelled has its own input and the stability is investigated for the whole "composite subsystem".

All theoretical considerations have been checked by simulation and the obtained results are given. The model of general anthropomorphic system is formed using a software package, developed for this purpose, in which nominal dynamics is synthesized by prescribed synergy method. Afterwards, all suggested control laws have been applied and the system's behaviour is studied by simulation.

Finally, a numerical example of stability analysis, with special attention paid to the composite subsystems, has been presented.

## 3.2. Survey of Results on Biped Posture and Locomotion Control and Stability Analysis

How to control an artificial locomotion activity, especially the biped locomotion, it has always been a challenging problem. Bernstein [1] was the first to describe some global feedback and overall control philosophy of such systems. According to him, basic elements of a control cycle are (Fig. 3.1): actuator, a programming device which prescribes the necessary value of the regulated parameter to the system, receptor (sensor), a discriminator which gives the value and the sign of the difference between the actual and desired parameter value, a coder which converts the discriminator data into correcting signals that are transmitted to the regulator by means of the feedback loop, and a controller which controls the effector operation. The spontaneous motion due to the redistribution of tension in muscle groups modifie  the relation between forces, bringing these relations to equilibrium, or taking them away from the equilibrium position. Then, variations in muscle tension cause motion and the motion causes further change in muscle tension by changing the degree of contraction or extension. Bernstein has named the described interconnection of action the peripheral cycle of interactions. Mathematical analysis of the relationship between force and movement indicates that this form of interaction does not have a

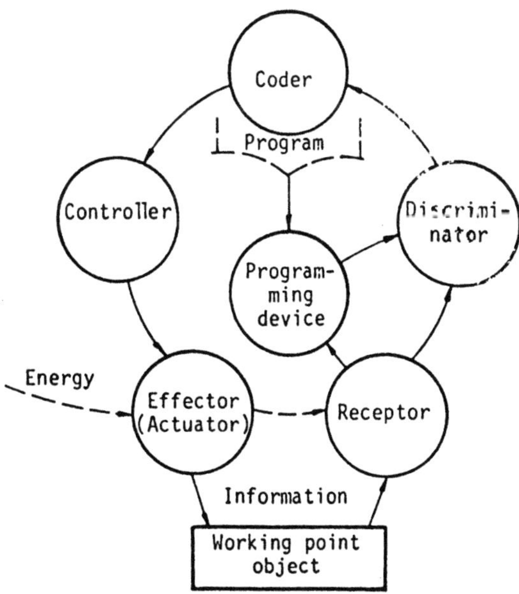


Fig. 3.1. Motion control cycle (after [1])

unique dependence since its mathematical description is based on a second-order differential equation whose solution requires two initial values (initial position of the link, and its initial velocity). Depending on particular value of these integration constants, quite different effects may be initiated by the same innervation. A logical conclusion follows that a proper coordination and agreement of movements under the direction of the living organism is possible only under the condition that the central nervous system has already working information available on these independent parameters of integration, and that it generates its actuator pulses in dependence on variation of these parameters. The appropriate information is provided by the proprioceptive system, and it forms the second feedback in Bernstein's terminology. In that loop, the actuator pulses change the muscle tension, thus causing acceleration in the system's parts and, consequently, in the whole system. The accelerations lead to changes in positions and velocities, and these values cause changes in muscle tension, thus generating the proprioceptive signals. These signals affect the flow of effector pulses by introducing appropriate corrections and exciting the effector's central mechanism to adapt to the modified peripheral operating conditions. From these facts it is possible to obtain, to a certain extent, an insight into the problems that have to be solved if a stable motion is to be synthesized.

From the viewpoint of dynamics, the two-leg locomotion belongs to a class of very complex problems. Modelling a mechanism which can perform such a motion, if use is not made of some specially dedicated computer software package (e.g. like the one described in the previous chapter), naturally requires significant simplification. Depending on the degree of simplification, the models vary from a single inverted pendulum (with fixed base, or with prescribed base trajectory) through multi-linked planar models up to the multi-linked spatial models.

Lower-order and linearized models enable the application of well-known methods of control synthesis. These models are also suitable for an easy handling and solving in analytical form, allowing thus the analysis of the influence of different parameters on a particular solution value. However, some important features of the actual system behaviour can be lost by simplification, and only the main motion characteristics can be investigated. More complex nonlinear models are closer to the actual system, but in general, their solution cannot be obtained in analytical form; they are only solvable by numerical methods.

### 3.2.1. <u>Postural control</u>

The simplest model that can be used as an approximation of the locomotion mechanism is a single-link model.

Chow and Jacobson [2] studied spatial motion of a massive link having its supporting point at its base. It is assumed the torso base follows a trajectory prescribed in advance, and which should be realized by appropriate motion of legs. The trajectory is synthesized to satisfy only kinematic constraints. Thus, the problem is reduced to the modelling of inverted pendulum, and the control problem to the maintenaning of the vertical upright position. The body angles are generated by the three succesive rotations $\theta$, $\psi$, and $\phi$. This choice of body angles differs from the usual Eulerian description in rigid body dynamics in that the Euler angles become ill-conditioned as they become small. The motion equations are derived as Lagrange's second-order equations under assumption that the angle of rotation about the longitudinal axis ($\phi$) is small as compared to the angles in the sagittal ($\theta$) and in the frontal plane ($\psi$). On assuming that the torso deviations from vertical position are small, and the point $\theta = \psi = \dot{\theta} = \dot{\psi} = 0$ corresponds to the position of unstable equilibrium, the authors linearized the system of equations and derived the linear feedback law which succesfully stabilized the torso. An important role in stabilization has the effect analogous to that of hard cubic spring. For the purpose of stability investigation, Chow and Jacobson generalized the hard spring analogy and use it to derive suitable Lyapunov's function. In this way they proved the torso stability under all initial conditions of the angular displacement and velocities.

As an illustration of the proposed approach, a numerical example of the cyllindrically-shaped torso is presented. An initial angular displacement is assumed and the corresponding feedback gains are derived. Finally, a stable behaviour of the system is shown by simulation.

The significance of this work is mainly theoretical since it demonstrates how the general Lyapunov theory can be used for stability investigation of locomotion systems. However, the simplification introduced by splitting up the system into two "independent" subsystems, i.e., the "trunk" and the "legs", is rather rough and does not correspond to the reality. The trunk, in any case, influences the legs motion as well as vice versa; a correct separation of the two effects can be only done by

taking into account their dynamic reactions. In spite of the fact that
the authors suppose this model can be used in the walking process simulation, its actual contribution is to the posture maintenance control.

Other authors have also used simple inverted pendulum model to investigate  some aspects of the human posture. Hemami et al. used it in [3,
4, 5]. In [3], it was the torso modelled by inverted pendulum and its
stability was investigated by Lyapunov's function and by simulation.
In [4], different control laws were synthesized and applied, and their
effectiveness was tested by simulation, whereas in [5] the validity of
the adopted control structure and feedback gains was verified by two
experiments (TV and force platform system measurement).

A very detailed piece of research, based on multi-linked models (both
planar and spatial) has been also carried out by Hemami et al. [7-14].
They paid special attention to the type of constraints imposed. The
problem investigated in [7] is how to control a dynamic system in such
a way that the holonomic constraints are maintained and the forces of
constraints are a priori specified. Two cases have been considered:
constant forces of constraint and forces that are functions of the
state. In [8], the constraint forces are assumed to be the explicit
functions of the state and the input, while in [11], the mechanism is
subjected to certain on-off constraints. In [12], in the presence of
large external forces acting on the model, connection constraints are
maintained by the ligaments and other soft-tissue structures. In [14],
the dynamics of a multi-linked spatial model with arbitrary holonomic
and non-holonomic constraints at the joints is formulated.

An example [10] of stability analysis using Lyapunov's functions will
be briefly presented. The authors investigated stability of the system
by determining the region of stability of the original nonlinear system with a proposed feedback. A nonlinear system can be written in the
state space form

$$\dot{x} = f(x) \tag{3.2.1}$$

where

$$\frac{\partial f}{\partial x} = A \tag{3.2.2}$$

and where A is stable (all eigen-values are in the left half of the
complex plane). A positive definite quadratic function

$$V = x^T H x \qquad (3.2.3)$$

is used as a Lyapunov function, constructed in such a way that

$$x^T (A^T H + HA) x \qquad (3.2.4)$$

is negative definite. One possible estimate of the region of stability of the original system (3.2.1) is to select the largest ellipsoid

$$x^T H x = \ell^2 \qquad (3.2.5)$$

inside which

$$\dot{V} = \frac{dV}{dt} \qquad (3.2.6)$$

is negative:

$$\dot{V} = f^T H x + x^T H f \leq 0 \qquad (3.2.7)$$

As an illustration is used an example of a two d.o.f. biped powered by two torque actuators, one at the ankle, the other at the hip joint. A variety of control structures is applied and for each case is selected the ellipsoid of stability, which represents an estimate of the region in state space, inside which the system is stable. A similar approach to stability analysis has been applied in [2]. In [6], postural stability about vertical stance is investigated, and the Lyapunov second method (like in [10]) is used to estimate the region where the biped, with corresponding control structure, is stable. Two biped models used are double inverted pendulum and the three-link biped model with motions restricted to the sagittal plane. In [18], a two-link biped system consisting of the massive body and massive legs is modelled with addition of ligaments at the joints. Stability in two cases (unconstrained and constrained system) is investigated by the second Lyapunov method. For each case, the corresponding Lyapunov function is constructed and simulation results of mechanism's behaviour presented. In [11], stability is investigated by movement simulation during the observed time interval. In [15], the impact effect on a five-linked biped is considered. Two links are used to model feet, two for leg and one for torso. Such planar biped is subjected to the instant velocity change (landing on the heels and toes, or toes first) and its behaviour is studied using computer simulation.

Especially interesting is the work of Hill [19] in which he formed a seven-linked planar anthropomorphic model with all massive links. The links represent trunk, thigh, shank, upper arm, forearm, and head. Each link is considered to have its mass and rotational inertia. The muscle systems are assumed to produce torques about joints and it is assumed that both the heel and toe can be in contact with ground. The nonlinear differential equations are derived to describe the system's motion. The generalized forces (acting at each joint) arise from muscle torques and ground reaction forces. The ground reaction forces are expressed in terms of heel and toe coordinates and their derivatives. Based on such complete nonlinear model, the simplification is carried out for three physical situations: free fall with no muscle torques, stand effects and leap. In the first case, the model is in free fall until it touches the ground, after which it collapses under the action of ground reaction forces and gravity. In the second case, muscle torques proportional to the linear combination of joint angle and joint angular velocity are applied so as to attempt to drive all joint angles to zero, resulting in stable upright position. In the third case, the control laws specifying the muscle torques are modified to cause the hip and thigh to extend and the foot to deflect in such a way to propel the back into the air as in an imitation of "graceful leap". Hill linearized this model around vertical equilibrium position. Simulation results for linearized models show large deviations from those obtained for a nonlinear model. Thus, some important features inherent to nonlinear mathematical models can be lost by linearization.

By investigating the linearized and nonlinear equations of one link taken separately, Hill showed that the use of linearization is not admissible even for somewhat larger angles. The results of this analysis definitely prove that for large-scale active spatial mechanisms, a linearized model is rather far from the actual system. Thus the control synthesis based only on the linearized model equations, i.e., neglecting the nonlinear effects, cannot satisfy the requirements of speed and efficiency of stabilization of the system.

All the above references are related only to planar models. Gubina[20] considered a spatially placed massive body supported on two massless legs of variable length (Fig. 3.2). The body of mass $m_t$, has main inertia moments $J_x$, $J_y$, $J_z$. The variable length of legs replaces the knee function, and the body positing in space is achieved by leg extension (it is supposed that the "knee" can generate force F). The body orientation is maintained by three torques $P_1$, $P_2$ and $P_3$ applied at the mechanism's hip which rotate the body in the frontal and sagittal plane, as well as around the vertical body axis (self-rotation). Only single-support phase is considered. A spherical coordinate frame has been chosen because it is closest to that used by humans.

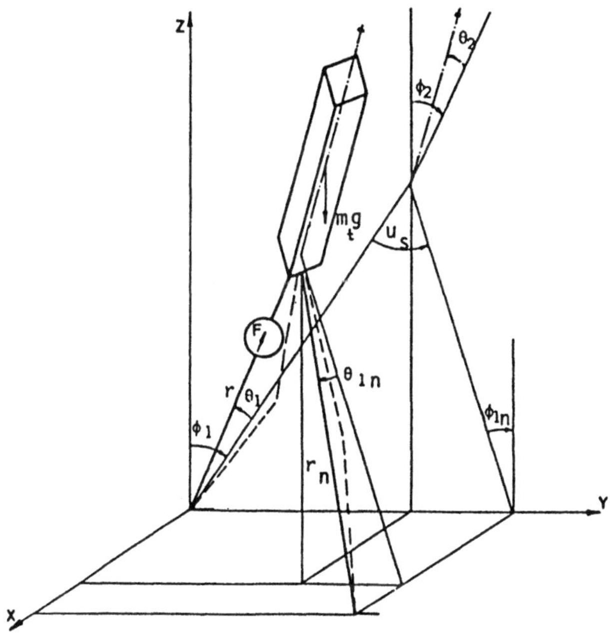


Fig. 3.2. Spatial biped model with massless legs (after [20])

The non-disturbed motion is realized in the sagittal plane; the leg position is defined by its length r and angle $\phi_1$; the body position is determined by the angle $\phi_2$ and the distance between the hip and mass centre of the body $\ell$. In the case of disturbance in the frontal plane, the position of the body is defined by the angles $\theta_1$ and $\theta_2$.

The equations of motion are nonlinear with one second-order differential equation for each d.o.f.

$$f(q, \dot{q}, \ddot{q}, Q) = 0 \tag{3.2.8}$$

where the vectors q and Q contain the generalized coordinates and generalized forces, respectively (the generalized forces are reduced to unit mass)

$$q = (r, \phi_1, \phi_2, \theta_1, \theta_2, \psi)^T$$

$$Q = (F, P_1, P_2, P_3)^T \tag{3.2.9}$$

The point of unstable equilibrium is defined by $q=[r_o, 0, 0, 0, 0, 0]^T$ and $\dot{q} = 0$ represents the upright mechanism position. The linearized equations of motion (3.2.8) can be written in the following form

$$\dot{x} = \begin{bmatrix} A_1 & & & & & \\ & A_2 & & & 0 & \\ & & A_1 & & & \\ & & & A_2 & & \\ & 0 & & & A_1 & \\ & & & & & A_1 \end{bmatrix} x + \begin{bmatrix} B_1 & & & & \\ & B_2 & & & \\ & B_3 & & & \\ & & B_2 & & \\ & & B_3 & & \\ & & & B_1 \end{bmatrix} u \tag{3.2.10}$$

with x as the state vector

$$x = (r-r_o, \dot{r}, \phi_1, \dot{\phi}_1, \phi_2, \dot{\phi}_2, \theta_1, \dot{\theta}_1, \theta_2, \dot{\theta}_2, \psi, \dot{\psi})^T \tag{3.2.11}$$

and u as the control vector

$$u = Q - Q^* \tag{3.2.12}$$

where $Q^*$ is the value of generalized forces at the point of unstable equilibrium. The matrix blocks in (3.2.10) are

$$B_1 = [0 \ 1]^T \quad B_2 = [0 \ -c]^T \quad B_3 = [0 \ d]^T \tag{3.2.13}$$

$$A_1 = \begin{bmatrix} 0 & 1 \\ 0 & 0 \end{bmatrix} \quad A_2 = \begin{bmatrix} 0 & 1 \\ b_2 & 0 \end{bmatrix}$$

with $b_2 = g/(\ell+r)$, $c = (J_x+r_o\ell+\ell^2)/J_xr_o^2$, $d = (r_o+\ell)/J_xr_o$

It is evident from (3.2.10) that the linearized system can be partitioned into six modes, each of which is associated with one d.o.f., owing to favourable selection of the state coordinates. The matrix of control distribution, however, indicates the coupling between some of the modes. In spite of this, four completely uncoupled subsystems appear, describing the motion of the body along its leg, in the sagittal plane, in the lateral direction, and around the body major axis. The independent subsystems suggest the independent control loops, each containing only the state variables associated with the corresponding mode. Furthermore, it is clear that, in addition to the four control components of u, it should be added the step length control which is discrete, in contrast to the body attitude and altitude control which are continuous. The step length may be changed after each change of the supporting leg, so that sampling period T/2 is equal to the duration of one half-step cycle.

Linear feedback laws have been derived according to the parameters of the linearized model using pole - placement techniques for both continuous and time-discrete controls. For example, the body altitude control is maintained by the leg length control which supplies the system with the energy, lost at the moment of leg switching. From the equations of the first mode

$$\dot{x}_r = A_1 \cdot x_r + B_1 \cdot u_1 \qquad (3.2.14)$$

it can be derived the linear feedback $(x_r = (r-r_o, \dot{r})^T = (x_1, x_2)^T)$:

$$u_1 = k_1 \cdot x_1 + k_2 \cdot x_2 \qquad (3.2.15)$$

which gives the poles of the closed-loop subsystem $\lambda_1$ and $\lambda_2$ with the following selection of feedback gains

$$k_1 = -\lambda_1 \lambda_2, \quad k_2 = \lambda_1 + \lambda_2 \qquad (3.2.16)$$

In the same way are defined the feedback gains for the other three subsystems, as well as the gains for discrete control of the step length in the sagittal and frontal plane. In such a way, stable locomotion on a certain path in three-dimensional space can be achieved.

For larger excursions from the point of unstable equilibrium, such linear control of each mode is not effective any more. The autonomous action of continuous controllers has to be supported by additional terms $Q^*$ in the nonlinear region (eq. (3.2.12)). The task of this additional non-linear control is to keep the body in upright position, $\phi_2 = \dot{\phi}_2 = \theta_2 = \dot{\theta}_2 = 0$ independent of legs motion. Using such dynamic compensation and linear continuous control, the stable motion of the body supported on massless legs may be achieved at each step.

The effectiveness of the derived control concept and the biped body
stability of gait have been tested in the following example. A model
with the parameters similar to those of the human body started its motion with a large disturbance of the body position ($\phi_2$ = 3 [rad]). After two subsequent steps, the body reached almost the normal vertical
position.

The body path stability of the six d.o.f. nonlinear model was tested
in the case when the model was forced to walk on a line under the initial lateral disturbance. The body was bent in the lateral direction
for $\theta_2$ = 57$^\mathrm{O}$ with the remaining components of the state vector equal
to zero. The required velocity was $v_o$ = 1.5 [m/s], T = 0.5 [s]. The
body attitude control brought the body into the upright position within two steps. By the foot action, the lateral disturbance was elimina-
ted almost completely in twelve successive steps. The model was brought on line, and continued locomotion with a stable gait.

In spite of the fact that the author has supported the mechanism's
torso on a pair of massless legs, this work actually deals with the
problem of postural control. In the suggested stabilization procedure,
the torso is subjected to external loads, which in the case of locomotion mechanisms cannot be accepted at all. It is obvious that the driving torques of active mechanisms of this type belong to the category
of internal forces, which will obviously happen if the legs masses are
introduced. A change of the massless legs position has as a consequence only the change of the direction of the generated "knee" force,
which is characteristic of the external load application. However,
with internal driving torques arises the question: how to establish
the correspondence between the external coordinates (Euler's angles,
spherical coordinates, or any others) and the internal ones, when the
driving torques are not known? In other words, how to define the driving torques which would realize the internal synergy, or, how to define the internal synergy by means of which the driving torques can be
obtained? The approach proposed cannot offer the acceptable answers to
these questions.

A further progress has been made in the work by Vukobratović et al.
[22, 23] in which the problem of postural stabilization of a biped was
considered. In [23], the locomotion mechanism has been modelled as a
multi-linked spatial mechanism with all links massive, so that the model was quite close to the actual system. The authors considered both
small and large disturbances.

In the case of small disturbances, an approximate method for posture preservation via ground reaction force measurement was used. It is clear that there is a direct relationship between the overall system force vector and the ground reaction on the supporting foot. If the system is in static equilibrium, the ground reaction should act at a point which is within the stable region, i.e., within the area covered by the foot. Any violation of the posture causes a displacement of ZMP from the previous (nominal) position. Now, with respect to the nominal position of ZMP an additional moment is acting which can be used as a measure of the overall deviation of the system from its nominal position. The values of resultant ground reaction force and its position are used for the biped stabilization. For this purpose three force sensors, at least, are to be used at each sole.

In the case of large perturbations, the system has to enlarge stride and change the initial posture; the system stabilization can be achieved using another control strategy.

Consider the system in its static equilibrium. If a certain disturbance occurs, the phase point will move from the equilibrium position. If it is still within the region of stability reserve for the considered posture, the phase point will automatically return to the equilibrium position. If the disturbance carries the phase point out of the stable zone, the system has to switch to another algorithm for stabilization by striding. Naturally, striding enlarges the zone of stability reserve, thus, if the system state can be transferred to a new adequate equilibrium position, the initial disturbance will be automatically compensated for.

The suggested approach has been checked by computer simulation (Fig. 3.3). Initially, the model was in the double-support phase with parallel feet (posture 1). The system inclination in the sagittal plane was $\phi_o = 0.08$ [rad] and it rotated with $\dot{\phi} = 0.4$ [rad/s].

While the left foot remains at the same place, the right leg moves to enlarge stride and acquire a new stable posture. In this way, the problem of large perturbations is reduced to stabilization in the presence of small perturbations which can be solved by the force measurement method.

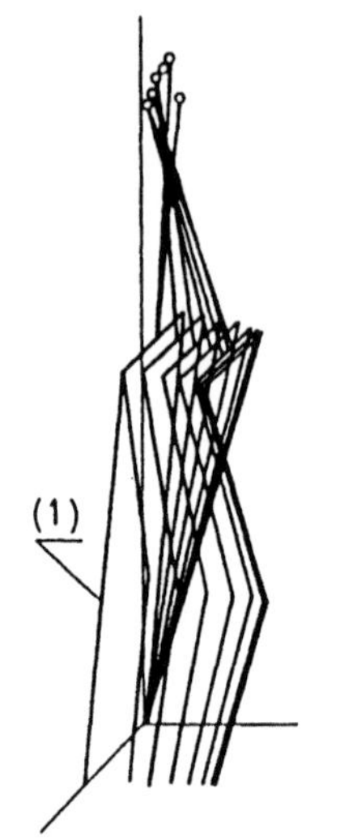



Fig. 3.3. Biped stabilization process

A further development of the approach was made in [24]. Mathematical modelling of the mechanism is done automatically on the computer using the originally developed software. The overall system model consists of two parts: mechanical configuration and actuators i.e. models of actuators with its own dynamics have been included in the model of the complete system. Also, a different approach to the problem of control synthesis has been adopted.

The control concept used here is based on the procedure of two-stage control synthesis. While the first stage comprises the synthesis of programmed (nominal) regimes under ideal conditions i.e. without perturbations, the second stage solves the problem of realizing the programmed (nominal) motions under perturbed operating regimes.

Therefore, in the stage of perturbed regimes, the nominal trajectory $x^o(t)$ should be realized under the assumption that the actual state vector $x(t_o)$ differs from the nominal initial conditions of the mechanism $x^o(t_o)$, $x(t_o) \neq x^o(t_o)$.

At the level of perturbed regimes, the control is synthesized in two steps. First, the subsystems have to be stabilized as free of coupling, and, in the second step it is necessary to apply an additional control which has to take care of the system as a whole.

Thus, the task of local control is to stabilize the decoupled subsystem, i.e. the subsystem which is supposed not to be influenced by the motion of the rest of the mechanism.

If the control thus synthesized is not capable to stabilize the system, an additional stabilization is required to compensate for destabilizing influences of both the coupling and unpowered d.o.f.

This approach has been illustrated by a simplified biped model (Fig. 3.4). The mechanism parameters are chosen to correspond to the anthro-

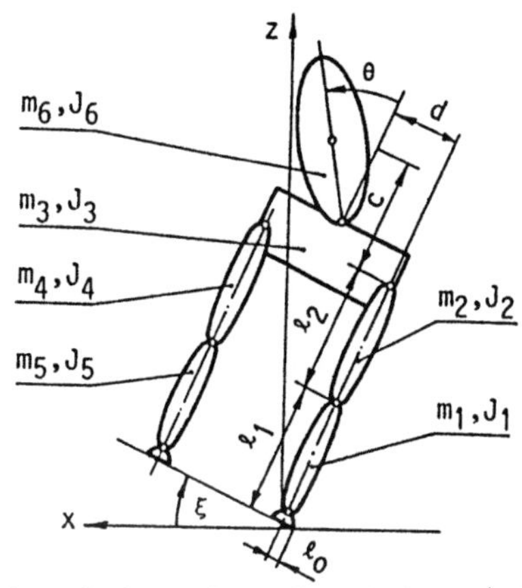



Fig. 3.4. A biped model in the
frontal plane

pomorphic biped system in the frontal plane. The model is assumed to have two d.o.f.: one between the foot and support ($\xi$) and the other the motion of the trunk around the horizontal axis $\theta$. Thus, the biped can perform a planar motion in the x-z plane. For the trunk d.o.f. a linear feedback is introduced in such a way to ensure that poles of free subsystem (body) are in the left part of the complex plane. But, with linear feedback the unpowered d.o.f. is not stabilized and a global feedback with respect to coupling is introduced. However, for $\xi > 0.12$ [rad] and $\dot{\xi} \geq 0$, such control is not sufficient to stabilize the system. Hence, the additional control which use the information about intensity and position of ground reaction force is introduced and then the system is capable of returning to the static equilibrium position.

### 3.2.2. Gait control

A problem more serious than the posture control of a biped is the problem of controlling its gait. A lot of research effort in this field has been devoted to mathematical modelling of mechanical systems and studying their behaviour using different control strategies. The problem arises with the complexity of the mechanism model used to represent the actual system, and then, with the complexity of the mathematical model which describes the mechanism. The fidelity of the obtained solution, and its applicability to the actual system, is a direct consequence of simplifications introduced at both stages (when a sufficiently simple mechanical model is chosen and when, in addition, the mathematical model is simplified). Our further discussion will be concerned with the results on the control of locomotion of anthropomorphic systems.

We shall review first some of the results of Beletskii [25], whose systematic studies encompassed the control of planar models of locomotion mechanisms. The models were mostly simplified up to the level allowing analytic solutions, which led to detailed analyses and discussions of the results.

A model, Beletskii used most frequently was the model of massive body
supported on massless legs. Such a system in three-dimensional space
is shown in Fig. 3.5.a).

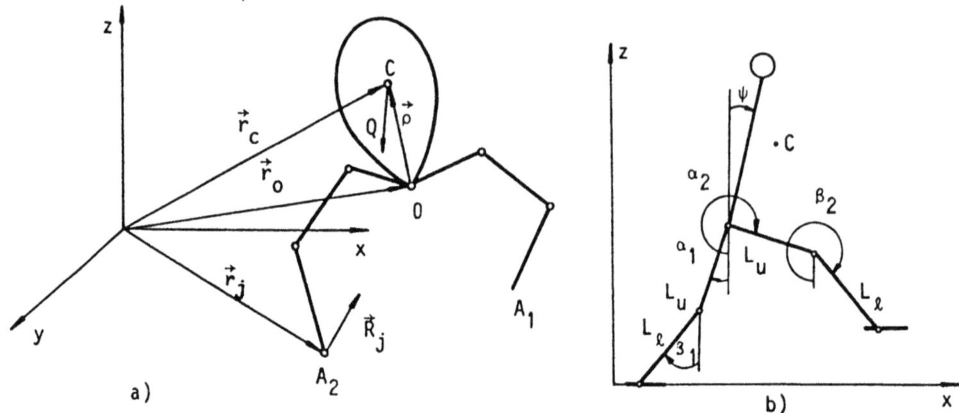


Fig. 3.5. Biped mechanism a) general spatial case, b) planar case
(after [25])

Point C is the mass centre of the body (and, in this case, of the whole
system), while 0 is the point at which legs are connected to the torso.
The $\vec{r}_c$, $\vec{r}_o$, and $\vec{r}_j$ are the radius vectors from the external coordinate
frame origin to points C and 0, and to the supporting point of the j-th
leg (j=1,2) on the ground, respectively; $\vec{\rho}$ is distance vector from the
point 0 to C; Q is the body weight; and $\vec{R}_j$ (j=1,2) is the reaction for-
ce of the supporting leg. In general, $\vec{R} = \sum\limits_{j=1}^{2} \vec{R}_j$ corresponds to the support phase; only one $\vec{R}_j$ component is characteristic of the single-sup-
port phase. For the single-support phase it will be supposed the leg
whose end is point $A_1$ is in the swing phase, while the other one en-
ding at $A_2$, is the supporting leg.

According to the theorem of motion of the system's mass centre it can
be written

$$\vec{R} = -\vec{Q} + M[\ddot{\vec{r}}_o + \dot{\vec{\omega}} \times \vec{\rho} + \vec{\omega} \times (\vec{\omega} \times \vec{\rho})]$$ (3.2.17)

where M is the system mass, $\vec{\omega}$ is the angular velocity of the rigid
body.

Theorem on the rate of change of moment of momentum relative to point
0 could be written as

$$\{I\}\dot{\vec{\omega}} + \vec{\omega} \times \{I\}\vec{\omega} + M\cdot\vec{\rho}\times\ddot{\vec{r}}_o = \vec{\rho}\times\vec{Q} - \sum_{j=1}^{2}(\vec{r}_o - \vec{r}_j)\times\vec{R}_j$$

where $\{I\}$ denotes the inertia tensor of the body for point 0.

If a single-support phase is considered, the ground reaction force is defined by the right-hand side of (3.2.17), and the previous expression may be written as

$$\{I\}\dot{\vec{\omega}} + \vec{\omega}\times\{I\}\vec{\omega} = [\vec{\rho} + (\vec{r}_o - \vec{r}_j)]\times(\vec{Q} - m_t\cdot\ddot{\vec{r}}_o) - (\vec{r}_o - \vec{r}_j)\times[\dot{\vec{\omega}}\times\vec{\rho} + \vec{\omega}\times(\vec{\omega}\times\vec{\rho})]m_t$$

$$(3.2.18)$$

where $m_t$ is the torso mass, which is equal to M for the case of massless legs.

This equation describes the motion of a spatially placed massive body supported on massless legs. If, for example, the planar motion (Fig. 3.5.b)) of such system is considered, (3.2.18) reduces to

$$[J_t + m_t\cdot\rho(z - z_{A_2})\cos\psi + m_t\cdot\rho(x - x_{A_2})\sin\psi]\ddot{\psi} \;+$$

$$+ m_t\cdot\rho\cdot\dot{\psi}^2[\cos\psi(x - x_{A_2}) - (z - z_{A_2})\sin\psi] - m_t\cdot\rho(g + \ddot{z})\sin\psi \;+$$

$$+ m_t\cdot\rho\cdot\ddot{x}\cos\psi = m_t(g + \ddot{z})(x - x_{A_2}) - m_t\ddot{x}(z - z_{A_2}) \qquad (3.2.19)$$

The angle $\psi$ is defined by the vertical direction and the longitudinal axis of body. Further, $\vec{r}_o(x, y, z)$ defines a trajectory of point 0, while the equality $\vec{r}_j = \vec{r}_{A_2}$ corresponds to the single-support phase. $J_t$ is the torso moment of inertia with respect to the axis perpendicular to the plane of motion and passes through point 0, and $\rho = |\vec{\rho}|$.

If the system walks upon level ground, point 0 moves monotonously with speed V with the distance between 0 and the ground being constant z=h, equation (3.2.19) is transformed into

$$[J_t + m_t\cdot\rho\cdot h\cdot\cos\psi + m_t\cdot\rho(V\cdot t - s)\sin\psi]\ddot{\psi} \;+$$

$$+ m_t\cdot\rho\cdot\dot{\psi}^2[(V\cdot t - s)\cos\psi - h\sin\psi] \;-$$

$$- m_t\cdot\rho\cdot g\sin\psi = m_t\cdot g(V\cdot t - s) \qquad (3.2.20)$$

where $s = x^o_{A_2} - x^o$, and the superscript o denotes initial values.

Equation (3.2.20) describes the case that will be used in further considerations of the nominal motion synthesis and stabilization of the system after perturbation. It is the planar motion of a massive body supported on two legs (Fig. 3.5.b)), each composed of two links whose lengths are $L_u$ and $L_\ell$, for the thigh and shank, respectively. $\alpha_1$ and $\alpha_2$ are the angles between the thighs and vertical and $\beta_1$ and $\beta_2$ between shanks and vertical. The subscript 1 corresponds to the leg being in swing phase and 2 to the supporting leg.

A consequence of the assumption of massless legs is that legs movements do not influence the body motion and their task is only to provide the desired trajectory of point 0. If this trajectory is known (i.e. prescribed in advance) the legs trajectories are defined by a sequence of supporting points of the legs ends on the ground. After that, it is possible to solve the equation of body motion and choose such a solution which will satisfy the additional requirement of repeatability on the one step period. Now, when the complete system dynamics is known, it is possible to compute the force of ground reaction and driving torques at joints, necessary to enable such motion.

Let us consider the case given by (3.2.20) with the purpose of computing the nominal torso trajectory. Point 0 moves along the horizontal straight line on z=h above the ground with a constant speed V, in such a way that $\ddot{x}=0$, $\ddot{z}=0$ and $x=Vt+x^o$. Then, expressions for the ground reactions $R_x$ and $R_z$ (only the single-support phase is considered) and joint torques are

$$R_x = m_t \cdot \rho \, (\ddot{\psi}\cos\psi - \dot{\psi}^2 \cdot \sin\psi)$$

$$R_z = m_t \cdot g - m_t \cdot \rho \, (\ddot{\psi}\sin\psi + \dot{\psi}^2 \sin\psi)$$

$$P_t = R_x \cdot h - R_z \, (V \cdot t - s)$$

$$P_k = L_\ell \, (R_x \cos\beta_2 - R_z \cdot \sin\beta_2)$$

where $P_k$ and $P_t$ are the driving torques at knee and hip, respectively. Now, the task may be formulated as a requirement to find out such initial values of the torso angle $\psi(0)$ and angular velocity $\dot{\psi}(0)$ which will ensure that in integration of (3.2.20), the repeatibility conditions are fulfilled, that is

$$\psi(0) = \psi(T), \qquad \dot{\psi}(0) = \dot{\psi}(T)$$

In this way it is possible to define the system motion under ideal conditions, i.e., the so-called nominal trajectories.

If it is supposed that the body angle $\psi$ during the system's motion is small, (3.2.20) can be linearized around the vertical position, and the equation of body motion acquires the form

$$(J_t + m_t \cdot \rho \cdot h)\ddot{\psi} - m_t \cdot \rho \cdot g \cdot \psi = m_t \cdot g \cdot (V \cdot t - s) \qquad (3.2.21)$$

and its solution is

$$\psi = \frac{L}{2 \cdot \rho}\left[\frac{e^{\alpha T \iota} - e^{-\alpha T(\tau-2)}}{e^{2\alpha} - 1} - (\tau - \sigma)\right] \qquad (3.2.22)$$

where $L = V \cdot T$ is the step length, $T$ is the step duration, and

$$\alpha = \frac{T}{2}\sqrt{\frac{m_t \cdot \rho \cdot g}{J_t + m_t \cdot \rho \cdot h}} \ , \qquad \sigma = \frac{2s}{L}, \qquad \tau = \frac{2}{T}t, \qquad \tau \in [0,2]$$

Hence, the initial conditions corresponding to the periodic solution are

$$\psi_0 = \frac{L}{2\rho}(\sigma - 1), \qquad \dot{\psi}_0 = \frac{V}{\rho}\left(\alpha\,\frac{e^{2\alpha}+1}{e^{2\alpha}-1} - 1\right) \qquad (3.2.23)$$

From the whole family of solutions one is singled out by choosing one particular value of s.

In Fig. 3.6. are shown three locomotion mechanism configurations at the

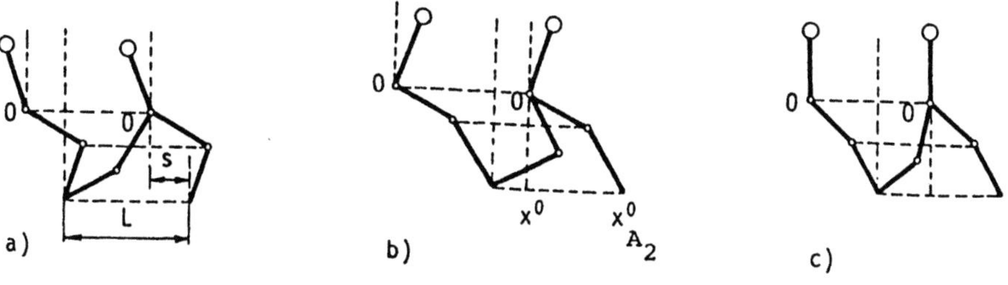


Fig. 3.6. Mechanism configuration at the moment of
changing supporting leg (after [25])

moment of changing supporting leg. Cases a), b), and c) correspond to solutions where s<L/2, s>L/2 and s=L/2, respectively. Fig. 3.7. shows the phase portrait of the planar body motion.

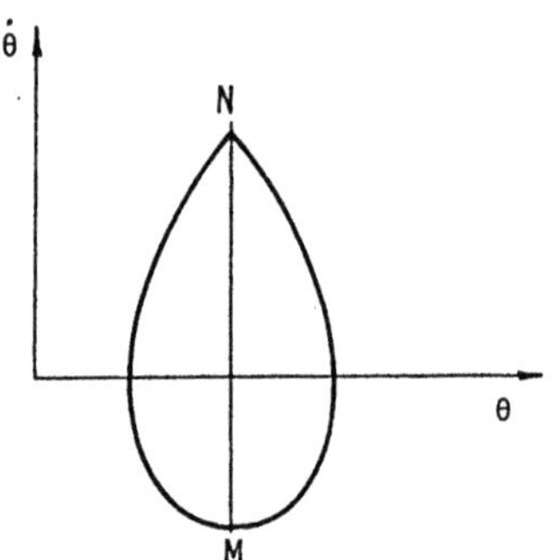

Fig. 3.7. Phase portrait of the planar torso motion (after [25])

The problem becomes more complex when a disturbance occurs, and Beletskii has offered two types of stabilization. The first one is related to the case when the system is forced to any appropriate periodical regime, differing from the original one, whereas the second type of stabilization is employed when the system is forced to return to its original regime that was performed before the disturbance.

The first case is illustrated by a five-link planar mechanism with massless legs whose motion is described by (3.2.21). The problem to be solved can be described as follows. Let in the beginning of the n-th step the system phase coordinates be $\psi_o$, $\dot{\psi}_o$, $x_o$, V, $z_o$ = h and $\dot{z}_o$ = 0. For the same step, the remaining parameters $L_N$, L, and T should be defined in such a way that the system, after completing the n-th step for time T and step length L, starting from the (n+1)-th step onward performs the periodical motion with $\dot{x}$=V, $z_o$=h, $\dot{z}_o$=0, step length $L_N$ and $T_N$ = $L_N$/V. (Note that $T_N$ is also the period of the body compensating motion).

If point 0 moves along a horizontal line with a constant speed, and, if z=h, the torso motion is described by (3.2.21). Then, the ground reactions are

$$R_x = m_t^2 \cdot g \cdot \rho \, \frac{x - x_{A_2} + \rho \cdot \psi}{J_t + m_t \cdot \rho (h - z_{A_2})} \, , \qquad R_z = m_t \cdot g,$$

while the disturbed motion is described by

$$x = x_o + V \cdot t, \qquad z = h$$

$$\rho \cdot \psi = c_1 \cdot e^{\alpha t} + c_2 e^{-\alpha t} + x_{A_2} - x$$

(3.2.24)

where constants $c_1$ and $c_2$ are

$$c_1 = \frac{1}{2}[\rho \cdot \psi_o - s + \frac{1}{\alpha}(\rho \dot{\psi}_o + V)],$$

$$c_2 = \frac{1}{2}[\rho \cdot \psi_o - s - \frac{1}{\alpha}(\rho \dot{\psi}_o + V)]$$

where $\psi_o$ and $\dot{\psi}_o$ are the body angle and angular velocity at the initial moment, respectively. The parameters $\alpha$ and $s$ are defined as

$$\alpha^2 = \frac{m_t \cdot \rho \cdot g}{J_t + m_t \cdot \rho (h - z_{A_2})}, \qquad s = x_{A_2} - x_o$$

The length of the n-th step, L, and the time period T should be such to ensure that the system, starting from the (n+1) step, moves periodically with the period $T_N$ and step length $L_N = V \cdot T_N$. These conditions may be written in analytic form as

$$\rho \cdot \psi(T) \equiv c_1 \cdot e^{\alpha T} + c_2 e^{-\alpha T} + x_{A_2} - x(T) = \rho \cdot \psi_N(T_N)$$

$$\rho \cdot \dot{\psi}(T) \equiv c_1 \alpha \cdot e^{\alpha T} - c_2 \alpha e^{-\alpha T} - V = \rho \cdot \dot{\psi}_N(T_N)$$

where $\psi_N(T_N)$ and $\dot{\psi}(T_N)$ are the initial values for periodical regime, which, according to (3.2.23), can be written as

$$\psi_N(T_N) = \frac{V \cdot T_N}{2\rho}\left[\frac{2x_{A_2}(n+1) - x_o - V \cdot T}{L_N} - 1\right]$$

$$\dot{\psi}_N(T_N) = \left[\frac{T_N}{2} \alpha \cdot \mathrm{cth}(\frac{T_N}{2} \cdot \alpha) - 1\right]\frac{V}{\rho}$$

$$\tag{3.2.25}$$

where $s = s_N = x_{A_2}(n+1) - (x_o + V \cdot T)$, where $x_{A_2}(n+1)$ denotes $x_{A_2}$ at the (n+1) step.

Equations (3.2.25) contain three variables: T, L, and $L_N$, which gives the possibility of introducing the additional conditions and thus of transforming them into

$$L(L - L_N) - \frac{L_N^2}{4\mathrm{sh}^2(\frac{\alpha \cdot L_N}{2V})} = D, \qquad D = 4c_1 c_2$$

$$T = \frac{1}{\alpha} \ln \frac{1}{4c_1}\left[2L + L_N(\mathrm{cth} \frac{\alpha \cdot L_N}{2V} - 1)\right]$$

$$\tag{3.2.26}$$

It can be concluded that the area of solution existence (which is, also, the area in which the system is controlable) is limited by

$$D + G^2 \geq 0$$

$$G + \sqrt{G^2 + D} > 2c_1$$

where $G = \dfrac{1}{2} L_N \cdot \text{cth}\left(\dfrac{\alpha \cdot L_N}{2V}\right)$

System (3.2.26) solves, in general, the problem of defining L, T, and $L_N$. In addition, it can be used for solving uniquely the problem when one of the three quantities is prescribed in advance, what in the author's opinion gives more freedom in the control realization.

Such a concept has been illustrated by computer simulation. The model parameters are close to those of a man performing nominal motion with V = 1[m/s], L = 0.6[m], and h = 0.7[m]. At the beginning of the third step, disturbance of $\psi$ and $\dot{\psi}$ was applied, and the mechanism transferred to a new state

$$\psi_o = 0.1[\text{rad}], \qquad \dot{\psi}_o = 0.8[\text{rad/s}], \qquad x_o = 0.8[\text{m}],$$

$$V = 1[\text{m/s}], \qquad z_o = 0.7[\text{m}] \quad \text{and} \quad \dot{z}_o = 0$$

The nonlinearities of the original differential equations of motion are considered as an additional disturbance. The process of returning the system to the nominal regime is given in Table 3.1., from which it can be seen that the mechanism attained a periodical regime in the 6-th step. The process of angular stabilization is illustrated in Fig. 3.8.

The second type of stabilization operates when the system is to return to the nominal regime. Again, the consideration will be concerned with the model of massive body on massless legs, and it will be limited to the single-support phase. The system motion is described by the following set of equations

$$m_t \cdot \ddot{x}_C = R_x$$

$$m_t \cdot \ddot{z}_C = R_z - M \cdot g$$

$$(J_t - m_t \rho^2)\ddot{\psi} = (x_C - x_{A_2})R_z - (z_C - z_{A_2})R_x$$

$$(3.2.27)$$

$$x - x_{A_2} = x_C - x_{A_2} - \rho \sin\psi, \qquad z - z_{A_2} = z_C - z_{A_2} - \rho \cos\psi$$

where $J_t = J_c + m_t \rho^2$, $J_c$ is the central moment of inertia.

It should be noted that equations (3.2.27) are written with respect to the mass centre of the whole system, C, unlike the previous case when they were written with respect to point 0.

Table 3.1. Simulation results (after [25])

| STEP k PARAMETERS | 1 | 2 | 3 | 4 | 5 | 6 | 7 | 8 |
|---|---|---|---|---|---|---|---|---|
| $\psi_0$ [rad] | 0.0 | 0.0 | 0.1 | 0.011 | -0.001 | -0.002 | -0.002 | -0.002 |
| $\dot{\psi}_0$ [rad/s] | 0.57 | 0.57 | 0.8 | 0.544 | 0.621 | 0.621 | 0.621 | 0.621 |
| $x_0$ [m] | -0.3 | 0.3 | 0.8 | 1.511 | 2.128 | 2.755 | 3.381 | 4.007 |
| $V_0$ [m/s] | 1.0 | 1.0 | 1.0 | 1.008 | 1.007 | 1.007 | 1.007 | 1.007 |
| $z_0$ [m] | 0.7 | 0.7 | 0.7 | 0.688 | 0.687 | 0.687 | 0.687 | 0.687 |
| $\dot{z}_0$ [m/s] | 0.0 | 0.0 | 0.0 | -0.002 | -0.001 | -0.001 | -0.001 | 0.000 |
| T [s] | 0.6 | 0.6 | 0.689 | 0.610 | 0.622 | 0.622 | 0.622 | 0.622 |
| L [m] | 0.6 | 0.6 | 0.614 | 0.627 | 0.626 | 0.626 | 0.626 | 0.626 |
| *)$\varepsilon$ | 0.0 | 0.0 | 0.345 | 0.089 | 0.001 | 0.000 | 0.000 | 0.000 |

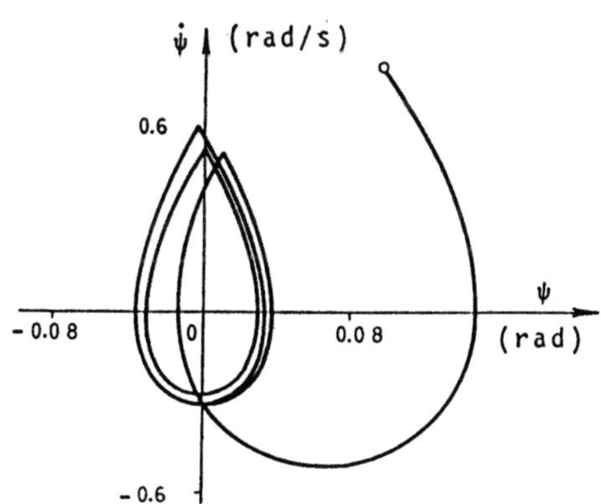


Fig. 3.8. Angular stabilization (after [25])

The control torques at the knee and hip of the supporting leg are

$$P_k^1 = L_\ell (R_x \cdot \cos\beta_1 - R_z \cdot \sin\beta_1), \quad P_t^1 = R_x (z - z_{A_2}) - R_z (x - x_{A_2}) \quad (3.2.28)$$

*) $\quad \varepsilon = |\psi_0(k+1) - \psi_0(k)| + |\dot{\psi}_0(k+1) - \dot{\psi}_0(k)|$

whereas the torques at joints of the leg in swing phase are equal to zero.

As the disturbance are adopted deviations of $\psi$, $\dot{\psi}$, $x$, $\dot{x}$ and $z$, $\dot{z}$ from their nominals, as well as, the deviation of $z_{A_2}$ from $z = 0$. Deviation of $z_{A_2}$ defines the change of ground profile (if it differs from the level surface).

It can be accepted with no loss in generality, that disturbance acts at the beginning of each step. The accepted control values will be the ground force reactions $R_x$ and $R_z$, the step length L, and its duration T; the moments $P_k$ and $P_t$ are the functions of the ground reaction force and geometrical characteristics of the mechanism position.

The procedure of mechanism's returning to its nominal trajectory is as follows.

Let the mechanism move along an undisturbed trajectory, and let the height $(h_N)$ and speed $(V_N)$ of point 0 be constant. The body moves periodically with respect to the angle $\psi$ and the angular velocity $\dot{\psi}$ (with the step length $L_N$). So,

$$\dot{x} = V_N = \text{const}, \qquad z = h_N = \text{const}, \qquad L = L_N$$

$$\psi(0) = \psi(T), \quad \dot{\psi}(0) = \dot{\psi}(T)$$

$$(3.2.29)$$

At the beginning of the next step, the mechanism is subjected to disturbance, and the disturbed state is characterized by $(\psi_0, \dot{\psi}_0, \ldots, \dot{z}_0)$. Then, the task is to return the system to (3.2.29).

The synthesis is carried out in two phases. In the first phase, the required height of point 0, $h_N$, has to be achieved (vertical stabilization), while in the second phase it should be ensured the constant translatory horizontal speed $V_N$ (longitudinal stabilization) and step length $L_N$, as well as, the periodical motion of the trunk (angular stabilization).

Let consider the first phase. Let the ground reactions $R_x$ and $R_z$ be

$$R_x = m_t \cdot \alpha^2 (x_C - x_{A_2}),$$
$$R_z = m_t \cdot g - m_t \cdot \alpha \cdot \ddot{z}_C - m_t \cdot \beta^2 (z_C - H)$$

$$(3.2.30)$$

where $\alpha$ and $\beta$ are positive constants, and H is the mass centre height, prescribed in advance. Reaction forces (3.2.30) are expressed by coordinates of the mass centre of the whole system C. By introducing them into (3.2.27), we can obtain solutions of (3.2.27) as

$$x(t) = x_{A_2} - \rho \cdot \sin\psi(t) + c_1 \cdot e^{\alpha t} + c_2 \cdot e^{-\alpha t}$$

$$z(t) = H - \rho \cdot \cos\psi(t) + d_1 \cdot e^{k_1 t} + d_2 e^{k_2 t} \qquad (3.2.31)$$

$$\psi(t) = \sum_{i=1}^{2} [a_i \cdot e^{\pm \alpha t} + b_i \cdot e^{(k_i + \alpha)t} + f_i \cdot e^{(k_i - \alpha)t}] + a_3 t + a_4$$

Solution (3.2.31) is derived for an arbitrarily chosen n-th step. The constants $a_i$, $b_i$, $c_i$, $d_i$, $f_i$, $k_i$ (i=1,2), $a_3$ and $a_4$ are expressed by the initial values of $(\psi_o, \dot\psi_o, x_o, V_o, z_o, \dot z_o)$, corresponding to the n-th step, using the following expressions

$$c_i = \frac{1}{2}[\rho \cdot \sin\psi_o + x_o - x_{A_2} \pm \frac{1}{\alpha}(V_o - \rho \cdot \dot\psi_o \cdot \cos\psi_o)],$$

$$k_i = -\frac{\alpha}{2} \pm \sqrt{\frac{\alpha^2}{4} - \beta^2},$$

$$d_i = \frac{(-1)^{i+1}}{k_2 - k_1}[k_j(z_o - \rho\cos\psi_o - H) - (\dot z_o - \rho\dot\psi_o \sin\psi_o)],$$

$$a_i = \frac{m_t \cdot c_i}{J_C \cdot \alpha^2}[g - \alpha^2(h - z_{A_2})], \qquad b_i = \frac{m_t \cdot c_1 \cdot d_i (k_i^2 - \alpha^2)}{J_C (k_i + \alpha^2)},$$

$$f_i = \frac{m_t \cdot c_2 \cdot d_i (k_i^2 - \alpha^2)}{J_C (k_i - \alpha)^2}$$

$$a_3 = \dot\psi_o - \alpha(a_1 - a_2) - \sum_{i=1}^{2} [b_i(k_i + \alpha) + f_i(k_i - \alpha)]$$

$$a_4 = \psi_o - \sum_{i=1}^{2} (a_i + b_i + f_i)$$

where $[i=1,2; \; j=i+(-1)^{i+1}]$.

According to (3.2.30), the vertical stabilization appears to be an independent task.

If $R_z$ is known (prescribed), according to (3.2.30) the vertical coordinate of the whole system centre of gravity, $z_C$, approaches assumptotically the prescribed value H, $z_C = z + \rho \cdot \cos\psi = H$. If small body inclinations are supposed, it follows that the height of point 0 is practically constant, $z = H - \rho = h_N$.

In the control equations (3.2.30) there are two unspecified parameters, $\alpha$ and $\beta$, as well as, L and T. The task of vertical stabilization is solved in the following way.

In the beginning of the step, $\alpha$ and T should be chosen such to satisfy the boundary conditions

$$\psi(\alpha, T) = 0, \quad \dot{x}(\alpha, T) = V_N \tag{3.2.32}$$

The parameter $\beta$ is chosen so as to ensure fast convergence of $z_C$ to H; the step length being defined as $L = x(T) - x(0)$. In the first phase, it is not required the realization of either torso periodical motion or constant gait speed V. This phase is finished when the condition $|z(T) - h_N| < \varepsilon$ is satisfied. The step duration T is determined from (3.2.32), step length from the condition that the periodical motion of the body is initiated from the vertical position $\psi_0 = 0$.

In the second phase it is necessary to ensure the mechanism motion with the constant speed $V_N$, periodic motion of torso with the prespecified step length $L_N$, and with permanent requirement that point 0 is being kept within the $\varepsilon$ area with respect to $h_N$.

If in (3.2.30) we set $\beta^2 \equiv 0$, while $\alpha$ is, as before, a free parameter, and if the modified system (3.2.30) is introduced into the first two expressions of (3.2.27), then integration leads (because of the third equation of (3.2.27)) to

$$x(t) = x_{A_2} - \rho \cdot \sin\psi(t) + c_1 e^{\alpha t} + c_2 e^{-\alpha t}$$

$$z(t) = \tilde{H} - \rho\cos\psi(t) - d_3 e^{-\alpha t} \tag{3.2.33}$$

$$\psi(t) = k(c_1 \cdot e^{\alpha t} + c_2 e^{-\alpha t}) + c_3 \cdot t + c_4$$

where

$$\tilde{H} = z_0 + \rho\cos\psi_0 + \frac{1}{\alpha}(\dot{z}_0 - \rho\dot{\psi}_0\sin\psi_0), \quad d_3 = \frac{1}{\alpha}(\dot{z}_0 - \rho\dot{\psi}_0\sin\psi_0)$$

$$\tag{3.2.34}$$

$$k = \frac{m_t}{J_C\alpha^2}(g + \alpha^2(z_{A_2} - \tilde{H})), \quad c_3 = \dot{\psi}_0 - \alpha k(c_1 - c_2), \quad c_4 = \psi_0 - k(c_1 + c_2)$$

while the values for $c_1$ and $c_2$ are the same as in (3.2.31).

It is also supposed that the angle $\psi$ is small, and the adopted control will be applied to the original equations (3.2.27), while the influence of nonlinearities will be taken as distrubance. From (3.2.33) (provided that $\psi$ is small) it follows

$$\dot{x}(t) = \alpha(1-\rho k)(c_1 e^{\alpha t} - c_2 e^{-\alpha t}) - \rho\dot{\psi}_o(1-\rho k) + \rho k V_o$$

Let us choose $\alpha$ such to ensure a constant walking speed. For this purpose, let $\rho \cdot k = 1$. Then, we have

$$\dot{x}(t) = V_o = V_N = \text{const}$$

From the condition $\rho \cdot k = 1$, and according to (3.2.34), $\alpha$ can be determined as

$$\alpha^2 = \frac{m_t \cdot \rho \cdot g}{J_t + m_t \cdot \rho(\tilde{H} - \rho - z_{A_2})} \tag{3.2.35}$$

Consider the expressions for $\tilde{H}$ in (3.2.34) and (3.2.35) as equations (for small $\psi$ and $\dot{\psi}$) with respect to $h = H - \rho$ and $\alpha$. If we express these parameters in terms of $z_o$, $\dot{z}_o$, and $z_{A_2}$, it will follow

$$h = z_o + \frac{\dot{z}_o^2}{2g} + \frac{\dot{z}_o}{2g}\sqrt{\dot{z}_o^2 + \frac{4g^2}{\alpha_*^2}}$$

$$\tag{3.2.36}$$

$$\alpha = \alpha_* \left[\frac{1}{2g}\left(\sqrt{\dot{z}_o^2 + \alpha_*^2 + 4g^2} - \dot{z}_o \cdot \alpha_*\right)\right]$$

where

$$\alpha_*^2 = \frac{m_t \cdot \rho \cdot g}{J_t + m_t \cdot \rho(z_o - z_{A_2})}$$

If it is assumed that $\psi$ is small, system (3.2.33) will become

$$x(t) = x_o + V_N \cdot t, \qquad z(t) = h - d_3 \cdot e^{-\alpha t}$$

$$\rho\psi(t) = c_1 e^{+\alpha t} + c_2 e^{-\alpha t} + x_{A_2} - x(t)$$

where $\alpha$ should be determined from (3.2.36).

In this way, using the control according to (3.2.30) with $\beta^2 \equiv 0$ and selecting $\alpha$ according to (3.2.36), the system motion in the second phase with required walking speed $V_N$ will be ensured. At the same time, the vertical coordinate of point 0 which is in the $\varepsilon$ - region of its nominal value, approaches h asymptotically according to (3.2.36). This is achieved by choosing a correct value for $\beta$ in the first phase of stabilization.

The suggested stabilization algorithm was checked by simulation. The model used has the same characteristics as in the previous case. The system was supposed to walk with $V_N$ = 1 [m/s], $h_N$ = 0.7 [m/s], $L_N$ = 0.6 [m] and $\psi_0$ = 0. Disturbance to the coordinates $\psi$, $\dot{\psi}$, $\dot{x}$, $\dot{z}$ was applied at the beginning of the third step. Deviation of $\psi$ from its nominal value was $6^o$ ($\approx$0.1 [rad]), angular speed $\dot{\psi}$ = 0.8 [rad/s], and walking speed $V_0$ = 1.2 [m/s]. The disturbance to the vertical speed of point 0 was $\dot{z}$ = 0.2 [m/s].

The results are presented in Table 3.2.

The first phase of stabilization was performed on the interval from the third to the eleventh step, and the second phase between the twelvth (i.e., when $|z-h_N| < 0.01$ is satisfied) and the forteenth step. In the fifteenth step, the remainder of the vertical component of speed of point 0 was eliminated and z stabilized at z = 0.76 [m]. The described stabilization process is illustrated in Fig. 3.9. The first two steps of undisturbed motion correspond to line I, and the beginning of disturbed state (in the third step) corresponds to point B. The first phase of stabilization is described by the line along which the system state is transferred from B to line II, and the second phase from II to I.

Such approach to the control synthesis of locomotion systems is characterized by some specificities. First, the whole procedure is performed on a system which is significantly simpler if compared to the real system. The assumption of massless legs, whose motion does not influence the body motion, is not acceptable in general case. In addition, if the four massive links of the two legs move, the inertia

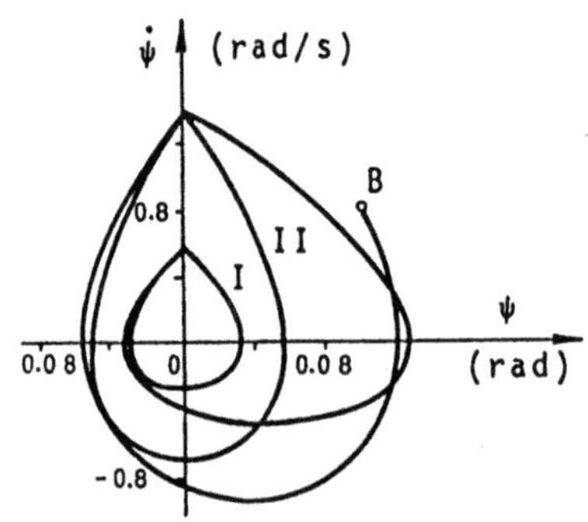


Fig. 3.9. Stabilization process in the phase plane (after [25])

Table 3.2. Simulation results (after [25])

| PARAMETER \ STEP k | 1 | 2 | 3 | 4 | 5 | ... | 10 | 11 | 12 | 13 | 14 | 15 |
|---|---|---|---|---|---|---|---|---|---|---|---|---|
| $\psi_0$ [rad] | 0.0 | 0.0 | 0.1 | 0.0 | 0.0 | ... | 0.0 | 0.0 | 0.0 | 0.0 | 0.0 | 0.0 |
| $\dot{\psi}_0$ [rad/s] | 0.57 | 0.57 | 0.8 | 1.41 | 1.41 | ... | 1.38 | 1.37 | 1.37 | 0.57 | 0.57 | 0.57 |
| $x_0$ | -0.30 | 0.30 | 0.90 | 1.50 | 2.10 | ... | 5.09 | 5.69 | 6.29 | 7.09 | 7.69 | 8.29 |
| $V_0$ [m/s] | 1.0 | 1.0 | 1.2 | 1.0 | 1.0 | ... | 1.0 | 1.0 | 1.0 | 1.0 | 1.0 | 1.0 |
| $z_0$ [m] | 0.7 | 0.7 | 0.7 | 0.74 | 0.75 | ... | 0.72 | 0.71 | 0.708 | 0.706 | 0.706 | 0.706 |
| $\dot{z}_0$ [m/s] | 0.0 | 0.0 | 0.2 | 0.06 | 0.01 | ... | -0.01 | -0.01 | -0.01 | -0.001 | 0.0 | 0.0 |
| T [s] | 0.6 | 0.6 | 0.39 | 0.41 | 0.41 | ... | 0.41 | 0.42 | 0.80 | 0.6 | 0.6 | 0.6 |
| L [m] | 0.6 | 0.6 | 0.605 | 0.599 | 0.598 | ... | 0.599 | 0.77 | 0.63 | 0.6 | 0.6 | 0.6 |
| $z_{A_2}$ [m] | 0.0 | 0.0 | 0.1 | -0.1 | 0.0 | ... | -0.1 | 0.0 | 0.0 | 0.0 | 0.0 | 0.0 |
| *) $\varepsilon$ | 0.0 | 0.33 | 0.708 | 0.003 | 0.006 | ... | 0.002 | 0.002 | 0.8 | 0.0 | 0.0 | 0.0 |
| $\alpha$ [s$^{-1}$] | 2.82 | 2.82 | 2.33 | 2.18 | 2.29 | ... | 2.26 | 2.37 | 2.81 | 2.81 | 2.81 | 2.81 |
| $\beta$ [s$^{-1}$] | 0.0 | 0.0 | 1.16 | 1.09 | 1.14 | ... | 1.13 | 1.18 | 0.0 | 0.0 | 0.0 | 0.0 |

*) $\varepsilon = |\psi_0(k+1) - \psi_0(k)| + |\dot{\psi}_0(k+1) - \dot{\psi}_0(k)|$

moments make the system equation much more complex. Then, a question arises on whether such, relatively simple, analysis is valid at all. Besides, all stabilization algorithms have been formed for a linear model and then applied to the nonlinear one. If a more complex model is used, the influence of nonlinearities will be significantly greater. Also, it should be noted that only the system states at the ends of steps are considered, i.e., the transient regimes are not observed. In the case of locomotion mechanisms, this may cause some undesirable consequences.

Beletskii also considered some special cases of locomotion activities, like the underwater locomotion, which may be of importance for studying the characteristics of human walk in the diving suit, or, for constructing some underwater locomotion machines. The area of his interest is also the energy consumption in walking processes; he has investigated the optimization of the energy needs for different types of walk ("knees back" or "knees before", body above point 0, or beneath it).

In considering massive legs, Beletskii adopts a different approach to motion synthesis, known as semi-inverse, or, prescribed synergy method, previously introduced by Vukobratović and Juričić [22].

A very interesting approach to the control of biped locomotion systems was proposed by Formal'skii [27]. It will be presented on the example of a five-linked planar mechanism whose all links are massive.

It is usually accepted that the control signals are acting on the system all over the regulation period. Of course, their intensity changes (even, the sign), but they are permanently acting.

In the approach suggested by Formal'skii, the control signals are applied only at the beginning of each half-step. Between two actions, the system is not subjected to any driving force or torque, and it is driven only by the inertial forces. This type of control has been named the impulse control.

The suggested approach was tested by simulation on a planar, five-linked model which consists of body and two legs, and performes the walk upon level ground that consists of two phases: single-support phase, and double-support phase, for which is supposed that lasts a very short time. At the very instant when the leg being in the swing phase makes contact with the ground, the other leg (which till that moment was the supporting leg) deploys from the ground and passes to swing phase. Such control strategy was tested by simulation on digital computer. A planar model of five massive links, with parameters close to those of man, was considered.

The corresponding system of equations was solved in two ways. The first one included the linearization around the point $q^i = 0$, $\dot{q}^i = 0$, $(i = 1,...,5)$, where, $[q^i] = [\psi, \alpha_1, \alpha_2, \beta_1, \beta_2]^T$ is state vector. Here, $\psi$ is trunk deviation angle from vertical, while $\alpha_2$, $\beta_2$ and $\alpha_1$, $\beta_1$ are the deviation angles from vertical of thigh and shank for supporting leg in swing phase, respectively. The initial position was defined by $\psi(0) = 0$, $\alpha_1(0) = \beta_1(0) = -0.25$ [rad]; $\alpha_2(0) = \beta_2(0) = 0.25$ [rad]. Fig. 3.10. shows the stick-diagram of the system motion during one half-step.

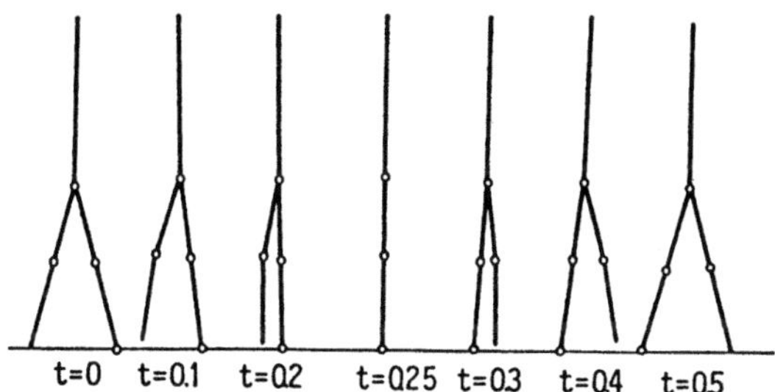


Fig. 3.10. Solution of the linearized model (after [22])

The second approach was concerned with solving the original nonlinear model. Obviously, since there was no analytic solution, a computer iterative procedure had to be employed. The problem was reduced to find out the vector $\dot{q}(0)$ which would transfer the system from the initial state $q(0)$, $\dot{q}(0)$ to the terminal state $q(T)$. Obviously, the vector $\dot{q}(0)$ minimizes the deviation of the system solution from the state $q(T)$ for $t = T$. As a measure of quality of the solution, served the sum of squared deviations of the solution from the given state

$$I = \sum_{i=1}^{5} k_i [\Delta q^i(T)]^2$$

where $k_i$ are the weighting coefficients, greater than one.

Two solutions were obtained:

- symmetric, satisfying $q(0) = [0, \alpha, -\alpha, \alpha, -\alpha]^T$ and $q(\frac{T}{2}) = [0, 0, 0, 0, 0]^T$ (T denotes duration of one half-step), and

- asymmetric, if the previous conditions are not satisfied.

Fig. 3.11. shows the stick-diagram of the solutions of the nonlinear model. The half-step length was 0.45 [m] and its duration 0.5 [s].

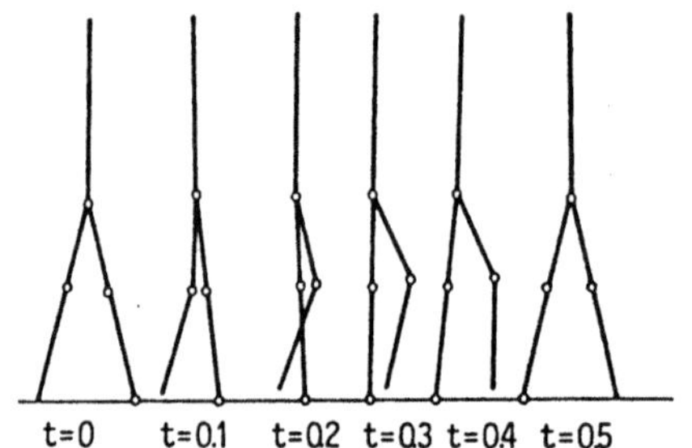

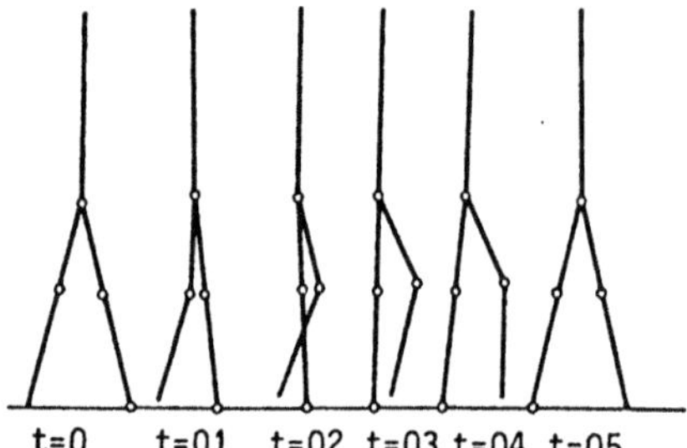


Fig. 3.11. Symmetric (a) and asymmetric (b) solutions
of the nonlinear model (after [27])

In all previous papers there is no any experiment and validity of theoretical results is checked mostly by computer simulation.

It is characteristic of some recent works that they contain experimental checking of theoretical considerations. Experiments have been usually carried out on simple mechanical structures for legs, while the trunk with arms is represented by a single massive link. The legs may be either single - or multi-linked.

Two similar mechanical structures, BIPER-3 and BIPER-4 have been considered in [28]. We shall describe in more detail the structure BIPER-3. The system consists of two single-linked legs without feet (the contact between the leg and the ground is realized at one point) with two d.o.f. at the hip joint. When forming the equations of motion, the following assumptions have been adopted:

1. the motion around all three axes is independent;

2. the rotation around vertical axis is neglected;

3. there is no friction at joints;

4. linearization of motion equations is possible;

5. the contact of the leg with ground is considered as a two d.o.f. joint, and slippage is not permitted;

6. the duration of the double-support phase is much shorter than of the single-support one. (So, it can be assumed that the posture of each link and, accordingly, of the whole mechanism is constant during the double-support phase);

7. collision of the leg with the ground surface is plastic, and the condition $\phi = \dot{\phi} = 0$ corresponds to the case when both feet are in contact with the ground.

Fig. 3.12. shows a scheme of the mechanical structure BIPER-3.

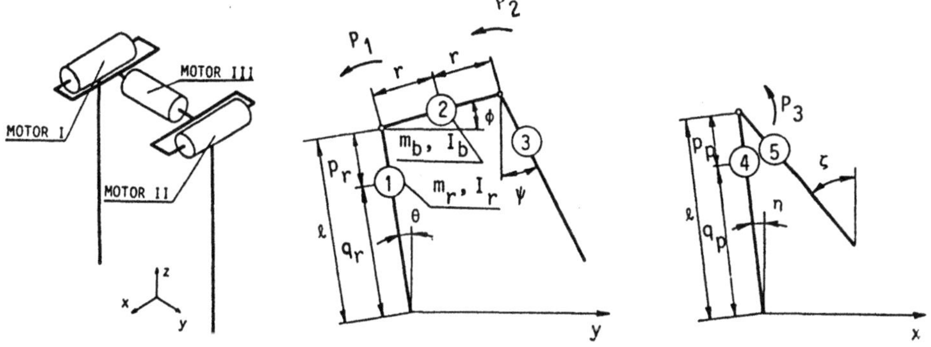


Fig. 3.12. Locomotion mechanism BIPER-3 (after [28])

All data about masses, moments of inertia, lengths, and angles are clear from the figure.

The equations of motion are given as

$$\alpha_1 \ddot{\theta} + \alpha_8 \ddot{\psi}\sin(\phi-\theta) - \alpha_2 \ddot{\psi}\cos(\psi-\theta) =$$

$$= -\alpha_8 \dot{\phi}^2 \cos(\phi-\theta) - \alpha_2 \dot{\psi}^2 \sin(\psi-\theta) + \alpha_3 \sin\theta - P_1$$

$$\alpha_8 \ddot{\theta}\sin(\phi-\theta) + \alpha_4 \ddot{\phi} + \alpha_9 \ddot{\psi}\sin(\psi-\phi) =$$

$$= \alpha_8 \dot{\theta}^2 \cos(\phi-\theta) - \alpha_9 \dot{\psi}^2 \cos(\psi-\phi) - \alpha_5 \cos\phi + P_1$$

214

$$-P_2 - \alpha_2 \ddot{\theta}\cos(\psi-\theta) + \alpha_9 \ddot{\phi}\sin(\psi-\phi) + \alpha_6 \ddot{\psi} =$$

$$= \alpha_2 \dot{\theta}^2\sin(\psi-\theta) + \alpha_9 \dot{\phi}^2\cos(\psi-\theta) - \alpha_7\sin\psi + P_2$$

$$\beta_1 \ddot{\eta} - \beta_2 \ddot{\zeta}\cos(\zeta-\eta) = -\beta_2 \dot{\zeta}^2\sin(\zeta-\eta) + \beta_3\sin\eta - P_3$$

$$\beta_2 \ddot{\eta}\cos(\zeta-\eta) - \beta_4 \ddot{\zeta} = -\beta_2 \dot{\eta}^2\sin(\zeta-\eta) + \beta_5\sin\zeta - P_3$$

After linearization they become

$$\alpha_1 \ddot{\theta} - \alpha_2 \ddot{\psi} = \alpha_3\theta - P_1$$

$$\alpha_4 \ddot{\phi} = -\alpha_5 + P_1 - P_2$$

$$\alpha_6 \ddot{\psi} - \alpha_2 \ddot{\theta} = -\alpha_7\psi + P_2 \qquad\qquad (3.2.37)$$

$$\beta_1 \ddot{\eta} - \beta_2 \ddot{\zeta} = \beta_3\eta - P_3$$

$$\beta_2 \ddot{\eta} - \beta_4 \ddot{\zeta} = \beta_5\zeta - P_3$$

where

$$\alpha_1 = I_r + m_r(q_r^2+\ell^2) + m_b\ell^2, \qquad \alpha_2 = m_r p_r \ell,$$

$$\alpha_3 = (m_r(q_r+\ell) + m_b\ell)g, \qquad \alpha_4 = I_b + 4m_r r^2 + m_b r^2,$$

$$\alpha_5 = (2m_r+m_b)gr \qquad \alpha_6 = I_r + m_r p_r^2$$

$$\alpha_7 = m_r g p_r, \qquad \alpha_8 = (m_b+2m_r)r\ell,$$

$$\alpha_9 = 2m_r p_r r,$$

$$\beta_1 = I_p + m_p(q_p^2+\ell^2), \qquad \beta_2 = m_p p_p \ell,$$

$$\beta_3 = m_p(q_p+\ell)g, \qquad \beta_4 = I_p + m_p p_p^2$$

$$\beta_5 = m_p g p_p$$

Using simple transformations, system (3.2.37) may be written in the matrix form

$$\dot{x}_r = Ax_r + BP_r + E,$$

$$\dot{x}_p = Cx_p + DP_3 \tag{3.2.38}$$

where

$$x_r = \begin{bmatrix} \theta \\ \dot{\theta} \\ \phi \\ \dot{\phi} \\ \psi \\ \dot{\psi} \end{bmatrix}, \quad A = \begin{bmatrix} 0 & 1 & 0 & 0 & 0 & 0 \\ a_1 & 0 & 0 & 0 & a_2 & 0 \\ 0 & 0 & 0 & 1 & 0 & 0 \\ 0 & 0 & 0 & 0 & 0 & 0 \\ 0 & 0 & 0 & 0 & 0 & 1 \\ a_3 & 0 & 0 & 0 & a_4 & 0 \end{bmatrix}, \quad B = \begin{bmatrix} 0 & 0 \\ b_1 & b_2 \\ 0 & 0 \\ b_3 & -b_3 \\ 0 & 0 \\ -b_2 & b_4 \end{bmatrix}$$

$$E = \begin{bmatrix} 0 \\ 0 \\ 0 \\ e_1 \\ 0 \\ 0 \end{bmatrix}$$

$$P_r = \begin{bmatrix} P_1 \\ P_2 \end{bmatrix}, \quad x_p = \begin{bmatrix} \eta \\ \dot{\eta} \\ \zeta \\ \dot{\zeta} \end{bmatrix}, \quad C = \begin{bmatrix} 0 & 1 & 0 & 0 \\ c_1 & 0 & c_2 & 0 \\ 0 & 0 & 0 & 1 \\ c_3 & 0 & c_4 & 0 \end{bmatrix}, \quad D = \begin{bmatrix} 0 \\ d_1 \\ 0 \\ d_2 \end{bmatrix}$$

$$\Delta_r = \alpha_1 \alpha_6 - \alpha_2^2, \quad \Delta_p = \beta_1 \beta_4 - \beta_2^2, \quad a_1 = \frac{\alpha_3 \alpha_6}{\Delta_r}, \quad a_2 = \frac{\alpha_2 \alpha_7}{\Delta_r}$$

$$a_3 = \frac{\alpha_2 \alpha_3}{\Delta_r}, \quad a_4 = -\frac{\alpha_1 \alpha_7}{\Delta_r}$$

$$b_1 = -\frac{\alpha_6}{\Delta_r}, \quad b_2 = \frac{\alpha_2}{\Delta_r}, \quad b_3 = \frac{1}{\alpha_4}, \quad b_4 = \frac{\alpha_1}{\Delta_r},$$

$$e_1 = -\frac{\alpha_5}{\alpha_4}, \quad c_1 = \frac{\beta_3 \beta_4}{\Delta_p}, \quad c_2 = -\frac{\beta_2 \beta_5}{\Delta_p}, \quad c_3 = \frac{\beta_2 \beta_3}{\Delta_p}$$

$$c_4 = -\frac{\beta_1 \beta_5}{\Delta_p}, \quad d_1 = \frac{\beta_2 - \beta_4}{\Delta_p}, \quad d_2 = \frac{\beta_1 - \beta_2}{\Delta_p}$$

After introducing numerical values of parameters into (3.2.38), we obtain

$$a_1 = 34.8, \qquad a_2 = -2.45, \qquad a_3 = 50.9, \qquad a_4 = -50.1,$$

$$b_1 = -6.77, \qquad b_2 = 9.91, \qquad b_3 = 201, \qquad b_4 = 203,$$

$$e_1 = -132, \qquad c_1 = 37.4, \qquad c_2 = -4.55, \qquad c_3 = 57.7,$$

$$c_4 = -56.1, \qquad d_1 = 4.06, \qquad d_2 = 131$$

Since some of these values are significantly smaller than others, (3.2.38) can be simplified

$$\ddot{\theta} = a_1\theta,$$

$$\ddot{\phi} = b_3 P_1 - b_3 P_2 + e_1,$$

$$\ddot{\psi} = a_3\theta + a_4\psi + b_4 P_2 \qquad\qquad (3.2.39)$$

$$\ddot{\eta} = c_1\eta$$

$$\ddot{\zeta} = c_3\eta + c_4\zeta + d_2 P_3$$

Now, behaviour of (3.2.38) can be investigated using the solution of (3.2.39).

Nominal trajectories of the system have to be defined to satisfy (3.2.38) and the repeatability conditions. If the subscripts i and f correspond to the beginning and the end of single-support phase, respectively, the repeatability conditions may be written as

$$x_{ri} = \psi_r x_{rf}, \qquad x_{pi} = \psi_p x_{pf} \qquad\qquad (3.2.40)$$

where

$$x_{ri} = [\theta_i \;\; \dot{\theta}_i \;\; \phi_i \;\; \dot{\phi}_i \;\; \psi_i \;\; \dot{\psi}_i],$$

$$x_{rf} = [\theta_f \;\; \dot{\theta}_f \;\; \phi_f \;\; \dot{\phi}_f \;\; \psi_f \;\; \dot{\psi}_f],$$

$$x_{pi} = [\eta_i \;\; \dot{\eta}_i \;\; \zeta_i \;\; \dot{\zeta}_i],$$

$$x_{pf} = [\eta_f \ \dot{\eta}_f \ \zeta_f \ \dot{\zeta}_f],$$

$$\psi_r = \begin{bmatrix} 0 & 0 & 0 & 0 & -1 & 0 \\ 0 & -1 & 0 & 0 & 0 & 0 \\ 0 & 0 & 0 & 0 & 0 & 0 \\ 0 & 0 & 0 & 0 & 0 & 0 \\ -1 & 0 & 0 & 0 & 0 & 0 \\ 0 & -1 & 0 & 0 & 0 & 0 \end{bmatrix}, \qquad \psi_p = \begin{bmatrix} 0 & 0 & 1 & 0 \\ 0 & 1 & 0 & 0 \\ 1 & 0 & 0 & 0 \\ 0 & 1 & 0 & 0 \end{bmatrix}$$

Let us denote nominal trajectories by $\theta^o(t)$, $\phi^o(t)$, $\psi^o(t)$, $\eta^o(t)$, $\zeta^o(t)$ and the driving torques that should enable their realization by $P_1^o(t)$, $P_2^o(t)$, and $P_3^o(t)$. Then, the following conditions will be satisfied

$$\dot{x}_r^o = Ax_r^o + BP_r^o + E, \qquad \dot{x}_p^o = Cx_p^o + DP_3^o,$$

$$x_{ri}^o = \psi_r x_{rf}^o, \qquad x_{pi}^o = \psi_p x_{pf}^o \tag{3.2.41}$$

After subtracting (3.2.41) from (3.2.38), the deviation of the actual trajectories from their nominals will be described by the model

$$\Delta\dot{x}_r = A\Delta x_r + B\Delta P_r, \qquad \Delta\dot{x}_p = C\Delta x_p + D\Delta P_3, \tag{3.2.42}$$

$$\Delta x_{ri} = \psi_r \Delta x_{rf}, \qquad \Delta x_{pi} = \psi_p \cdot \Delta x_{pf} \tag{3.2.43}$$

where $\Delta$ denotes the difference between the actual and nominal value, i.e., $\Delta x_r \equiv x_r - x_r^o$, $\Delta\theta \equiv \theta - \theta^o$. Conditions (3.2.43) are approximately fulfilled.

We shall consider now the stabilization around y axis. Using (3.2.39), equation (3.2.42) may be simplified, so that

$$\Delta\ddot{\eta} = c_1\Delta\eta, \qquad \Delta\ddot{\zeta} = c_3\Delta\eta + c_4\Delta\zeta + d_2\Delta P_3 \tag{3.2.44}$$

If we denote with (n) all variables belonging to the n-th step, equations (3.2.43) and (3.2.44) can be written as

$$\begin{bmatrix} \Delta\eta_f^{(n)} \\ \Delta\dot{\eta}_f^{(n)} \end{bmatrix} = \exp\left( \begin{bmatrix} 0 & 1 \\ c_1 & 0 \end{bmatrix} T \right) \begin{bmatrix} \Delta\eta_i^{(n)} \\ \Delta\dot{\eta}_i^{(n)} \end{bmatrix} \tag{3.2.45}$$

218

$$
\begin{bmatrix} \Delta n_i^{(n+1)} \\ \Delta \dot{n}_i^{(n+1)} \end{bmatrix} = \begin{bmatrix} 1 & 0 \\ 0 & 1 \end{bmatrix} \begin{bmatrix} \Delta \zeta_f^{(n)} \\ \Delta \dot{n}_f^{(n)} \end{bmatrix}
\tag{3.2.45}
$$

If it is supposed that the leg being in swing phase can be accurately positioned at $\Delta \zeta_f^{(n)}$

$$
\Delta \zeta_f^{(n)} = k_{1p} \Delta n_f^{(n)} + k_{2p} \Delta \dot{n}_f^{(n)} .
\tag{3.2.46}
$$

where $k_{1p}$ and $k_{2p}$ are constants, equation (3.2.45) may be written in the form

$$
\begin{bmatrix} \Delta n_i^{(n+1)} \\ \Delta \dot{n}_i^{(n+1)} \end{bmatrix} = \begin{bmatrix} k_{1p} & k_{2p} \\ 0 & 1 \end{bmatrix} \begin{bmatrix} \Delta n_f^{(n)} \\ \Delta \dot{n}_f^{(n)} \end{bmatrix}
\tag{3.2.47}
$$

Then, from (3.2.45) and (3.2.47) we obtain

$$
\begin{bmatrix} \Delta n_i^{(n)} \\ \Delta \dot{n}_i^{(n)} \end{bmatrix} = \left( \begin{bmatrix} k_{1p} & k_{2p} \\ 0 & 1 \end{bmatrix} \begin{bmatrix} \cos\sqrt{c_1}T & \frac{1}{\sqrt{c_1}}\sinh\sqrt{c_1}T \\ \sqrt{c_1}\sinh\sqrt{c_1}T & \cosh\sqrt{c_1}T \end{bmatrix} \right)^n \begin{bmatrix} \Delta n_i^{(0)} \\ \Delta \dot{n}_i^{(0)} \end{bmatrix}
\tag{3.2.48}
$$

A characteristic equation of (3.2.48) is:

$$
\lambda^2 - \{ (k_{1p}+1)\cosh\sqrt{c_1}T + k_{2p}\sqrt{c_1}\sinh\sqrt{c_1}T \}\lambda + k_{1p} = 0
\tag{3.2.49}
$$

If the absolute values of the solutions of (3.2.48) are negative the quantities $\Delta n_i^{(n)}$ and $\Delta \dot{n}_i^{(n)}$ will converge to zero for arbitrary initial values of $\Delta n_i^{(0)}$ and $\Delta \dot{n}_i^{(0)}$. Because of that (according to (3.2.43)), $\Delta x_{p_i}^{(n)}$ will also converge to zero. Therefore, it can be concluded that the locomotion system can be stabilized by the accurate positioning of the free leg (i.e., controlled by step length). But, these conclusions are based on equations (3.2.44), which are only the approximations of the original equations, and on the assumption that the free leg can be accurately positioned in space. If, however, a linear feedback is introduced for controlling the position of the free leg

$$
\Delta P_3 = f_{1p}\{\Delta\zeta - (k_{1p}\Delta n_f^{(n)} + k_{2p}\dot{n}_f^{(n)})\} + f_{2p}\Delta\dot{\zeta} - \frac{c_3}{d_2}\Delta n
\tag{3.2.50}
$$

($f_{1p}$ and $f_{2p}$ are feedback gains), the positioning error may jeopardize the system stability, and therefore an estimation of $\Delta n_f^{(n)}$ is required.

But, if we accept that the system motion is described by (3.2.42), the corrective moment $\Delta P_3$ is given as

$$\Delta P_3 = f_{1p}\{\Delta\zeta - (k_{1p}\Delta n + k_{2p}\Delta\dot{n})\} + f_{2p}\Delta\dot{\zeta} - \frac{c_3}{d_2}\,\Delta n \tag{3.2.51}$$

Then, (3.2.42) transforms into

$$\Delta\ddot{n} = c_1\Delta n + c_2\Delta\zeta + d_1\Delta P_3,$$

$$\Delta\ddot{\zeta} = c_3\Delta n + c_4\Delta\zeta + d_2\Delta P_3 \tag{3.2.52}$$

By eliminating $\Delta P_3$ from (3.2.51) and (3.2.52) we obtain

$$\Delta\dot{x}_p = \begin{bmatrix} 0 & 1 & 0 & 0 \\ c_1 - \dfrac{d_1}{d_2}\,c_3 - f_{1p}k_{1p} & -d_1 f_{1p}k_{2p} & c_2 + d_1 f_{1p} & d_1 f_{2p} \\ 0 & 0 & 0 & 1 \\ -d_2 f_{1p}k_{1p} & -d_2 f_{1p}k_{2p} & c_4 + d_2 f_{1p} & d_2 f_{2p} \end{bmatrix} \Delta x_p \tag{3.2.53}$$

If the transition matrix of (3.2.53) is denoted by $\phi_p(T)$ and taking into account (3.2.41) it follows

$$\Delta x_{pi}^{(n)} = \{\psi_p \cdot \phi_p(T)\}^o \Delta x_{pi}^{(0)} \tag{3.2.54}$$

Stabilization around x axis is achieved in a similar way and under the assumption $\Delta P_1 = \Delta P_2$, so that

$$\Delta\ddot{\theta} = a_1\Delta\theta + a_2\Delta\psi + (b_1 + b_2)\Delta P_1$$

$$\Delta\ddot{\psi} = a_3\Delta\theta + a_4\Delta\psi + (-b_2 + b_4)\Delta P_1$$

$$\Delta\ddot{\phi} = 0$$

$$\Delta P_1 = \Delta P_2 = f_{1r}\{\Delta\psi - (k_{1r}\Delta\theta + k_{2r}\Delta\dot{\theta})\} + f_{2r}\Delta\dot{\psi} - \frac{a_2}{b_4 - b_2}\,\Delta\theta$$

The validity of the proposed control algorithm has been confirmed both

220

by simulation and by experiment. The control torques are realized as

$$P_r = P_r^O + \Delta P_r, \qquad P_3 = P_3^O + \Delta P_3.$$

Using the same approach, similar results have been obtained for the structure BIPER-4, possessing a number of d.o.f. which is much closer to that of man.

Furusho and Masubuchi [29] used a simplified mathematical model of the corresponding planar mechanism as the basis for building a five-link spatial mechanism and the suitable control synthesis (Fig. 3.13).

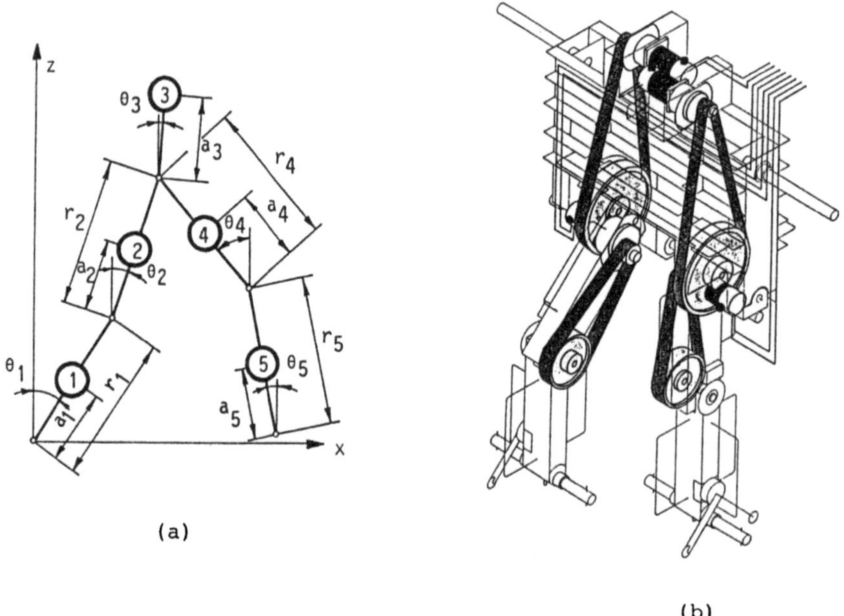



(a)

(b)

Fig. 3.13. Five-link planar biped model (a) and
corresponding mechanism (b) (after [29])

The equations of motion are

$$A(\theta) \cdot \ddot{\theta} + B(\theta) \cdot h(\dot{\theta}) + C \cdot g(\theta) = D \cdot P \qquad (3.2.55)$$

where

$$\theta = \begin{bmatrix} \theta_1 \\ \theta_2 \\ \vdots \\ \theta_5 \end{bmatrix} \qquad P = \begin{bmatrix} P_1 \\ P_2 \\ P_3 \\ P_4 \end{bmatrix} \qquad h(\dot{\theta}) = \begin{bmatrix} \dot{\theta}_1^2 \\ \dot{\theta}_2^2 \\ \vdots \\ \dot{\theta}_5^2 \end{bmatrix} \qquad g(\theta) = \begin{bmatrix} \sin\theta_1 \\ \sin\theta_2 \\ \vdots \\ \sin\theta_5 \end{bmatrix}$$

$$A(\theta) = \{q_{ij}\cos(\theta_i-\theta_j)+p_{ij}\}$$

$$B(\theta) = \{q_{ij}\sin(\theta_i-\theta_j)\}$$

$$C = \text{diag}\{-h_i\}$$

$$D = \begin{bmatrix} 1 & 0 & 0 & 0 \\ -1 & 1 & 0 & 0 \\ 0 & -1 & 1 & 0 \\ 0 & 0 & -1 & 1 \\ 0 & 0 & 0 & -1 \end{bmatrix}$$

The quantities $q_{ij}$, $p_{ij}$, $(i=1,\ldots,5,\ j=1,\ldots,5)$ and $h_i$, $(i=1,\ldots,5)$ depend on physical characteristics of the mechanism. The vector $P = [P_1\ P_2\ P_3\ P_4]^T$ denotes the driving torques at the knee and hip of the supporting leg and at the knee and hip of the leg in swing phase, respectively. After the linearization around the vertical equilibrium point $\theta = \dot{\theta} = 0$, (3.2.55) becomes

$$A_s \cdot \ddot{\theta} + C \cdot \theta = D \cdot P \qquad (3.2.56)$$

where $A_s = \{q_{ij}+p_{ij}\}$.

By introducing the state vector $x = [\theta^T,\ \dot{\theta}^T]^T$, (3.2.56) is transformed into

$$\dot{x} = F \cdot x + G \cdot P \qquad (3.2.57)$$

where

$$F = \begin{bmatrix} 0 & I \\ -A_s^{-1}C & 0 \end{bmatrix} \qquad G = \begin{bmatrix} 0 \\ A_s^{-1}D \end{bmatrix}$$

At each joint, the proportional and derivative local feedbacks are adopted. For example, at the knee joint of the supporting leg, it is

$$P_1 = f \cdot k_{p1} \{\theta_1^O - (\theta_2 - \theta_1)\} + f \cdot k_{d1} (\dot\theta_2 - \dot\theta_1)$$

Here f is a gain constant; $f \cdot k_{p1}$ and $f \cdot k_{d1}$ are the proportional and derivative gains, respectively. $\theta_1^O$ is a reference (i.e., nominal) value of the angle of the knee joint. Using matrix notation, it can be written

$$P = -f \cdot K_p \cdot \theta - f \cdot K_d \cdot \dot\theta - f \cdot K_r \cdot \theta^O \qquad (3.2.58)$$

where $\theta^O = [\theta_1^O \; \theta_2^O \; \theta_3^O \; \theta_4^O]^T$, $K_r = \mathrm{diag}\{k_{pi}\}$, $K_p = \mathrm{diag}\{k_{pi}\}D^T$, $K_d = \mathrm{diag}\{k_{di}\}D^T$.

Now, the closed loop system becomes

$$\dot{x} = \bar{F} \cdot x + \bar{G} \cdot \theta^O \qquad (3.2.59)$$

where

$$\bar{F} = \begin{bmatrix} 0 & I \\ A_s^{-1}(-C-fDK_p) & -fA_s^{-1}DK_d \end{bmatrix} \qquad \bar{G} = \begin{bmatrix} 0 \\ -fA_s^{-1}DK_r \end{bmatrix}$$

By investigating the locus of the eigen-values of $\bar{F}$ with respect to the constant f the authors concluded that there are two dominant eigen-values. So, they diagonalized the system (3.2.59) in such a way that the two dominant modes are diagonal elements of the matrix $\Lambda_1$, while the remaining fast modes are diagonal elements of the matrix $\Lambda_2$. Thus

$$\begin{bmatrix} \dot\xi_1 \\ \dot\xi_2 \end{bmatrix} = \begin{bmatrix} \Lambda_1 & 0 \\ 0 & \Lambda_2 \end{bmatrix} \begin{bmatrix} \xi_1 \\ \xi_2 \end{bmatrix} + \begin{bmatrix} \Gamma_1 \\ \Gamma_2 \end{bmatrix} \theta^O \qquad (3.2.60)$$

where $\xi_1$ and $\xi_2$ are the vectors of size two and eight, respectively. If the change of the reference input $\theta^O$ is sufficiently smooth (so that $\dot\xi_2 \approx 0$), system, (3.2.60) can be approximated as

$$\dot\xi_1 = \Lambda_1 \cdot \xi_1 + \Gamma_1 \cdot \theta^O \qquad (3.2.61)$$

$$\xi_2 = -\Lambda_2^{-1} \Gamma_2 \cdot \theta^O \qquad (3.2.62)$$

Thus, the reduced order model of biped is given by (3.2.61).

The authors have demonstrated that the reduced model approximates quite
well the original model of higher order.

Furthermore, the authors investigated stability of walk. They accepted
that locomotion is called stable if the deviation from the steady walking state is decreasing with time. For these considerations the reduced model (3.2.61) was used, to which particular values of the mechanism parameters were applied. Also, the reference inputs $\theta^o$ are selected to be appropriate for this purpose. For such a particular case, a
recursive relation is derived which proves that deviation from the
steady regime decreases with time.

This conclusion was proved in two ways: by computer simulation and by
experiment. In both cases it was confirmed that the considered mechanism can perform a continuous stable walk.

In [30], a real mechanism for the realization of artificial gait is
described. The mechanism has nine d.o.f. and is composed of two legs,
each having three d.o.f. (ankle, knee, and hip), and a trunk in the
form of the inverted massive pendulum having also three d.o.f. (the
motion in the frontal plane, the motion in the sagittal plane, and the
change of the pendulum length). The method of prescribed synergy is
adopted for the motion synthesis. For the given motion of the legs and
ZMP, such trunk motion is calculated which enables a stable walk of
the mechanism; the corresponding patterns are calculated for different
gait types.

The weight of the mechanism is 107 [kg] and its height when not moving
1.8 [m]. It is powered by an electro-hydro servo system. Apart from the
sensors to monitor the position and velocity at each joint, the mechanism's soles are equiped with two microswithes to monitor the contact
of the toe and the hill with the ground.

Different gait types upon level ground, including the stage of starting and stopping, are realized experimentally. The results show good
agreement of the preset and the realized walking patterns.

This review of results on biped posture and locomotion control has
shown a variety of models and control structures used in these studies.
Mechanical models varied from the single inverted pendulum with fixed
base, to the mechanisms with such a number of d.o.f. which is close to

that used by man in his locomotion activity. Mathematical modelling
has been mostly subjected to simplifications. Most often this was the
linearization in the vicinity of the operating point, but this was also
done in some other ways. These simplifications have enabled an easier
handling and solving the corresponding systems of equations. However,
simplifications take the model further away from the actual system on
which the results should be applied.

It is evident that there is no a general or predominanting approach to
the control synthesis. Locomotion systems are characterized by the presence of unpowered d.o.f., formed in contact of the foot and the gro-
und. They may be influenced only by appropriate movements of the rest
of the system. The models of locomotion mechanisms are highly nonlinear
and there are no methods for a convenient control synthesis. This is
the reason why most of controls have been synthesized on the models of
low order, and then applied to the original nonlinear equations, for
which nonlinearities are considered as disturbances.

Finally, it should be pointed out that there is no a generally accepted method for stability analysis. In the case this was achieved by
Lyapunov's second method, there was no way suggested how to construct
the Lyapunov function, and, the construction varied from case to case.
Another approach used was to simulate the desired motion and then ob-
serve the system's behaviour.

We may conclude that a general characteristic of the reviewed research
would be the lack of a systematic approach, independent of configuration and complexity of the system under study. In the  text  to follow,
we are going to present a unified approach to both the control synthesis and stability analysis.

## 3.3. Model of the System

The mathematical model of biped developed in Chapters 1 and 2 is concerned only with the mechanical part of the system, i.e., with the deriving relationships between the motion and generalized forces (driving torques for the revolute and forces for prismatic joints). However, the active spatial mechanisms are powered by actuators, whose dynamics should be also included into the system model. Therefore, a mathematical model of the overall system [31] consists of two parts: the model of mechanical configuration $S^M$ and the model of actuators $S_a^i$ located at mechanism's joints. The model of mechanical part (described in detail in Section 2.3) for the mechanism which can be represented as a set of open kinematic chains is presented as

$$S^M: \quad P = H(q) \cdot \ddot{q} + h(q, \dot{q}) \qquad (3.3.1)$$

where $q \in R^n$ is the vector of mechanism's generalized coordinates; $P \in R^n$ is the vector of generalized forces at mechanism's joints; $H(q): R^n \to R^{n \times n}$ is the inertial matrix of the system, and $h(q, \dot{q}): R^n \times R^n \to R^n$ is a vector comprising the gravitational, centrifugal, and Coriolis moments.

The mechanical configuration of n d.o.f. is powered by m (m<n)) actuators whose models are

$$S_a^i: \quad \dot{x}_c^i = A_c^i \cdot x_c^i + b_c^i \cdot N(u^i) + f_c^i \cdot P_c^i \qquad \forall i \in I_1 \qquad (3.3.2)$$

where $A_c^i \in R^{n_i \times n_i}$ is the subsystem matrix, whereas $b_c^i \in R^{n_i}$ and $f_c^i \in R^{n_i}$ are distribution vectors of the input control signal and force, respectively; $x_c^i \in R^{n_i}$ is the subsystem state vector; $u^i \in R^1$ and $P_c^i \in R^1$ are the scalar values of control input and generalized force of the i-th subsystem; $i \in I_1$, $I_1 = \{i, i=1,2,\ldots,m\}$; $N(u^i)$ is the nonlinearity of the amplitude saturation type:

$$N(u^i) = \begin{cases} -u_m^i & \text{if} \quad u^i \leq -u_m^i \\ u^i & \text{if} \quad -u_m^i \leq u^i \leq u_m^i, \quad i \in I_1 \\ u_m^i & \text{if} \quad u^i \geq u_m^i \end{cases} \qquad (3.3.3)$$

where $u_m^i$ is the maximum value of the input control to the i-th actuator. It is assumed that the parameters of the actuators' model are constant over the whole observation period T.

By uniting the actuators models $S_a^i$ and the model of mechanical part $S^M$, a model of the overall system can be formed. For this purpose, let us unite the models of all actuators in the form

$$\dot{x}_C = A_C \cdot x_C + F_C \cdot P_C + B_C \cdot N(u) \tag{3.3.4}$$

where $x_C \in R^{N_C}$ is the state vector of actuators system $x_C = (x_C^{1^T}, x_C^{2^T}, \ldots \ldots, x_C^{n^T})^T$; $N_C$ is the order of actuator system; $F_C = \text{diag}[f_C^i]$, $\forall i \in I_1$, $B_C = \text{diag}[b_C^i]$, $\forall i \in I_1$, $A_C = \text{diag}[A_C^i]$, $\forall i \in I_1$; $P_C = (P_C^1, P_C^2, \ldots, P_C^m)^T$ is the vector of generalized forces, whereas $N(u) = (N(u^1), N(u^2), \ldots \ldots, N(u^m))^T$ is the vector of system inputs.

Let, for the purpose of separating coordinates belonging to the powered d.o.f. from those for the unpowered ones, the matrix of transformation $T' \in R^{n \times n}$ be

$$T' \cdot q = (q_C^T \mid q_N^T)^T \tag{3.3.5}$$

where $q_C \in R^m$ and $q_N \in R^{n-m}$ denote the vectors of generalized coordinates of the powered and unpowered d.o.f., respectively. Introducing (3.3.5) into (3.3.1) yields[*]

$$P_C = H_{CC}\ddot{q}_C + H_{CN}\ddot{q}_N + h_C \tag{3.3.6}$$

$$P_N = H_{NC}\ddot{q}_C + H_{NN}\ddot{q}_N + h_N \tag{3.3.7}$$

where $T'HT'^{-1} = \begin{bmatrix} H_{CC} & H_{CN} \\ H_{NC} & H_{NN} \end{bmatrix}$ and $T'h = \begin{bmatrix} h_C \\ h_N \end{bmatrix}$, $H_{CC}$, $H_{CN}$, $H_{NC}$, $H_{NN}$, $h_C$ and $h_N$ are the matrices and vectors of appropriate dimensions, $T'P = [P_C^T \mid P_N^T]^T$, $P_C \in R^m$, $P_N \in R^{n-m}$ are the generalized forces of the powered and unpowered d.o.f., respectively. It is assumed that the $P_N$ values are known in advance (for locomotion systems they are usually zero).

The acceleration of unpowered d.o.f. may be expressed from (3.3.7) as

$$\ddot{q}_N = H_{NN}^{-1}(P_N - H_{NC} \cdot \ddot{q}_C - h_N) \tag{3.3.8}$$

---

[*] For simplicity $H$ and $h$ will be used instead of $H(q)$ and $h(q, \dot{q})$.

If the subsystem output is given by

$$y^i = C^i \cdot x^i_c = (q^i, \dot{q}^i)^T \qquad (3.3.9)$$

where $y^i \in R^2$ is the output of the subsystem $S^i$ and $C^i \in R^{2 \times n_i}$ is a constant matrix, such a matrix transformation can be introduced which will separate the velocities from the subsystem state vector, that is

$$\dot{q}_c = T \cdot C_c \cdot x_c \qquad (3.3.10)$$

where $C_c = \text{diag}[C^i]$, $C_c \in R^{2m \times N_c}$ is a constant matrix, $T = \text{diag}[T^i]$, $T \in R^{m \times 2m}$, $T^i \in R^{1 \times 2}$, $\forall i \in I_1$, $T^i$ are the constant row vectors $T^i = [0\ 1]$. From (3.3.4) and (3.3.10), one obtains

$$\ddot{q}_c = T \cdot C_c \cdot \dot{x}_c = TC_c[A_c \cdot x_c + F_c P_c + B_c \cdot N(u)] \qquad (3.3.11)$$

After introducing (3.3.11) and (3.3.8) into (3.3.6), the expression for generalized forces is obtained in the form

$$P_c = [H_{CC} - H_{CN}H_{NN}^{-1}H_{NC}]\{TC_c[A_c \cdot x_c + F_c P_c + B_c N(u)]\} +$$
$$+ H_{CN}H_{NN}^{-1}(P_N - h_N) + h_c \qquad (3.3.12)$$

or

$$P_c = (I_m - H_p TC_c F_c)^{-1}\{H_p[TC_c A_c \cdot x_c + TC_c B_c N(u)] +$$
$$+ H_{CN}H_{NN}^{-1}(P_N - h_N) + h_c\} \qquad (3.3.13)$$

where $H_p = (H_{CC} - H_{CN}H_{NN}^{-1}H_{NC})$. Substituting (3.3.13) into (3.3.4) yields

$$\dot{x}_c = A_c \cdot x_c + F_c(I_m - H_p TC_c F_c)^{-1}\{H_p[TC_c A_c x_c + TC_c B_c N(u)] +$$
$$+ H_{CN}H_{NN}^{-1}(P_N - h_N) + h_c\} + B_c N(u) \qquad (3.3.14)$$

If we take into account (3.3.10), (3.3.13), and (3.3.14), expression (3.3.8) will be transformed into

$$\ddot{q}_N = H_{NN}^{-1}\{P_N - H_{NC}TC_c[A_c \cdot x_c + F_c(I_m - H_p TC_c F_c)^{-1}(H_p TC_c A_c \cdot x_c +$$
$$+ H_p TC_c B_c N(u) + H_{CN}H_{NN}^{-1}(P_N - h_N) + h_c) + B_c N(u)] - h_N\} \qquad (3.3.15)$$

By uniting (3.3.14) and (3.3.15), the overall system model, S, is obtained as

$$S: \quad \dot{x} = \hat{A}(x) + \hat{B}(x) \cdot N(u) \tag{3.3.16}$$

where $x = (x_c^T, q_N^1, \dot{q}_N^1, \ldots, q_N^{n-m}, \dot{q}_N^{n-m})^T$, $x \in R^N$ is a state vector of the overall system S, $\hat{A}(x): R^N \to R^N$ and $\hat{B}(x): R^N \to R^{N \times m}$ are vector functions and matrices, according to

$$\hat{A}(x) = \begin{bmatrix} A_p \\ \{\hat{A}^j\} \end{bmatrix}, \qquad B(x) = \begin{bmatrix} F_c(I_m - H_p TC_c F_c)^{-1} H_p TC_c B_c + B_c \\ \{\hat{B}^j\} \end{bmatrix}$$

$$A_p = A_c x_c + F_c(I_m - H_p TC_c F_c)^{-1}[H_p TC_c A_c x_c + H_{CN} H_{NN}^{-1}(P_N - h_N) + h_c]$$

$$\hat{A}^j = \begin{bmatrix} \dot{q}^j \\ [H_{NN}^{-1}(P_N - H_{NC} TC_c A_p - h_N)]_j \end{bmatrix} \quad \forall j \in J$$

$$\hat{B}^j = \begin{bmatrix} 0 \\ -[H_{NN}^{-1} H_{NC} TC_c \{F_c(I_m - H_p TC_c F_c)^{-1} H_p TC_c B_c + B_c\}]_j \end{bmatrix}, \quad \forall j \in J$$

The set J is defined as $J = \{j, j=1,2,\ldots,n-m\}$. In the expressions for $\hat{A}^j$ and $\hat{B}^j$ $[\ ]_j$ denotes the j-th row of the vector or of the matrix. $\{\hat{A}^j\}$ denotes a vector whose block elements are given by the vectors $\hat{A}^j$; $\{\hat{B}^j\}$ denotes a matrix whose rows coincide with the matrices $\hat{B}^j$, $\{\hat{A}^j\} = (\hat{A}^{1T} \hat{A}^{2T} \ldots \hat{A}^{n-mT})^T \in R^{2(n-m)}$, $\{\hat{B}^j\} = (\hat{B}^{1T} \hat{B}^{2T} \ldots \hat{B}^{n-mT})^T \in R^{2(n-m) \times m}$.

The whole system order, N, is

$$N = \sum_{i=1}^{m} n_i + \sum_{j=1}^{n-m} k_j$$

where $k_j$ is the number of state vector coordinates of the j-th unpowered subsystem.

It is supposed that $\hat{A}(x)$ and $\hat{B}(x)$ satisfy the smoothness conditions, and solutions of (3.3.16) are unique and continuous with respect to time and initial conditions.

The mathematical model (3.3.16) represents a complete system i.e. the mechanical part powered by the actuators, and can be used either for the motion and control synthesis or for motion simulation. In Chapter 2, the synthesis of mechanism motion was discussed in detail. The motion, which satisfies all the requirements imposed (desired gait type, repeatibility conditions and required ZMP position) is named nominal motion. Let assume that the nominal trajectories of all state coordinates $x^o(t)$ be synthesized. Let us assume further that we may determine the nominal control signals $u^o(t)$ which satisfy

$$S^M: \quad P^o = H(q^o) \cdot \ddot{q}^o + h(q^o, \dot{q}^o)$$

$$S_a^i: \quad \dot{x}_c^{oi} = A_c^i \cdot x_c^{oi} + b_c^i \cdot (u^{oi}) + f_c^i \cdot P_c^{oi}$$

where the superscript "o" corresponds to nominal motion i.e. $x^{oi}(t)$ is the nominal trajectory or the i-th state vector, $q^o(t)$ is n×1 vector of nominal trajectories of the biped joint angles $q^o(t) = (q^{o1}, q^{o2}, \dots \dots, q^{on})^T$, $P_c^{oi}(t)$ is the nominal driving torque of the i-th actuator, and $P^o(t)$ is the n×1 vector of nominal generalized forces. In other words, we calculate the nominal programmed control signals which satisfy above relations in any instant of time.

The nominal programmed control $u^o(t)$ will drive the system along nominal trajectories $x^o(t)$ if the following conditions are satisfied:

1) The model of the system (3.3.1) and (3.3.2) is perfect, which means that all parameters of the system are ideally precisely identified.

2) The initial state of the system $x(0)$ coincides with the nominal initial conditions $x^o(0)$.

3) No perturbation is acting upon the system.

Obviously, none of these conditions are satisfied. Therefore, the actual system state deviates from the nominal trajectory. The additional control signals $\Delta u^i = u^i - u^{oi}$ must be applied to keep the system state as close as possible to the nominal trajectory.

For the purpose of control synthesis, we use a model of the system deviation from its nominal value. Again, two separate models are to be considered: the model of mechanical configuration, $S^M$, and the model of actuators, $S_a^i$. Thus,

230

$$S^M: \quad \Delta P = H^*(\Delta q) \cdot \Delta \ddot{q} + h^*(\Delta q, \Delta \dot{q}) \tag{3.3.17}$$

where $H^*: R^n \to R^{n \times n}$, $h^*: R^n \times R^n \to R^n$, and the asterisk indicates the model is related to the system's deviation from its nominal trajectory $x^o(t)$; $\Delta P = P(t) - P^o(t)$, $\Delta P \in R^n$, $P^o(t)$ are the generalized forces corresponding to the nominal trajectory $x^o(t)$, $\Delta q \in R^n$, $\Delta \dot{q} \in R^n$, $\Delta \ddot{q} \in R^n$, $\Delta q = q(t) - q^o(t), \dots$ . In the further text, we shall use a shorter notation $H^* = H^*(\Delta q)$ and $h^* = h^*(\Delta q, \Delta \dot{q})$.

The models of actuators deviations from nominal trajectories may be presented in the form

$$S_a^i: \quad \Delta \dot{x}_c^i = A_c^i \Delta x_c^i + f_c^i \Delta P_c^i + b_c^i N(\Delta u^i), \qquad \forall i \in I_1$$

$$\Delta y^i = c^i \cdot \Delta x_c^i \tag{3.3.18}$$

where $\Delta x_c^i = x_c^i(t) - x_c^{oi}(t)$, $\Delta x_c^i \in R^{n_i}$, $\Delta P_c^i \in R^1$, $\Delta P_c = (\Delta P_c^1, \Delta P_c^2, \dots, \Delta P_c^m)^T$, $\Delta y^i(t) = y^i(t) - y^{oi}(t)$, $\Delta y^i \in R^{k_i}$, $A_c^i \in R^{n_i \times n_i}$.

For $(\Delta u^i)$ we may write

$$N(\Delta u^i) = \begin{cases} -u_m^i - u^{oi}(t) & \text{for} \quad \Delta u^i \leq -u_m^i - u^{oi}(t) \\[2ex] \Delta u^i & \text{for} \quad -u_m^i \leq \Delta u^i + u^{oi}(t) \leq u_m^i \\[2ex] u_m^i - u^{oi}(t) & \text{for} \quad \Delta u^i \geq u_m^i - u^{oi}(t) \end{cases}$$

It is obvious that the deviation models (3.3.17) and (3.3.18) are of the same form as the basic models $S^M$ (3.3.1) and $S_a^i$ (3.3.2). Thus, the procedure applied to the equations (3.3.4) - (3.3.15) could be repeated again. Therefore, the deviations of generalized forces $\Delta P_c$ are expressed as

$$\Delta P_c = (I_m - H_p^* TC_c F)^{-1} \{ H_p^* [TC_c A_c \Delta x_c + TC_c B_c N(\Delta u)] +$$

$$+ H_{CN}^* H_{NN}^{*-1} (\Delta P_N - h_N^*) + h_c^* \} \tag{3.3.19}$$

where $\Delta P_c(t) = P_c(t) - P_c^o(t)$, $\Delta P_c \in R^m$, $\Delta P_N(t) = P_N(t) - P_N^o(t)$, $\Delta P_N \in R^{n-m}$, $\Delta P_N$ is the deviation of generalized forces of unpowered d.o.f. from their nominals. In the same way as before $T' \cdot \Delta P = [\Delta P_c \mid \Delta P_N]^T = [\Delta P_1, \Delta P_2, \dots, \Delta P_m, \Delta P_{m+1}, \dots, \Delta P_n]^T$, $\Delta x_c \in R^{N_c}$, $\Delta x_c = (\Delta x^{1T}, \Delta x^{2T}, \dots$

$$\ldots,\Delta x^{mT})^T, \quad A_c = \text{diag}\{A_c^i\}. \text{ Also,}$$

$$T'HT'^{-1} = \left[\begin{array}{c|c} H_{CC}^* & H_{CN}^* \\ \hline H_{NC}^* & H_{NN}^* \end{array}\right], \qquad T'h = \left[\begin{array}{c} h_C^* \\ \hline h_N^* \end{array}\right]$$

$$H_P^* = (H_{CC}^* - H_{CN}^* H_{NN}^{*-1} H_{NC}^*)$$

By analogy with equation (3.3.15), the model for the deviation of unpowered d.o.f. may be written in the form

$$\Delta\ddot{q}_N = H_{NN}^{*-1}\{\Delta P_N - H_{NC}^* TC_c[A_c\Delta x_c + F_c(I_m - H_p^* TC_c F_c)^{-1} \cdot (H_p^* TC_c \cdot A_c\Delta x_c +$$

$$+ H_p^* TC_c B_c N(\Delta u) + H_{CN}^* H_{NN}^{*-1}(\Delta P_N - h_N^*) + h_c^*) + BN(\Delta u)] - h_N^*\} \tag{3.3.20}$$

where $\Delta\ddot{q}_N(t) = \ddot{q}_N(t) - \ddot{q}_N^O(t)$, $\Delta\ddot{q}_N \in R^{n-m}$, $T'\Delta q = [\Delta q_C \mid \Delta q_N]^T$.

In state space representation, analogously to (3.3.16), the model of deviation of unpowered d.o.f. is

$$\Delta\dot{x}_N = A_N^*(\Delta x_N, \Delta x_c) + B_N^*(\Delta x_N, \Delta x_c)N(\Delta u) \tag{3.3.21}$$

where $\Delta x_N \in R^{2(n-m)}$, $\Delta x_N = (\Delta q_N^1, \Delta\dot{q}_N^1, \ldots, \Delta q_N^{n-m}, \Delta\dot{q}_N^{n-m})^T$, $A_N^*: R^{2(n-m)} \times R^{N_c} \to R^{2(n-m)}$, $B_N^*: R^{2(n-m)} \times R^{N_c} \to R^{2(n-m)\times m}$ and $A_N^*, B_N^*$ are of the form $A_N^* = [A_N^{*1T} \mid A_N^{*2T} \mid \ldots \mid A_N^{*(n-m)T}]^T$, $B_N = (B_N^{*1T} \mid B_N^{*2T} \mid \ldots \mid B_N^{*(n-m)T})^T$ and where

$$A_N^{*j} = \left[\begin{array}{c} \Delta\dot{q}^j \\ [H_{NN}^{*-1}(\Delta P_N - H_{NC}^* TC_c A_p - h_N^\wedge)]_j \end{array}\right], \quad \forall j \in J$$

$$B_N^{*j} = -\left[\begin{array}{c} 0 \\ [H_{NN}^{*-1}H_{NC}^* TC_c\{F_c(I_m - H_p^* TC_c F_c)^{-1}H_p^* TC_c B_c + B_c\}]_j \end{array}\right], \quad \forall j \in J$$

Here, $A_N^{*j}: R^{2(n-m)} \times R^{N_c} \to R^2$, $B_N^{*j}: R^{2(n-m)} \times R^{N_c} \to R^{2\times m}$.

On the basis of the form of $A_N^*$ and $B_N^*$, the model (3.3.21) can be written as a set of subsystems $S^j$:

$$S^j: \Delta\dot{x}_N^j = A_N^{*j}(\Delta x_N, \Delta x_c) + B_N^{*j}(\Delta x_N, \Delta x_c)N(\Delta u), \quad \forall j \in J, \tag{3.3.22}$$

If we want to investigate the subsystem $S^j$ independently of the rest of the system, we should suppose it is possible to separate from the vector functions $A_N^j$ those parts that are dependent only of $\Delta x^j$, i.e., to write the models $S^j$ in the form

$$S^j: \quad \Delta\dot{x}^j = a^{*j}(\Delta x^j) + B_N^{*j}(\Delta x_N, \Delta x_C)N(\Delta u) + \bar{A}_N^{*j}(\Delta x_N, \Delta x_C) \qquad (3.3.23)$$

where $\bar{A}_N^{*j} = A_N^{*j} - a^{*j}$, $\forall j \in J$, $a^{*j}: R^{n_j} \to R^{n_j}$, $\bar{A}_N^{*j}: R^{2(n-m)} \times R^{N_C} \to R^{n_j}$, $n_j = 2$ is the order of the subsystem $S^j$.

The partition of the subsystem model $S^j$ into the part which will be joined to the free (decoupled) subsystem (part dependent on $\Delta x^j$) and the part which will be joined to the coupling is not unique.

The model of deviation of the whole system from the nominals may be expressed in the form of subsystems $S_a^i$, $\forall i \in I_1$ from (3.3.18) and subsystems $S^j$, $\forall j \in J$ from (3.3.23).


## 3.4. Control Synthesis

### 3.4.1. <u>Some notes on biped control synthesis</u>

Our thinking about the synthesis of artificial locomotion activity of a biped is always burdened by the example of man who in the best way demonstrates how such a complex activity can be performed. Bernstein [32] expressed that in the form of a magnificent comparison: "As in the orchestra, each instrument plays its individual score, so in the act of walking, each joint reproduces its own curve of movements and each center of gravity its sequence of accelerations; each muscle produces its melody of efforts, full with regularly changing but stable details. And in like manner, the whole of this ensemble acts in unison with a single and complete rhythm, fusing the whole enormous complexity into a clear and harmonic simplicity. The consolidator and manager of the complex entity, the conductor and at the same time composer of the analyzed score, is of course the central nervous system".

The attempts of mere copying of the natural system have no chance to be successful for, at least, two reasons.

First, it would be a technical nonsense to realize such mechanical

structure which would correspond to the skeletal system, and to power
it by a number of actuators acting as muscles. Such a great number of
d.o.f. does not permit technical realization of the system. The second
reason is the unavailability of an artificial control unit which would
have potentialities of the nervous system acting as the control unit
in living organisms. Because of that, it is possible to adopt from humans only some main principles of the walk control organization, and,
on the basis of them, develop an approach which would be suitable for
various technical applications.

The locomotion activity, and gait in particular, belongs to the class
of motions which are the most automated. This fact was already recognized by Bernstein [1]. When a man is walking in a steady regime, no
central nervous system is involved, including even cases when small
disturbances occur. They are compensated automatically by the subject's
control system without violation of the anthropomorphic character of
the motion. The system behaves in the same way as if it followed a
prescribed nominal trajectory. In case of large disturbances, the system actions are directed only to preservation of the system overall
stability, i.e., to prevention of the system from falling down. This
requirement is of primary importance in locomotion.

Hierarchy is a basic principle on which control of large scale systems,
in general, is based. This also holds for robots. The hierarchical organization of the control system is most often vertical, so that each
control level deals with some wider aspects of the overall system behaviour than the lower level. A higher control level always refers to
the lower ones, and it controls those system parameters that vary more
slowly. A higher level communicates with a lower level, giving it in-
structions and receiving from it relevant information required for the
decision-making. After obtaining the information from a lower level,
each level makes decision taking into account general decisions obtained
from a higher level and forwards them to the lower level for execution.

In general, the control structure most often involved in robots is of
four levels [33, 34]: the highest (which recognizes the obstacles in
the operating space and the conditions under which a task is being
performed, and makes the decisions on how the task imposed is to be
accomplished), the strategic level (which divides the imposed operation into elementary movements), the tactical level (which performs the
distribution of an elementary movement to the motion of each d.o.f. of
the robot) and the executive level (which executes the imposed motion

234

of each d.o.f.) (Fig. 3.14). All robots and manipulators have the two
lowest control levels. The tactical level generates the trajectories
of each d.o.f. which perform the desired functional movement, while
the executive level executes these trajectories by means of appropriate actuators, incorporated in each d.o.f.

The complexity of the control structure for locomotion robots depends
primarily on whether the robot walks on the terrain of a known or unknown profile.

The walking on a terrain of unknown profile requires all four control
levels with certain elements of artificial inteligence involved, but
this is out of the scope of this book. If the walk is performed on a
terrain of the known profile, the situation is much simpler. The trajectories of each link can be defined off-line, as well as the corresponding torques and controls of each actuator. All these data may be
stored at the mechanism's controller and used when the walk is performed. This is a problem which is solved at the tactical level. Then,
the task imposed at the executive level is how to ensure the realization of the precalculated trajectories if a certain perturbation occurred. This problem reduces to the problem of eliminating the deviation of the actual system state from the precalculated one, and this
is the only part of the control task which should be solved in real
time (under the above assumptions).

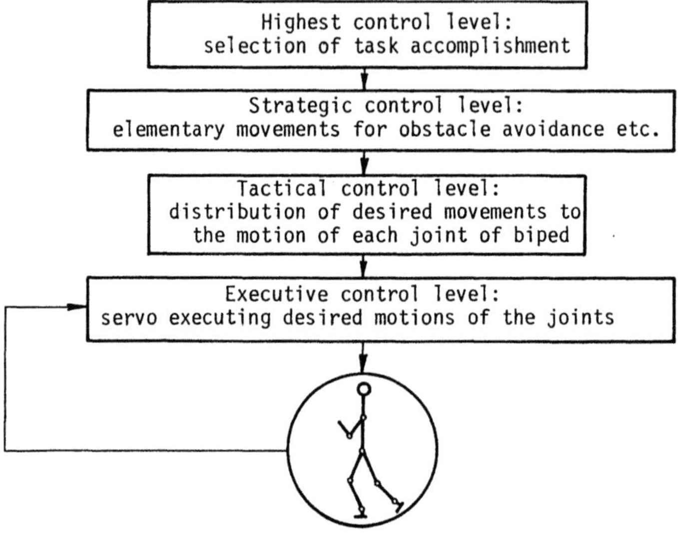


Fig. 3.14. Hierarchical architecture of control of biped locomotion robots

Let us consider the executive control level. Let us assume that desired nominal trajectories of the biped joints $x^o(t)$ have been computed at the tactical control level.

A control task at the executive control level for robots may be generally defined as in [31]. Suppose that the model of the overall system is formed by uniting the model of mechanical part and the models of actuators. Also, in all regions of the state space considered, the conditions of existence and uniqueness of the solution to the overall system are satisfied over the considered time interval $T=\{t, t\in(0,\tau)\}$, $\tau>0$. In other words, for each particular control $u(t)$, there is a unique solution $x(t)\in X^t(t)$, for $\forall t\in T$, where $X^t(t)$ is the bounded region $X^t(t)\subset R^N$, $\forall t\in T$.

It is characteristic of locomotion robots that precise tracking of the nominal state coordinates trajectories during a step cycle is not strictly required. As each disturbance is reflected on the ZMP position, and, as we already said, this should be kept within certain area (a much worse case is the single-support phase), it is allowed that state of the system belongs to the bounded regions of state space $X^t$.

Thus, the control task imposed on the locomotion system can be defined as the requirement of satisfying practical stability [31]. The practical stability of the biped system is defined in the following way. Let finite regions $X^I$ and $X^F$, $X^I\subset R^N$, $X^F\subset R^N$, in the state space be defined. Also, let the bounded region $X^t(t)\subset R^N$ be defined. The system is observed over the time interval T. The system is practicaly stable if for each initial state $x(0)$ belonging to the region $X^I$ (or the allowable initial states), a trajectory of the system satisfies $x(t)\in X^t(t)$, $\forall t\in T$ (i.e. the system state must belong to $X^t(t)$) and $x(t)\in X^F$, for $\tau_s<t\leq\tau$, where $\tau_s$ is time interval. In fact, it is required that the system state must be transferred from the region of the allowable initial states $X^I$ into the region $X^F$ of the allowable final states, within the time interval $\tau_s$, and during this transfer the system state must belong to finite region $X^t$. Here, it is assumed that $X^I\subset X^t(0)$ and $X^F\subset X^t(t)$, for $\tau_s<t\leq\tau$.

Here, our approach to synthesis of control at the executive control level will be explained.

We have adopted the two-stage control synthesis [31, 35,...,38], based on

a) nominal dynamics stage

b) perturbed dynamics stage.


Thus

$$u^i = u^{oi} + \Delta u^i$$

where $u^i \epsilon R^1$, $u^{oi} \epsilon R^1$, $\Delta u^i \epsilon R^1$, $(i=1,\ldots,m)$, are the total control input, nominal (programmed) control input and control defined at level of perturbed dynamic stage, respectively.


At the stage of nominal dynamics, the trajectories of all links should be determined in order to perform the desired motion $x^o(t)$, and then derive such a control which would transfer the subsystem S from the initial state $x^o(0) \epsilon X^I$ to a point in the region $X^F$ in the finite time $\tau_s < \tau$.


Namely, the nominal trajectory of the state vector is defined to satisfy $x^o(0) \epsilon X^I$, $x^o(t) \epsilon X^F$, for $\tau_s < t < \tau$, and $x^o(t) \epsilon X^t(t)$, $\forall t \epsilon T$. Practically, $x^o(t)$ corresponds to desired walk of the biped and its transfer from one place to another.


The motion of the links is defined using prescribed synergy method. The trajectories imposed on the legs should be such to ensure the desired gait type, and then, the trunk compensating movements are computed to preserve the system overall stability. The control synthesis at this stage is done on the basis of the complete centralized model S, without any approximation. This control (named the "nominal" or "programmed" control) is essentially centralized; at this stage one part of the coupling is taken into account so that the coupling at the state of perturbed regimes of control synthesis is reduced.


The synthesis of the nominal trajectories of the biped system (which ensure equilibrium of the system during the walk) has been considered in detail in Chapter 2. Therefore, we shall not consider it again. We assume that nominal trajectories of all biped joints are given, and that corresponding nominal diriving torques $P^o(t)$ and nominal centralized programmed control $u^o(t)$ are calculated, as explained in Section 3.3.


Here we shall consider the second stage, the stage of perturbed dynamics. We have explained that the nominal programmed control signals would drive the joints of the biped along the nominal trajectories only under ideal conditions (perfect modelling, no perturbations). Since these ideal conditions are never fulfilled, the system state always deviates from the nominal trajectories $x^o(t)$.

In the second stage, such a control should be defined which will force the actual system state to its nominal value under the condition of preserving the overall stability of the system. The presence of unpowered d.o.f., which is the most important characteristic of locomotion mechanisms, makes this task very complex. Therefore, a condition which should be fulfilled under all possible circumstances is that the acting point of the ground reaction force (ZMP) is within the area for which the system will be safe from rotating about the foot edge.

Let us consider the overall system model S defined as in (3.3.16)

$$S: \quad \dot{x} = \hat{A}(x) + \hat{B}(x) \cdot N(u)$$

Let it be assumed the part of the system corresponding to powered d.o.f. can be rearranged as a set of m subsystems $S_a^i$ (3.3.2) which are coupled through the term $(f_c^i \cdot P_c^i)$. So

$$S_a^i: \quad \dot{x}_c^i = A_c^i x_c^i + b_c^i N(u^i) + f_c^i P_c^i, \quad \forall i \in I_1$$

Let the nominal trajectory $x_c^o$, $x_c^o = (x_c^{o1^T}, x_c^{o2^T}, \ldots, x_c^{om^T})^T$ and the nominal control $u^o$, $u^o = (u^{o1}, u^{o2}, \ldots, u^{om})^T$ be introduced in such a way to satisfy

$$\dot{x}_c^{oi} = A_c^i x_c^{oi} + b_c^i N(u^{oi}) + f_c^i P_c^{oi}, \quad \forall i \in I_1 \tag{3.4.1}$$

Then, the model of subsystem deviation from the nominal is considered in the form

$$\Delta \dot{x}_c^i = A_c^i \cdot \Delta x_c^i + b_c^i \cdot N(\Delta u^i) + f_c^i \cdot \Delta P_c^i, \quad \forall i \in I_1 \tag{3.4.2}$$

The purpose of the synthesis of disturbance compensating control ($\Delta u$) is to force system deviations $\Delta x_c^i$, (i=1,...,m) to zero and to maintain the overall system stability.

The control at the level of perturbed regimes can be derived in various ways [31].

For the purpose of control synthesis at the level of perturbed regimes, a decentralized information structure is adopted at the first step, so that only the information about the state vector of corresponding d.o.f. serves for control of each actuator. The decoupling is accomplished in such a way that the system is considered as a set of subsystems corresponding to separate d.o.f., because such decomposition seems to be the most suitable for the mechanical systems of this type.

A basic scheme of adopted control structure is given in Fig. 3.15. The subsystems $S^1$ to $S^m$ which are powered consist of the appropriate part of the mechanism driven by the actuators. The control inputs to these subsystems are based on information of the own subsystem state coordinates, neglecting the influence of the rest of the system. In this way, each perturbation of subsystem state can be compensated to ensure that the subsystem state belongs to the allowable region of state space (but only if the coupling between the subsystems is neglected).

Thus, instead of the actual subsystem model (3.4.2), we consider approximative subsystems models in which the coupling with the rest of the system is neglected:

$$\Delta \dot{x}_c^i = A_c^i \Delta x_c^i + b_c^i N(\Delta u^i), \qquad \forall i \in I_1$$

If we compare this model with (3.4.2) we see that the coupling term $f_c^i \Delta P_c^i$ is now ignored. Therefore, the subsystem behaves as being decoupled from the rest of the biped system and we can adopted a local controllers which stabilize the decoupled subsystem, as denoted in Fig. 3.15. Obviously, such local controllers might include only local feedback loops (i.e. they use only local information on the subsystem state $\Delta x^i$). These local controllers may be synthesized in various ways. One possibility is to apply simple local servo system [31]. However, due to the specificity of the biped system a special form of control synthesis will be adopted (see Paragraph 3.4.2).

However, such local controllers would stabilize the local subsystems only if they were free of coupling. We have to examine the effects of the actual coupling $f_c^i \Delta P_c^i$ between the subsystems upon the system stability and, to check whether the actual system can be stabilized if only local controllers are implemented. This problem will be discussed in Section 3.5.

However, a specificity of the biped system is that, apart from powered

joints described by the model (3.4.2), it possesses unpowered degrees
of freedom. This part of the system is described by the model (3.3.23).
It can also be regarded as a set of subsystems $S^j$ (each associated to
one unpowered d.o.f.) in which coupling is neglected

$$S^j: \quad \Delta \dot{x}^j = a^{*j}(\Delta x^j), \qquad \forall j \in J$$

However, such free subsystems cannot be locally stabilized.

With the unpowered subsystems ($S^{m+1}$ to $S^n$), the situation is quite different from that with the powered subsystems. Their behaviour is to a
greatest extent dependent on and affected by the coupling of the other
subsystems. This influence of the powered on the unpowered subsystems

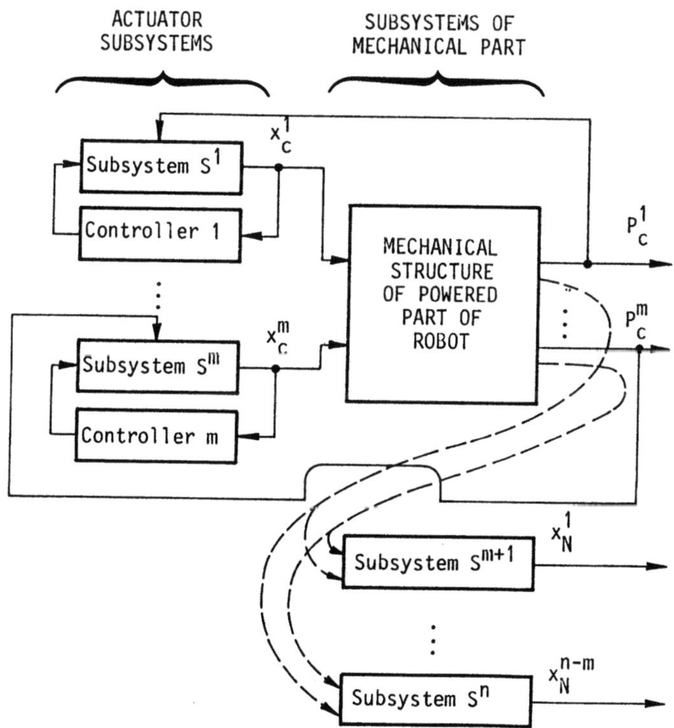


Fig. 3.15. Decentralized control structure for mechanisms
including unpowered d.o.f.

is in Fig. 3.15. represented by dashed lines. The synthesis of decentralized control and additional feedback loops which have to ensure
the required behaviour of unpowered degrees of freedom will be discus-

sed in detail in the coming sections. Namely, it is obvious that we must introduce some feedback loops from unpowered to powered subsystems in order to stabilize the unpowered ones.

A specificity of locomotion mechanisms is that the changes in mechanism's dynamics caused by compensating movements should be taken into account because they can affect significantly the mechanism behaviour. If the inertial forces, induced in the stabilizing process are not balanced or limited, the system may collaps (i.e., rotate around the foot edge) even if the internal nominal trajectories have been perfectly followed. In such a case, the displacement of ZMP from its prescribed position can be used as a measure of the whole system deviation from its nominal. For this purpose, the appropriate force sensors should be introduced at the sole of the mechanism's foot.

The problem of undesirable influence of inertial forces, induced as a side-effect in the stabilizing process, is solved in two steps. First, the admissible correctional accelerations are limited and, if necessary, an additional feedback with respect to the ZMP position is introduced.

Thus, the problem of realization of a stable gait is a problem related to the level of perturbed regimes. Because the biped performance is significantly characterized by the very strong coupling between the subsystems, one of the ways to minimize the problems arising at the level of perturbed regimes is to make different redistribution of the coupling $\Delta P_c^i$ [31]. If a decentralized control structure is adopted, the coupling $\Delta P_c^i$ can be redistributed in such a way that its larger part is associated with the free subsystem $S_a^i$, i.e. the model $S_a^i$ is formed with the larger inertia than the actuator inertia itself, and includes the corresponding link inertia. Thus, the local control takes into account not only the actuator dynamics, but also the overall system characteristics.

Now, the subsystems models $S_a^i$ can be written in the form

$$S_a^i: \quad \Delta \dot{x}_c^i = \bar{A}_c^i \cdot \Delta x_c^i + \hat{b}_c^i \cdot N(t, \Delta u^i) + \hat{f}_c^i \cdot \Delta P_c^{*i} \qquad (3.4.3)$$

where $\bar{A}_c^i = \bar{C}^i(A_c^i + f_c^i \bar{h}_i T_1^i)$, $\hat{b}_c^i = \bar{C}^i b_c^i$, $\hat{f}_c^i = \bar{C}^i f_c^i$, $\bar{C}^i = (I_{n_i} - f_c^i \bar{H}_{ii} T^i)^{-1}$, $\Delta P_c^{*i} = \Delta P_c^i - \bar{H}_{ii} T^i \Delta \dot{x}_c^i - \bar{h}_i T_1^i \Delta x_c^i$, $T^i$, $T_1^i \in R^{1 \times n_i}$, $T^i \Delta \dot{x}_c^i = \Delta \ddot{q}_c^i$ and $T_1^i \Delta x_c^i = \Delta q_c^i$.

If the system $S$ is considered as a set of subsystems $S_a^i$ (3.4.3) coupled through $\Delta P_c^{*i}$ instead through $\Delta P_c^i$, the model is distributed between the coupling and the subsystems in another way. By introducing $\bar{H}_{ii}$ into the subsystem $S_a^i$, each actuator is considered to have the inertia which includes not only its own inertia, but also the inertia of the corresponding mechanical part of the system which is powered by the actuator. The introduction of $\bar{h}_i$ into the subsystem model means that the actuator model is considered together with the gravitational load, i.e., the gravitational moment, which is a function of $\Delta q^i$, is associated with the subsystem model $S_a^i$.

The quantities $\bar{H}_{ii} > 0$ and $\bar{h}_i$ have to be chosen during the control synthesis.


### 3.4.2. <u>Synthesis of control with limited accelerations</u>

As we have explained above, we shall use the approach in which the control synthesis is done in two steps [39]. At the level of nominal regimes, the control is computed on the basis of the complete (nonlinear) model with the permanent requirement for satisfying dynamic equilibrium conditions of the overall mechanism. This control should enable the system (in the absence of disturbance) to follow the nominal trajectories. At the stage of perturbed regimes, control should force the actual state vector to its nominal value, i.e. to the nominal programmed trajectory. It is intuitively clear that the action should be "smooth", with no significant change in link acceleration, in order to keep its influence on unpowered degrees of freedom within an acceptable range.

Therefore, the acceleration can be accepted as a quantity that influences the system motion to the greatest extent, and because of that, it can be adopted as the control parameter. Now the problem can be redefined, as how to determine the trajectories of links accelerations that are capable of returning the system from a disturbed to the nominal trajectory under the condition that accelerations do not exceed a certain value, limited in advance.

We shall synthesize the local controller for the i-th actuator, i.e. for the $S_a^i$ subsystem whose model of state deviation around the nominal trajectory is given by (3.4.2). We want to define a controller for this subsystem which will reduce the state deviation $\Delta x_c^i(t)$ to zero, but in doing this we want to prevent the appearence of the too high accelerations. Therefore, we shall synthesize a controller which will ensure the acceleration of the corresponding joint $\Delta\ddot{q}^i$ is limited. To do this, we start from the simple problem of the second-order linear system with limited acceleration.

Let us consider the classical time-minimum problem. Let the system be described by

$$\Delta\dot{q}_1^i(t) = \Delta q_2^i(t)$$

$$\Delta\dot{q}_2^i(t) = u_i^*(t) \qquad |u_i^*(t)| \leq \Omega^i$$

(3.4.4)

with the initial conditions $\Delta q_1^i(0) = \alpha^i$ and $\Delta q_2^i(0) = \beta^i$ where $\Omega^i \in R^1$, $\Delta q_1^i \in R^1$ and $\Delta q_2^i \in R^1$. The value $u^*(t)$ should be computed in such a way to ensure that system (3.4.4) returns from the point $(\alpha^i, \beta^i)$ to the coordinate origin $(0, 0)$ in a minimal time interval.

Therefore, such a solution of (3.4.4) should be obtained that the functional

$$J = \int_0^{T^*} dt$$

is in the minimum, where $T^* \in R^1$ definies unspecified time interval. Such type of problem is well known [40], and for this particular case $(\Omega^i \neq 1)$ its solution is given by the expression

$$u_i^* = \begin{cases} +\Omega^i & \text{if} \quad \Delta q_1^i < \dfrac{-\Delta q_2^i|\Delta q_2^i|}{2\cdot\Omega^i} \quad \text{or} \quad \Delta q_1^i = \dfrac{+(\Delta q_2^i)^2}{2\Omega^i} \wedge \Delta q_2^i \leq 0 \\[4mm] -\Omega^i & \text{if} \quad \Delta q_1^i > \dfrac{-\Delta q_2^i|\Delta q_2^i|}{2\cdot\Omega^i} \quad \text{or} \quad \Delta q_1^i = -\dfrac{(\Delta q_2^i)^2}{2\Omega^i} \wedge q_2^i \geq 0 \end{cases}$$

(3.4.5)

We shall apply this solution to control one single actuator, i.e. the subsystem $S_a^i$ associated to the i-th joint.

Let us suppose the mechanical structure is powered by the DC motors whose models (3.3.18) are given in the form (the state vector $\Delta x_c^i$ is selected as $\Delta x_c^i = (\Delta q^i, \Delta\dot{q}^i, \Delta i_R^i)^T$):

$$\begin{bmatrix} \Delta\dot{q}^i \\ \Delta\ddot{q}^i \\ \Delta\dot{i}_R^i \end{bmatrix} = \begin{bmatrix} 0 & 1 & 0 \\ 0 & a_{22}^i & a_{23}^i \\ 0 & a_{32}^i & a_{33}^i \end{bmatrix} \begin{bmatrix} \Delta q^i \\ \Delta\dot{q}^i \\ \Delta i_R^i \end{bmatrix} + \begin{bmatrix} 0 \\ \bar{f}_c^i \\ 0 \end{bmatrix} \Delta P_c^{i*} + \begin{bmatrix} 0 \\ 0 \\ \bar{b}_c^i \end{bmatrix} \Delta u^i \qquad (3.4.6)$$

From the second equations of (3.4.6) we can write

$$\Delta i_R^{i*} = (u_i^* - a_{22}^i \Delta\dot{q}^i - \bar{f}_c^i \Delta P_c^i)/a_{23}^i \qquad (3.4.7)$$

where $\Delta\ddot{q}^i$ is replaced by the value of allowed link acceleration $u_i^*$ from (3.4.5). Then, $\Delta i_R^{i*}$ in (3.4.7) is the corresponding rotor current, and its derivative can be computed from

$$\Delta\dot{i}_R^i = (\Delta i_R^{i*} - \Delta i_R^i)/\Delta t \qquad (3.4.8)$$

where $\Delta t \in R^1$ is the control sampling period. Now, we can determine the control for (3.4.6).

Let us assume that we want to limit acceleration of the actuator (joint) to be within the limits $\Omega_{min}^i$ and $\Omega_{max}^i$. Starting from the time-minimum control (3.4.5) we can adopt the following control.

From the third equation of (3.4.6), the compensation control signal for the i-th actuator is

$$\Delta u^i = k_{i1}^L \Delta q_i + k_{i2}^L \Delta\dot{q}^i + k_{i3}^L \Delta i_R^i + k_{i4}^G \Delta P_c^{i*} + k_{i5}^G \qquad (3.4.9)$$

where the constant feedback gains are

$$k_{i1}^L = 0, \qquad k_{i2}^L = (-a_{22}^i - a_{23}^i \cdot a_{32}^i \cdot \Delta t)/d,$$

$$k_{i3}^L = -a_{23}^i(1 + a_{33}^i \Delta t)/d \qquad (3.4.10)$$

$$k_{i4}^G = -\bar{f}_c^i/d, \qquad k_{i5}^G = k_{i5}/d$$

$$k_{i5} = \begin{cases} \Omega_{max}^i & \text{if } \Delta q^i < \dfrac{-0.5\Delta\dot{q}^i|\Delta\dot{q}^i|}{\Omega^i} \quad \text{or} \quad \Delta q^i = \dfrac{+0.5(\Delta\dot{q}^i)^2}{\Omega^i} \ \wedge \ \Delta\dot{q}^i < 0 \\[4mm] \Omega_{min}^i & \text{if } \Delta q^i > \dfrac{-0.5\Delta\dot{q}^i|\Delta\dot{q}^i|}{\Omega^i} \quad \text{or} \quad \Delta q^i = \dfrac{-0.5(\Delta\dot{q}^i)^2}{\Omega^i} \ \wedge \ \Delta\dot{q}^i > 0 \end{cases}$$

where $d = a_{23}^i \bar{b}_c^i \Delta t$; $a_{jk}^i$, $\bar{b}_c^i$, $\bar{f}_c^i$ are the elements of the corresponding matrix and vectors of the actuator (3.4.6); $\Omega_{max}^i$ and $\Omega_{min}^i$ are the maximal and minimal values of the accelerations for the i-th link. The feedback gains synthesized in this way have to ensure the compensating movements such that the acceleration does not exceed a certain limit, fixed in advance. As a consequence, the induced inertial forces will not cause an undesirable motion of unpowered d.o.f., i.e. displacement of ZMP out of a prescribed area.

Obviously, the proposed control law (3.4.9) consists of two parts: the local control (concerning joint position feedback, velocity and rotor current) and global control (concerning feedback with respect to $\Delta P_c^{i*}$ from the rest of the system upon the i-th subsystem). The bang-bang term $k_{i5}^G$ might be conditionally associated to global control, although it is based upon local feedback information. Global control ($k_{i4}^G \Delta P_c^{i*}$) requires information on the coupling acting upon the i-th subsystem. This information $\Delta P_c^{i*}$ may be obtained in two ways: a) by measurement, or b) by on-line computation of coupling. Since the coupling upon the i-th subsystem $\Delta P_c^i$ represents the dynamic moment acting around the i-th joint, we can directly measure it by force transducers mounted on the joint shaft. This approach is very simple from the control point of view (since we have to add only one feedback loop from the force transducers to the actuator input), but it causes some additional technical problems. The second way for obtaining information on $\Delta P_c^i$ is to calculate it in the control microcomputer using dynamic model of the system. However, dynamic model of the biped system is very complex, and therefore, computing of the dynamic moments $\Delta P_c^i$ might require a lot of computation operations to be performed by computer. Since the dynamic moments must be computed during sampling interval which for biped systems must be less than 10-20[ms], this means that we have to apply some powerfull computer which will be capable to perform a large ammount of computation in such a short period of time. Therefore, instead complete dynamic model of the system we may use various approximative models of the system. However, the question is whether such approximative models will be sufficiently efficient to stabilize the overall system [31, 48].

A scheme of the control of a powered subsystem with control structure defined by (3.4.9) is presented in Fig. 3.16, while in Fig. 3.17. is given the control scheme for the whole system S consisting of the powered ($S^1 \div S^m$) and unpowered ($S^{m+1} \div S^n$) subsystems.

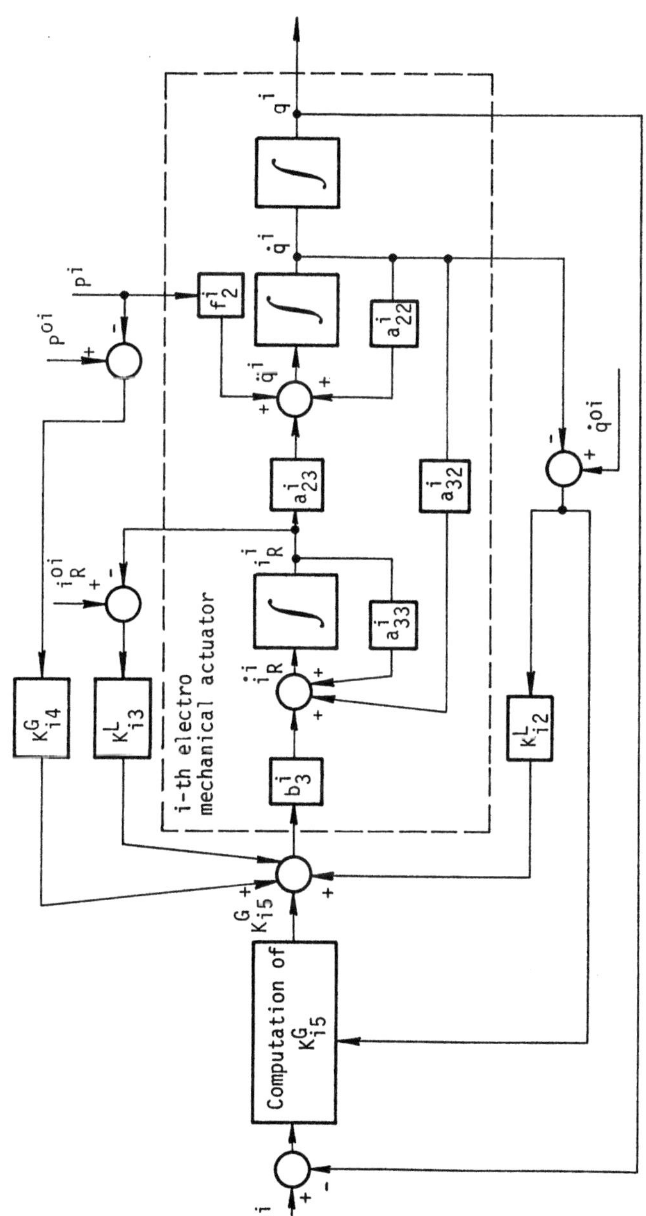


Fig. 3.16. Control scheme for the i-th powered subsystem if feedback gains defined by (3.4.10) are applied

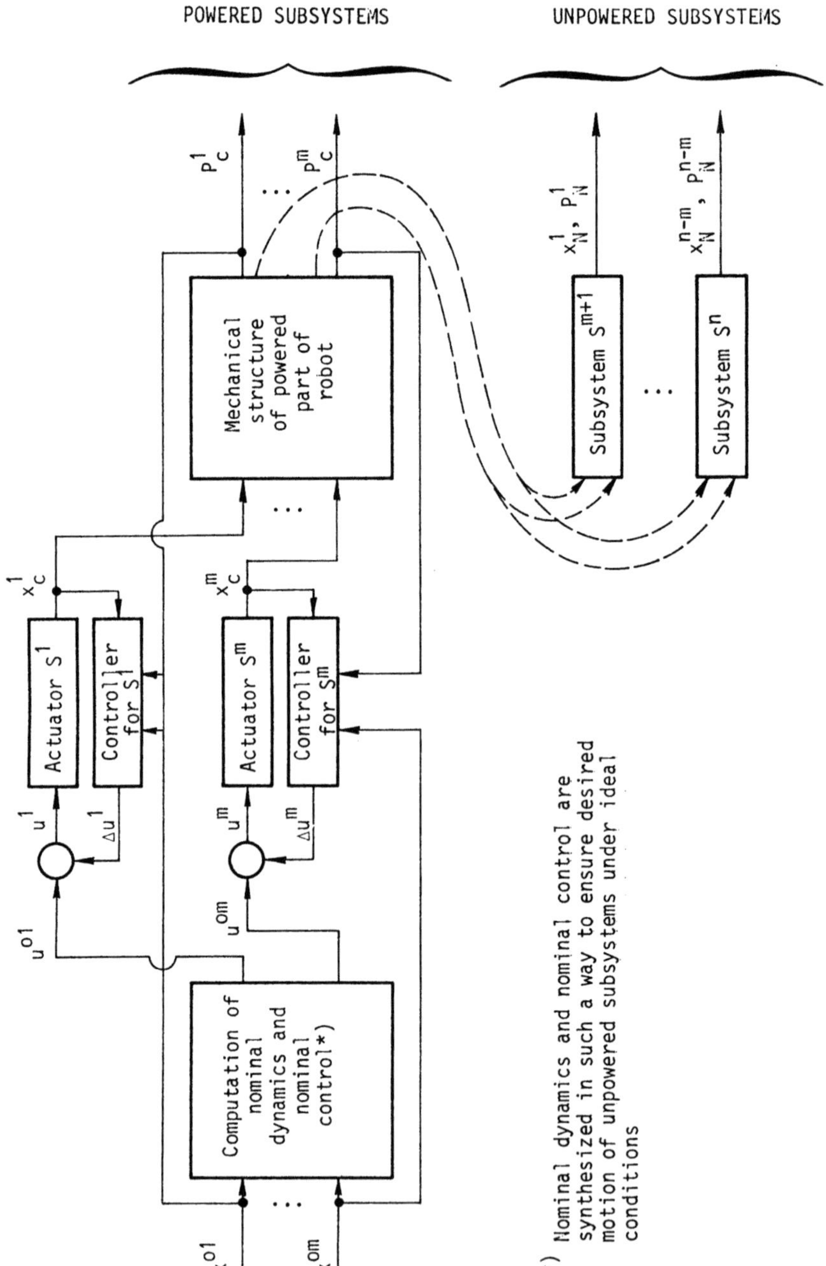


Fig. 3.17. Control scheme without feedback from unpowered subsystems

Obviously, it is an open question as for the amount of admissible accelerations, especially with respect to their disposition at joints. There are a number of very complex problems involved requiring detailed research and multidisciplinary studies. In Paragraph 3.4.4 we shall give some of the preliminary results based on simulation of walking processes.


### 3.4.3. Synthesis of global control with respect to ZMP position


If decentralized control defined by (3.4.9), applied at the mechanism's joints, is not sufficient to ensure the tracking of internal nominal trajectories with addition of the appropriate behaviour of unpowered subsystem, an additional feedback has to be introduced at one of powered joints to ensure a satisfactory motion of the complete mechanism. The task of this feedback is to reduce the destabilizing effect of the coupling acting upon the unpowered subsystems.

Again, it should be recalled that the locomotion mechanisms possess the d.o.f. powered by their own actuators ($S_a^i$) and the unpowered ones ($S^j$). The subsystems $S_a^i$ are coupled through $P_c^i$ and the measure of deviation of their actual value from the nominal ($\Delta P_c^i$) serves as the basis for introducing an additional feedback which has to reduce the destabilizing effect of the coupling between the subsystem $S_a^i$ and the rest of the system.

Because a predominant role in the system stability is played by the unpowered d.o.f., it is necessary to reduce the destabilizing coupling effects acting upon them. Because the unpowered subsystem cannot compensate for its own deviation from the nominal state, one of powered subsystems has to be chosen to accomplish it. As the coupling of subsystems $S^j$ is a function of control input to the i-th subsystem $S_a^i$, it is clear that a feedback from the subsystem $S^j$ to the inputs $\Delta u^i$ of the subsystems $S_a^i$ should be introduced.

Let us assume a measurable physical characteristic of the subsystem $S^j$ is a direct indicator of the destabilizing influence of the rest of mechanism. Let it be denoted by e [31]. Then, the complete global feedback can be introduced in the form

$$\Delta u_i^G = \phi_{1i}(\Delta x_c^i, \Delta P_c^i) + \phi_{2i}(\Delta x_c^i, e) \tag{3.4.11}$$

where $\phi_{1i}: R^{n_i} \times R^1 \to R^1$, $\phi_{2i}: R^{n_i} \times R^{n_e} \to R^1$, $e \in R^{n_e}$ and $n_e$ is the order of vector $e$.

If the control is synthesized according to (3.4.9) and (3.4.10), the first term of (3.4.11) is already defined by $(k_{i4}^G \cdot \Delta P_C^i)$. If the local control is synthesized in some other way, the first term of (3.4.11) should be defined in such a way that the coupling influence on $S_a^i$ is thus minimized.

The physical quantity $e$ can be measured and introduced into the feed-back with the purpose of minimization of coupling on subsystems $S^j$; in the case of active spatial mechanisms $e$ represents the forces acting on the chain ends that are in direct contact with the surroundings. With bipeds, where the unpowered d.o.f. are formed in contact of the feet and the ground, it is possible to measure the ground reaction force using force sensors (three at least) to determine actual acting point of total vertical reaction force.

For the known motion of the whole mechanism, the ground reaction force (or forces in double-support phase) is defined by its intensity, direction and the position of acting point on the foot. If force sensors A, B and C are introduced (Fig. 3.18), and the system is performing nominal gait, the measured values of vertical reaction forces $R_A$, $R_B$ and $R_C$ correspond to their nominal values, and the nominal position, of ZMP can be determined. Measurement of vertical reaction forces $R_A$, $R_B$ and $R_C$, when the mechanism is performing the gait in the presence of disturbances, enables determination of the actual position of ZMP.

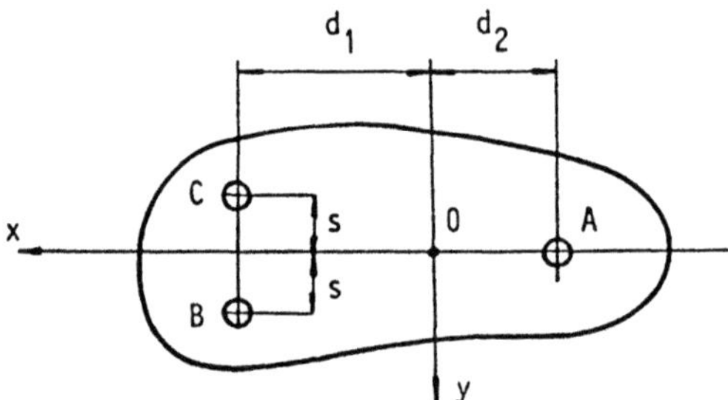


Fig. 3.18. Disposition of force sensors on the mechanism sole

If the nominal ZMP position corresponds to point 0, it can be written

$$s(\Delta R_B - \Delta R_C) = M_x = R_z \cdot \Delta y$$

$$d_1(\Delta R_B + \Delta R_C) - d_2 \cdot \Delta R_A = M_y = R_z \cdot \Delta x \qquad (3.4.12)$$

where $\Delta R_A$, $\Delta R_B$ and $\Delta R_C$ are deviations of the corresponding measured forces from their nominal values; $R_z$ is the total resultant vertical reaction force; $\Delta x$ and $\Delta y$ are the displacements of the actual position of ZMP from its nominal position. These displacements can be computed from (3.4.12), provided the sensors dispositions and vertical reaction forces are known. The actual position of ZMP is the best indicator of the overall biped behaviour, so that we are going to use it to achieve a stable motion.

Let us consider a mechanism performing the gait which deviaties from its nominal. In fact, it is not possible to realize the nominal biped gait in an exact way, because, any realized gait deviates, to a greater or lesser extent, from its nominal. As the resulting overall effect of all the system deviations, the total reaction force $\vec{R}$ differs from its nominal direction, intensity, and, what is the most important, the acting point 0 at the sole of the mechanism supporting foot is displaced. Let suppose all the values related to the $R_z$, vertical component of $\vec{R}$, be known (measured or calculated). The influence of the horizontal component is neglected because it is supposed the friction force is sufficiently large to prevent the slippage of supporting foot over the ground.

Our aim is to synthesize such control which will ensure a stable gait. The primary task of the feedback with respect to ZMP position is to prevent its excursion out of the allowable region, or, to prevent the system from falling down by rotation about the foot edge. If this is fulfilled, a further requirement imposed is to ensure that the actual ZMP position is as close as possible to its nominal.

Our further considerations will be restricted to biped motion in the sagittal plane, what means that the ground reaction force position will deviate only in the direction of x axis by $\Delta x$. Fig. 3.19. illustrates the case when the vertical ground reaction force $R_z$ deviates from the nominal position 0 by $\Delta x$; thus, the moment $R_z \cdot \Delta x = M_{ZMP}^x$ is a measure of the mechanism overall behaviour.

In the same way, we can consider the mechanism motion in the frontal plane, and $R_z \cdot \Delta y = M_{ZMP}^y$ is a measure of the mechanism behaviour in the

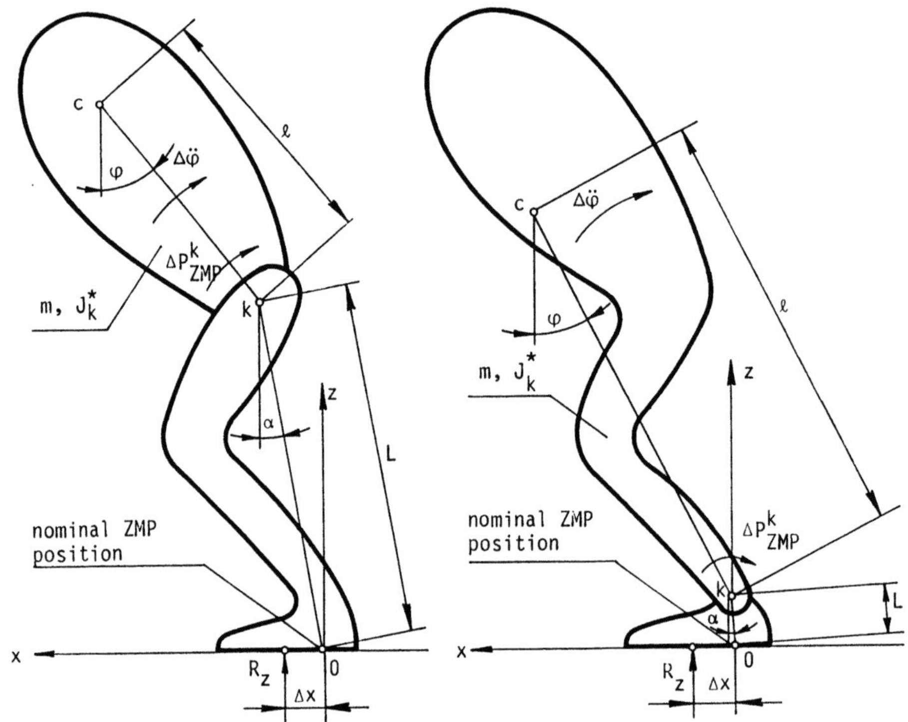



Fig. 3.19. Compensation of ZMP displacement by:
a) hip joint, b) ankle joint

direction of y axis.

Let us assume the correction of the $R_z$ acting point in one direction is done by the action at only one joint, arbitrarilly selected in advance. A basic assumption introduced for the purpose of simplicity is that the action at the chosen joint will not cause a change in the motion at any other joint. If we consider only this action, the system will behave as if composed of two rigid links connected at joint k, as presented in Fig. 3.19. In other words, the servo systems are supposed to be sufficiently stiff. In Fig. 3.19, two situations are illustrated, when the joint which has to compensate the ZMP displacement is the hip (case a) and the ankle joint (case b) of supporting leg. In both cases, this joint is denoted by k, and all links above and below it are considered as a single rigid body. The upper link is of total mass m and inertia moment $J_k^*$ for the axis of joint k. Of course, numerical values are different for case a) and case b). Furthermore, the distance from the ground surface to k is denoted by L, from k to c (c is the mass centre of the upper link) by $\ell$, whereas $\Delta P_{ZMP}^k$ stands for the correctional additional actuator torque, applied at joint k. In Fig. 3.19. the upper (compensating) link is presented as a single link above joint k. In fact, in both cases presented, the compensating link includes also the other leg which is in swing phase, and which is not drawn in the figure. The inertia moment $J_k^*$ has to be calculated in such a way to include all the links which are found further onward with respect to the selected compensating joint. In this analysis, all the joints except the k-th one are considered "frozen", and, as the consequence, the lower link, representing the sum of all the links below the k-th joint, is also considered as rigid body, which is standing on the ground surface and does not move.

The procedure by which the correctional amount of global control with respect to ZMP position is synthesized is as follows.

Assume the mechanism performs the gait such that a displacement of the ground reaction force $\vec{R}$ in the x direction occurs, so $M_{ZMP}^x = R_z \cdot \Delta x$. Then, the quantity $\Delta P_{ZMP}^k$ is to be determined on the basis of the value $M_{ZMP}^x$ and of the known mechanism and gait characteristics. It is supposed that the additional torque $\Delta P_{ZMP}^k$ will cause change in acceleration of the compensating link $\Delta\ddot{\varphi}$, while velocities will not change due to the action of $\Delta P_{ZMP}^k$, $\Delta\dot{\varphi}\approx0$. From the equations of planar motion of considered system of two rigid bodies (Fig. 3.19) which is driven by $\Delta P_{ZMP}^k$, and under assumption that terms $(\dot{\varphi}\Delta\dot{\varphi})$ and $(\Delta\dot{\varphi})^2$ from expression for normal component of angular acceleration of the upper link are neglected, it follows that

$$\Delta P_{ZMP}^k = \frac{M_{ZMP}^x}{1 + \dfrac{m \cdot \ell \cdot L \cdot \cos\varphi\cos\alpha}{J_k^*} + \dfrac{m \cdot \ell \cdot L \cdot \sin\varphi\sin\alpha}{J_k^*}} \qquad (3.4.13)$$

The control input to the actuator of the compensating joint which has to realize $\Delta P_{ZMP}^k$ can be computed from the model of the actuator deviation from the nominal. Thus,

$$\begin{bmatrix} \Delta \dot{q}^k \\ \\ \Delta \ddot{q}_T^k \\ \\ \Delta \dot{i}_R^k \end{bmatrix} = \begin{bmatrix} 0 & 1 & 0 \\ \\ 0 & a_{22}^k & a_{23}^k \\ \\ 0 & a_{32}^k & a_{33}^k \end{bmatrix} \begin{bmatrix} \Delta q^k \\ \\ \Delta \dot{q}^k \\ \\ \Delta i_R^k \end{bmatrix} + \begin{bmatrix} 0 \\ \\ \bar{f}_c^k \\ \\ 0 \end{bmatrix} (\Delta P_c^k + \Delta P_{ZMP}^k) + \begin{bmatrix} 0 \\ \\ 0 \\ \\ \bar{b}_c^k \end{bmatrix} (\Delta u^k + \Delta u_{ZMP}^k)$$

$$(3.4.14)$$

This model differes from (3.4.6) by terms $\Delta P^k_{ZMP}$ and $\Delta u^k_{ZMP}$. From the second equation of (3.4.14) the change of rotor current is

$$\Delta i^k_R = \frac{\Delta \ddot{q}^k_T - a^k_{22} \cdot \Delta \dot{q}^k - \bar{f}^k_c (\Delta P^k_c + \Delta P^k_{ZMP})}{a^k_{23}} \qquad (3.4.15)$$

Here, the subscript "T" is used for the acceleration $\Delta \ddot{q}^k$ from (3.4.14). It denotes the total change of link acceleration which consists of two parts. The first part is the "regular" change of acceleration due to the control already applied to each powered joint defined by (3.4.9), and corresponds to $\Delta P^k_c$. The second part is a direct consequence of the compensation torque $\Delta P^k_{ZMP}$. Thus

$$\Delta \ddot{q}^k_T \approx \Delta \ddot{q}^k + \Delta \ddot{q}^k_{ZMP} \approx \Delta \ddot{q}^k + \frac{\Delta P^k_{ZMP}}{J^*_k} \qquad (3.4.16)$$

where $\Delta \ddot{q}^k(t) \approx (\Delta \dot{q}^k(t) - \Delta \dot{q}^k(t-\Delta t))/\Delta t$. Then, from the third equation of (3.4.14) we have

$$\Delta u^k_{ZMP} = \frac{\Delta \dot{i}^k_R - a^k_{32} \cdot \Delta \dot{q}^k - a^k_{33} \cdot \Delta i^k_R}{\bar{b}^k_c} - \Delta u^k \qquad (3.4.17)$$

Here $\Delta u^k$ is the control defined by (3.4.9), while $\Delta \dot{i}^k_R$ stands for $\Delta \dot{i}^k_R(t) \approx (i^k_R(t) - i^k_R(t-\Delta t))/\Delta t$. Equation (3.4.17) defines the control input to the k-th actuator which has to produce $\Delta P^k_{ZMP}$. Taking into account that $\Delta P^k_{ZMP}$ is derived by introducing certain simplifications, an additional feedback gain $k^{Gk}_{ZMP} \epsilon R^1$ has to be introduced into (3.4.17). Thus, (3.4.17) becomes

$$\Delta u^k_{ZMP} = k^{Gk}_{ZMP} (\frac{\Delta \dot{i}^k_R - a^k_{32} \Delta \dot{q}^k - a^k_{33} \cdot \Delta i^k_R}{\bar{b}^k_c} - \Delta u^k) \qquad (3.4.18)$$

In this way, the additional feedback introduced and the correctional input to a selected powered mechanism's subsystem have only the purpose of maintaining the ZMP position. It is quite possible that the feedback introduced could even spoil the tracking of the internal nominal trajectory of joint k, but the stability of the overall system would be preserved, what is the most important task for a locomotion system. Which joint (ankle, hip,...) is the most suitable for this purpose it cannot be stated in advance, because the answer is dependent on the particular task imposed. Some of these questions, based on simulation of the walk under influence of disturbances, will be discussed in the next paragraph.

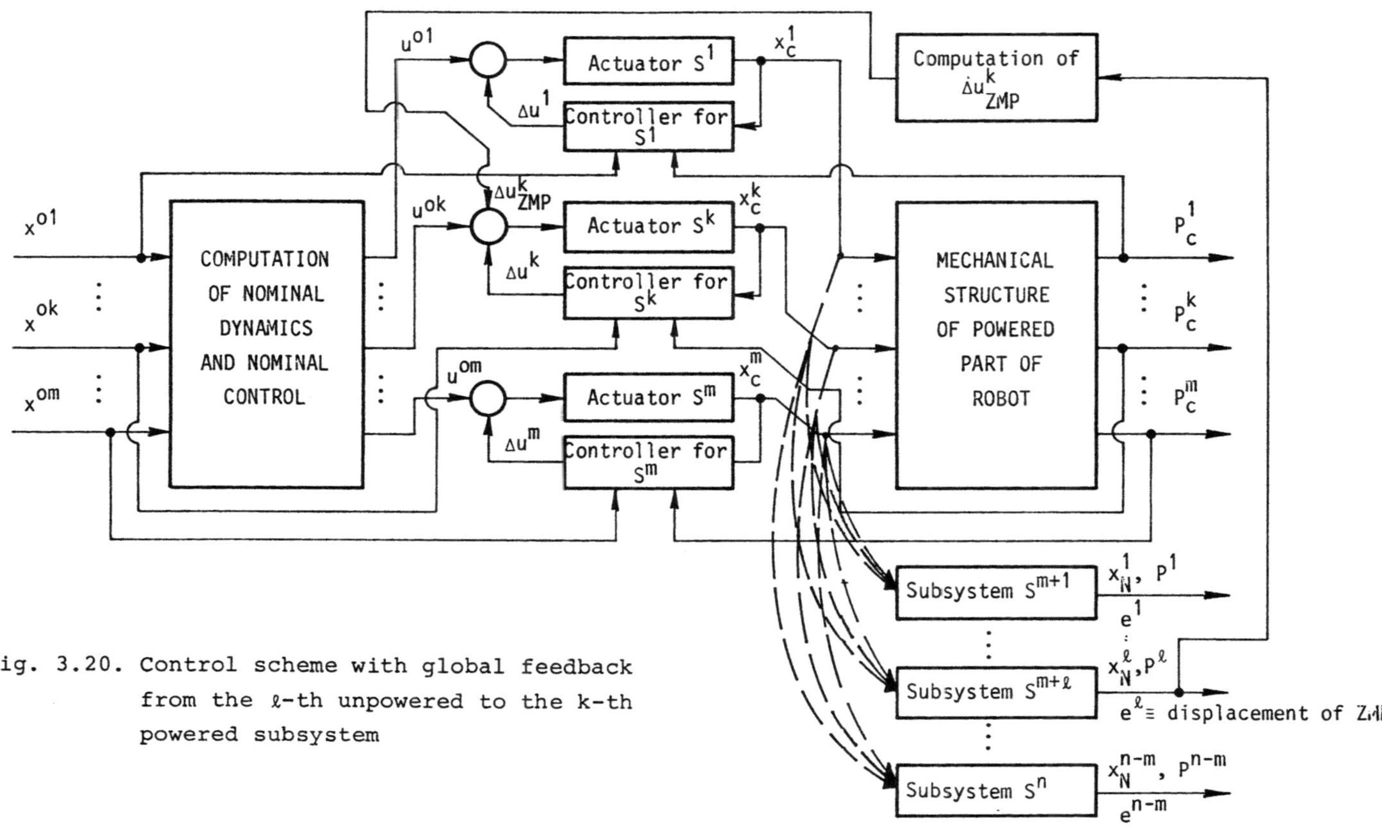


Fig. 3.20. Control scheme with global feedback from the $\ell$-th unpowered to the k-th powered subsystem

In Fig. 3.20. is given a scheme of the control with a feedback introduced with respect to the ZMP position.


### 3.4.4. <u>Simulation of walk for a specific biped structure</u>

The scheme of biped structure used for the walk simulation is presented in Fig. 3.21. and its mechanical parameters are given in Table 2.4. The mechanism is identical to that presented in Fig. 2.44., Chapter 2. The whole mechanism had eleven joints, each of them having either one, or two or three d.o.f. The joint with more than one d.o.f. has been modelled as a set of corresponding numbers of simple rotational joints connected with light links, and with no lenght[*)], $\vec{r}_{i,k}$ = 0 (because of certain specificities of the simulation programme, $|r_{i,k}|$=$\varepsilon$, $\varepsilon$=0.0001 [m] is used instead of $\vec{r}_{i,k}$=0). These links are called fictious links, and in Fig. 3.21. they are represented with dashed line. Thus, simple rotational joints with the unit rotational axes $\vec{e}_3$ and $\vec{e}_4$ connected to the fictious link 3, constitute the ankle joint of the right leg. The hip joint of the same leg had three d.o.f. represented by rotations about $\vec{e}_6$, $\vec{e}_7$ and $\vec{e}_8$, while the knee joint had only one d.o.f. In the same way are formed the joints for the left leg: $\vec{e}_9$, $\vec{e}_{10}$ and $\vec{e}_{11}$ for hip, $\vec{e}_{12}$ for knee, and $\vec{e}_{13}$ and $\vec{e}_{14}$ for ankle. The trunk joint had two d.o.f.: $\vec{e}_{15}$ and $\vec{e}_{16}$. The shoulder and elbow joint possessed only one d.o.f.: $\vec{e}_{17}$ and $\vec{e}_{18}$ for the right and $\vec{e}_{19}$ and $\vec{e}_{20}$ for the left arm, respectively.

As in the single support phase the mechanism as a whole can rotate about foot edges in the frontal and sagittal plane, these two d.o.f. were also modelled as simple rotational joints with axes $\vec{e}_1$ and $\vec{e}_2$ below the supporting foot. The mechanism rotation about z axis is supposed to be prevented by sufficiently large friction between the mechanism sole and ground surface.

Nominal motion is synthesized using prescribed synergy method. Only seven joints are included in the walking process with totally eight d.o.f. At each leg joint, only one simple rotational joint having one d.o.f. is participating in the motion: at the ankle joints it is

---

[*)] In the text below, the vectors $\tilde{\vec{r}}_{i,k}$ and $\tilde{\vec{e}}_i$ will be denoted by $\vec{r}_{i,k}$ and $\vec{e}_i$.

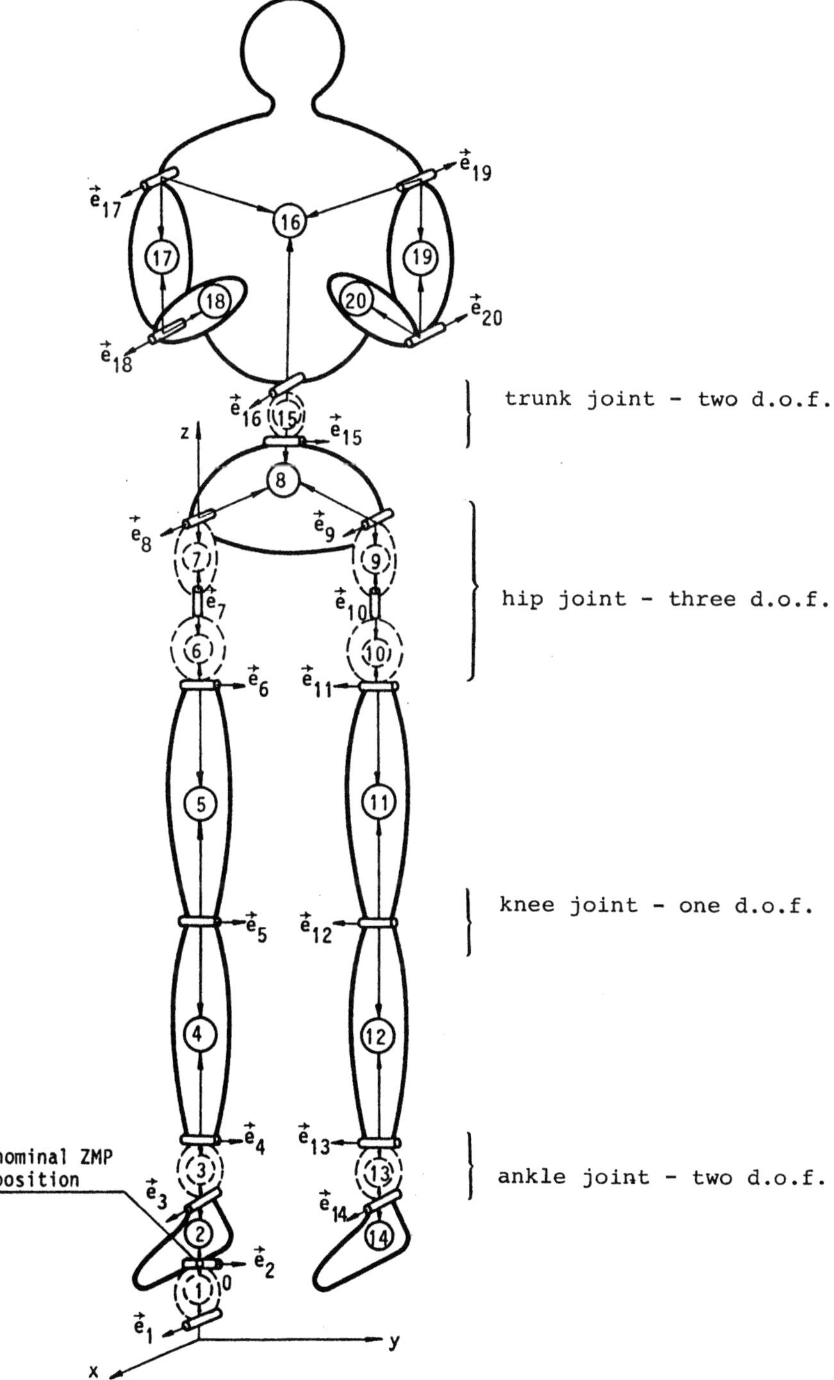



Fig. 3.21. Scheme of mechanical biped structure

256

only $\vec{e}_4$ and $\vec{e}_{13}$, at the knees $\vec{e}_5$ and $\vec{e}_{12}$, and at the hip joints $\vec{e}_6$ and $\vec{e}_1$. At the trunk joint, both d.o.f., represented by $\vec{e}_{15}$ and $\vec{e}_{16}$ are involved in the motion. The rest of them are inactive and locked, and do not change their relative position during the mechanism motion. The hands are crossed on the chest. Such choice of the active legs joints enables their motion in the planes parallel to the frontal plane, only. Thus, the compensating movements are executed by the trunk: in the frontal plane about $\vec{e}_{15}$, in sagittal about $\vec{e}_{16}$. Fig. 3.22. shows the stick diagram of the nominal biped gait in the sagittal plane. Here 0

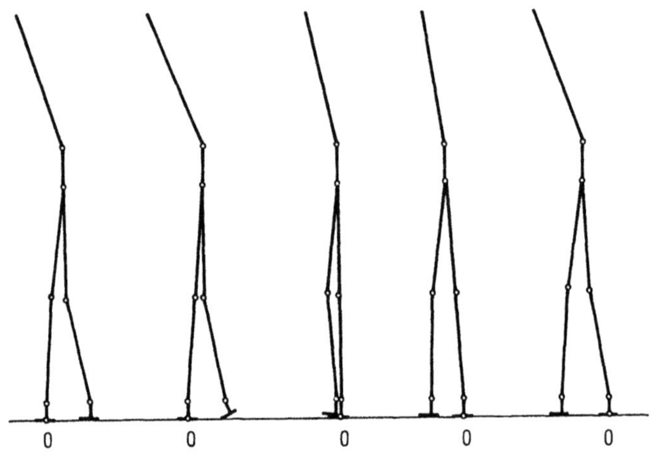

0        0        0        0        0

Fig. 3.22. Stick diagram of the nominal mechanism motion
in the sagittal plane during half-step period

denotes the nominal position of ZMP under the supporting foot, the other leg is being in swing phase. In Figs. 3.23. and 3.24. are given the nominal trajectories of all the internal active angles and of the corresponding driving torques, respectively.

Each active joint is powered by the corresponding actuator. The actuators used were DC motors of two types: Moore - Reed PCP 50 for actuator $M_1$, TRW GLOBE 102A 200-8 for actuator $M_2$. Mathematical model of each powered subsystem is given by (3.3.2), with matrix $A_c^i$ and vectors $b_c^i$ and $f_c^i$, as well as their distribution per joints and saturation voltage being given in Table 3.3. The motion was simulated for one half-step period, and for the single support phase, only. Duration time of simulated motion was T = 0.75[s]. The perturbed motion of the system around the nominal trajectory has been simulated, where, trunk angular displacement from the nominal trajectory in the frontal plane, of

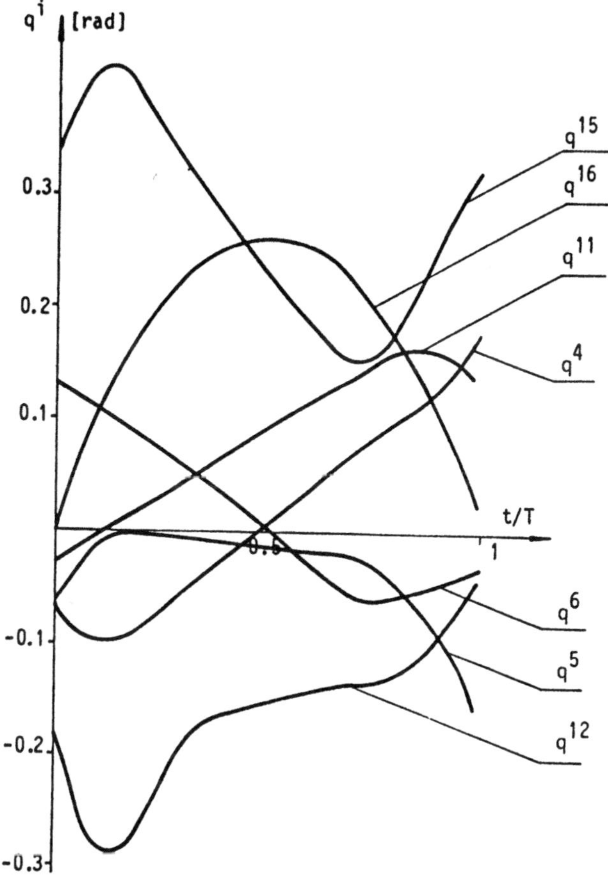


Fig. 3.23. Nominal trajectories of active internal angles

Table 3.3. Actuator parameters

| Actuator \ Term | $a^i_{22}$ | $a^i_{23}$ | $a^i_{32}$ | $a^i_{33}$ | $b^i_3$ | $f^i_2$ | Used at joint | Saturation [V] |
|---|---|---|---|---|---|---|---|---|
| $M_1$ | -0.198 | 15.96 | -2000 | -695.6 | 434.8 | -3.45 | 5, 6, 11, 12, 15, 16 | 24 |
| $M_2$ | -1 | 2.83 | -700 | -100. | 100 | -0.65 | 4, 13 | 27 |

$\Delta q^{15}(0) = +0.2$ [rad] at the initial moment was adopted as disturbance. In addition, to compare the mechanism behaviour, an initial displacement from the nominal of $\Delta q^4(0) = 0.1$ [rad] at the ankle joint of the supporting leg has also been considered.

Simulation was performed for different control schemes applied. These cases were:

1) Only local control defined by (3.4.9) was applied at each joint, while $k_{15}^G$ was introduced only at the joint where an initial angular displacement occurred (trunk or ankle).

2) The control structure was identical to the previous one, with addition of force feedback with respect to joints driving torques.

3) To improve the ZMP behaviour, an additional feedback was introduced from force sensors at mechanism's sole of the supporting foot to the compensating subsystem, selected in advance.

To enable an easy comparison of different situations, simulation results are presented in the same way for each case considered. Each gait is simulated using three different quantities of $\Omega_{max}^i$ (i=4 for ankle joint, i=15 for trunk) defined by $k_{15}^G$. They are $|\Omega_{max}^i|=3[rad/s^2]$, $|\Omega_{max}^i|=5[rad/s^2]$, $|\Omega_{max}^i|=7[rad/s^2]$.

The results belonging to one simulated gait case are given as a set of five diagrams. The first diagram shows angular deviation from nominal value for trunk, the second one gives the same for the ankle and hip joints of the supporting leg. Each diagram includes the results for all three values of $\Omega_{max}^i$. The next two diagrams present the example (for $|\Omega_{max}^i|=7[rad/s^2]$ only) of the control input to the actuator and driving torque deviation for the ankle and hip joints of the supporting leg, as well as for the trunk.

The last diagram of each set of results shows the deviation of ZMP from the nominal position.

The first set of simulation results describes the case when feedback is defined as in 1), and the trunk displacement for $\Delta q^{15}(0) = 0.2[rad]$ is adopted as a disturbance. Simulation results are given in Fig. 3.25. Fig. 3.25.a) shows the trunk position deviation from the nominal angle, and Fig. 3.25.b) shows the same for the hip and ankle joint. Fig. 3.25.c) illustrates control inputs to actuators, while diagram of driving torque deviations is ommitted because $\Delta P^i \approx 0$ (i=4, 6, 15). Fig. 3.25.d) shows the ZMP position deviation from its nominal position. The nominal ZMP position in the simulated gait, fixed to the origin of the

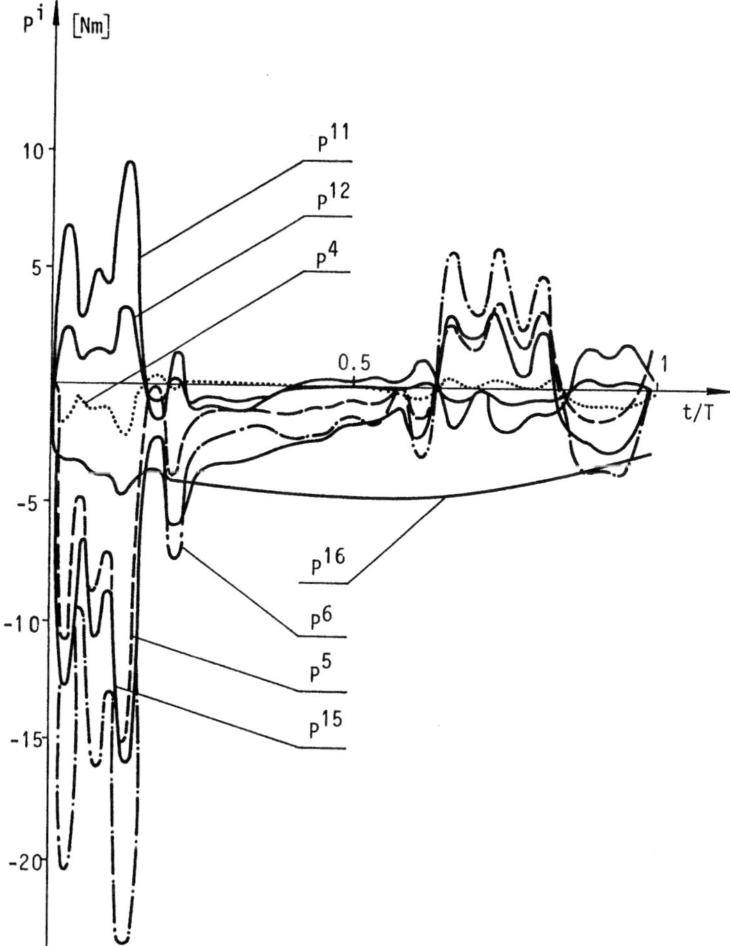



Fig. 3.24. Nominal driving torques of active joints

coordinate frame, is in Fig. 3.21 denoted by 0. During the whole time period considered the nominal ZMP remains at the same position.

From these simulation results we may conclude that such control scheme does not satisfy the requirements imposed on the locomotion mechanism. For the $|\Omega^{15}_{max}| = 3$ [rad/s$^2$], the trunk deviates from its nominal position, for $|\Omega^{15}_{max}| = 5$ [rad/s$^2$] trunk position deviation is almost constant, while for $|\Omega^{15}_{max}| = 7$ [rad/s$^2$] trunk position converges to the nominal position. Contrary to the trunk, the ankle and hip joint of

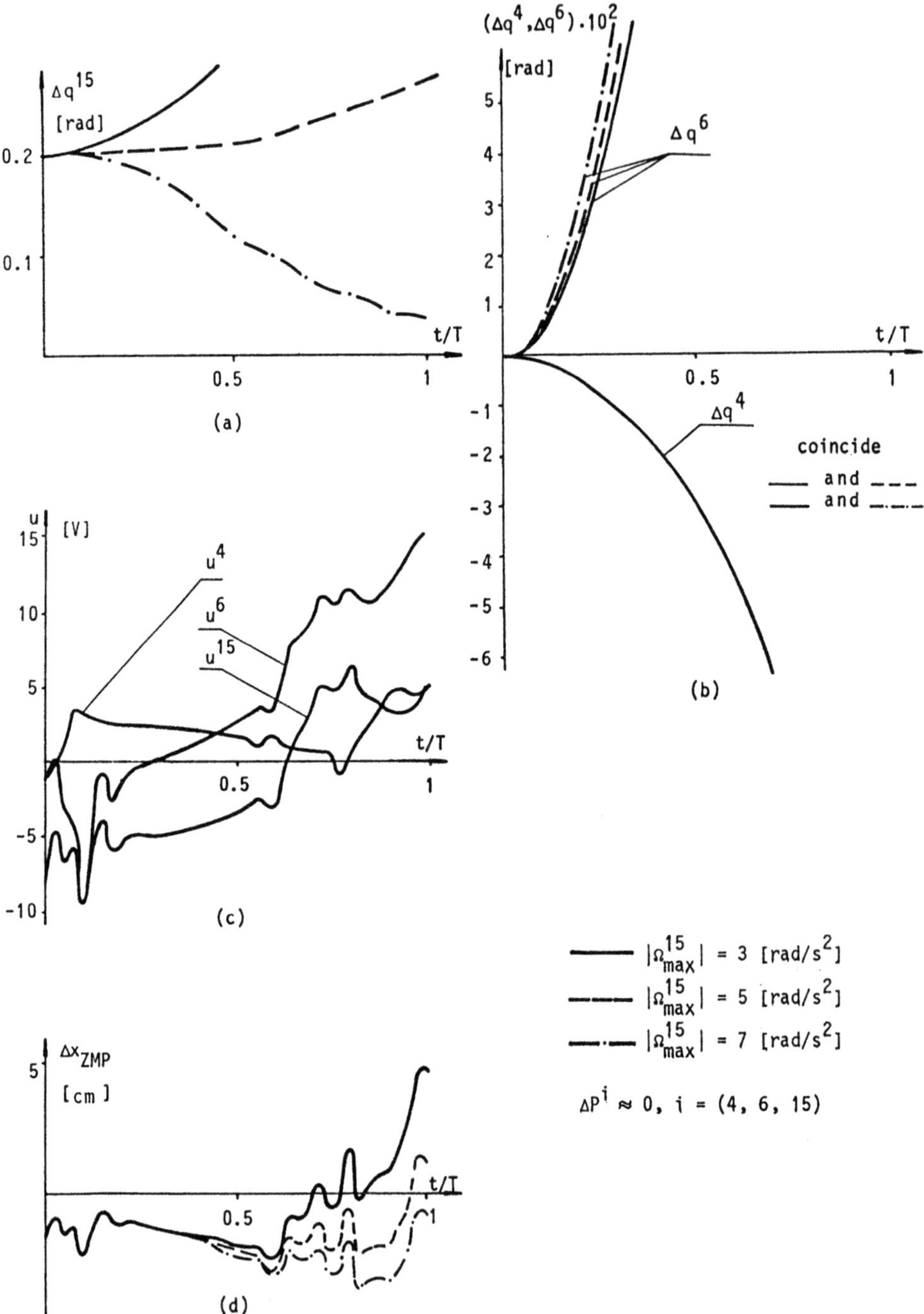



Fig. 3.25. Walk simulation without feedback with respect to $P^i$ and ZMP position, $\Delta q^{15}(0) = 0.2$ [rad]

the supporting leg repidly diverge for any value of $\Omega_{max}^{15}$, while the hip angle diverges somewhat faster. The behaviour of ZMP position is better at the beggining of the gait. In the second half of the step, ZMP position moves more rapidly.

The second example of gait simulation includes feedbacks with respect to driving forces at the mechanism joints. This case is illustrated in Fig. 3.26. It can be seen that the behaviour of all biped d.o.f. is improved: for each of them the diference between nominal and real position has been decreased. However, the behaviour of ZMP position has deteriorated, what could be expected, because the ZMP position was not taken into account in compensating the link angular deviations.

In both previous examples the trunk inclination for 0.2 [rad] was adopted as initial disturbance. Fig. 3.27. presents the case when the initial angular displacement assigned to the ankle joint, was $\Delta q^4 = 0.1$ [rad]. The complete feedback (3.4.9) was employed. In 3.27.a) is illustrated the behaviour of the trunk angle, and in 3.27.b) of the hip and ankle angle. In each case, the angular deviation is increased with time. In 3.27.c) is presented a diagram of the control input (voltage) to the actuators at joints 3, 6, and 15 (ankle, hip and trunk), and 3.27.d) shows the torque deviations from the nominals. The most illustrative is Fig. 3.27.e) where ZMP displacement progresses very rapidly. Such mechanism's behaviour is quite unacceptable, and additional feedback has to be introduced to prevent it from falling down. In all these diagrams, the lines representing the cases $|\Omega_{max}^4| = 3$ [rad/s$^2$], $|\Omega_{max}^4| = 5$ [rad/s$^2$] and $|\Omega_{max}^4| = 7$ [rad/s$^2$] coincide.

To improve behaviour of ZMP, an additional feedback with respect to its deviation from the nominal position is introduced. The main question is to which joint the task of its compensation should be assigned. Three different choices have been considered: the ankle joint, hip and trunk. In all cases the compensating control is computed according to (3.4.18). Another question is whether the global feedback is to be active permanently, or only when the ZMP position is out of a certain region. Namely, there are several possible strategies to correcting the ZMP position. One of them assumes the absence of any additional action aimed at decreasing this deviation while this is below a certain value fixed in advance. Because the ZMP position is a consequence of the complete mechanism motion, its displacement will be reduced to zero as the total mechanism state vector approaches its nominal. But, if the ZMP

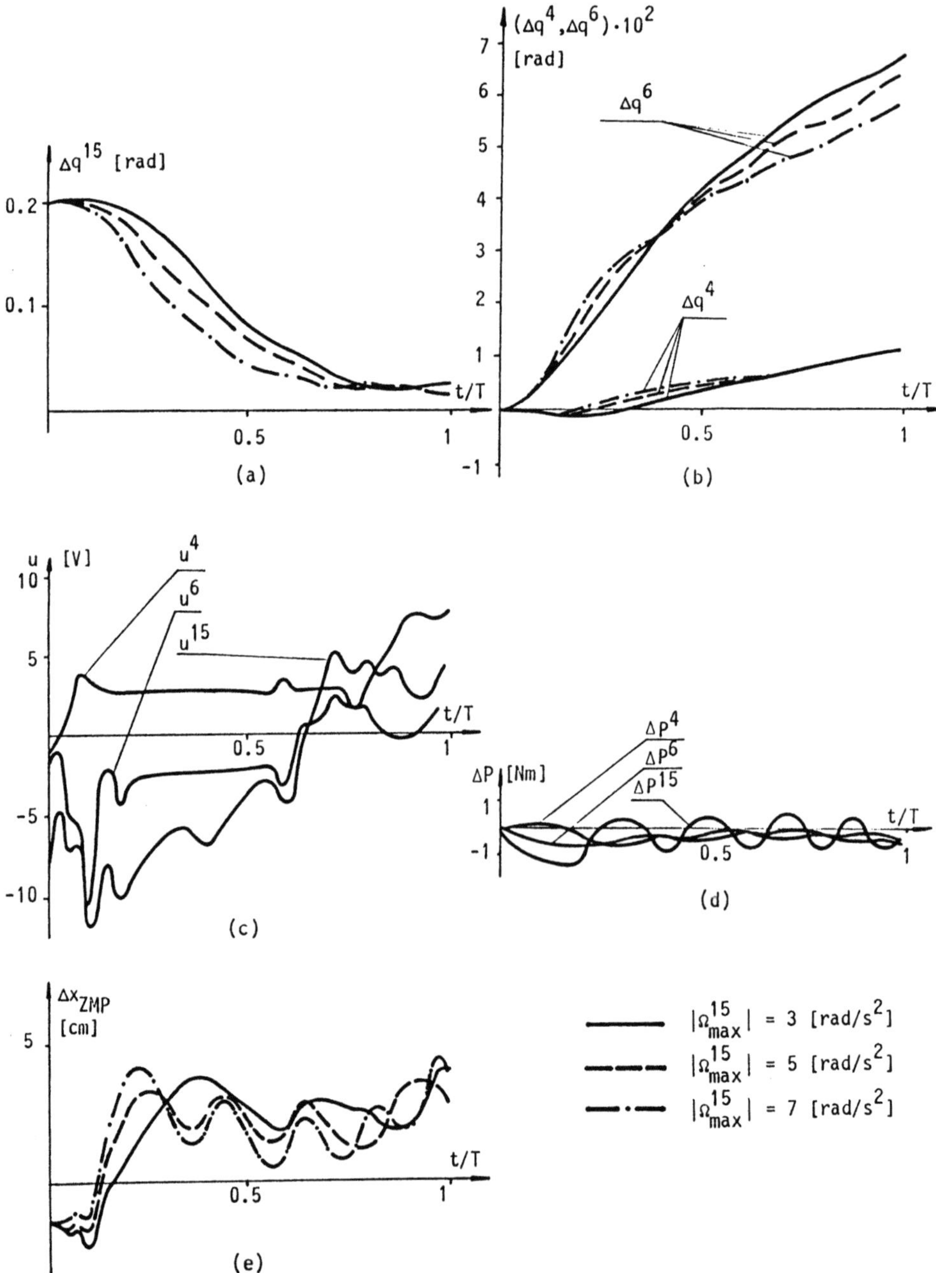

Fig. 3.26. Walk simulation without feedback with respect
to ZMP position, $\Delta q^{15}(0) = 0.2$ [rad]

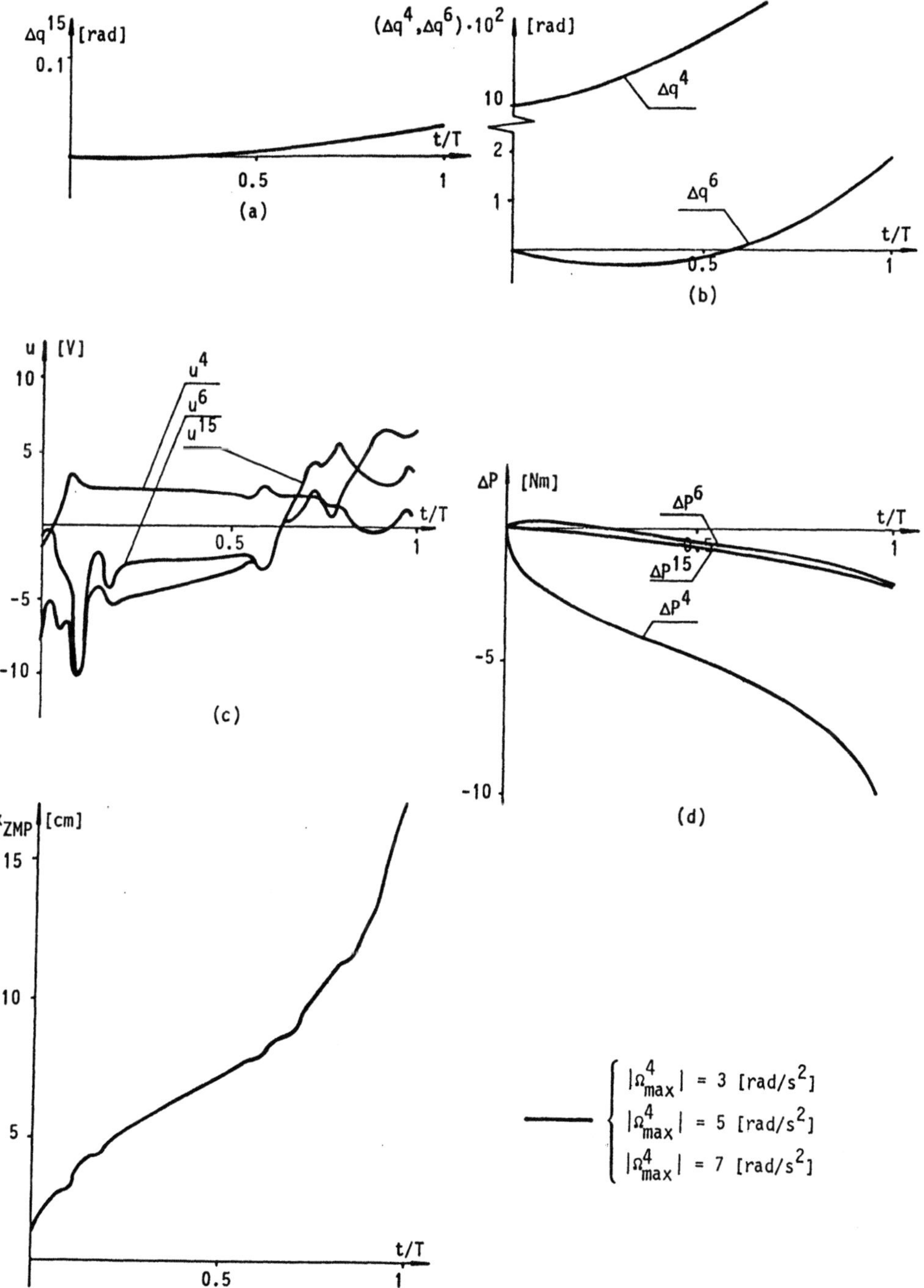

Fig. 3.27. Walk simulation, without feedback with respect to
ZMP position, $\Delta q^4(0) = 0.1$ [rad]

excursion is too large, an additional mechanism action has to bring it back to the allowed region. Also, the question of the $k_{ZMP}^{Gk}$ gain value has to be answered. The answers to all these questions will be proposed, on the basis of simulation results.

Let us start with the example when the trunk is selected to compensate ZMP displacement.

Fig. 3.28. presents the case when the initial displacement assigned to trunk is $\Delta q^{15}(0) = 0.2$ [rad], and $k_{ZMP}^{Gk} = 0.1$. In addition, a region of length ±3 [cm] under supporting foot where $k_{ZMP}^{Gk} = 0$ is introduced. In other words, in the zone $\Delta x_{NPER} = \pm3$ [cm], the action of global feedback with respect to ZMP is not permissible. In Fig. 3.28.a) is presented the angular trunk displacement for three values of $\Omega_{max}^{15}$. As is seen, the larger $\Omega_{max}^{15}$ is, the worse is the trunk behaviour, and for $|\Omega_{max}^{15}| = 7$ [rad/s$^2$] it becomes even oscillatory. The same is valid for the hip and ankle, (Fig. 3.28.b)). Fig. 3.28.c) shows an example of the actuators control input, and Fig. 3.28.d) of torque deviation, both for $|\Omega_{max}^{15}| = 7$ [rad/s$^2$]. Fig. 3.28.e) shows the behaviour of ZMP. The motion is oscillatory, with the frequency and amplitude increasing with $\Omega_{max}^{15}$.

Fig. 3.29. illustrates the gait under the same condition as in Fig. 3.28, except $k_{ZMP}^{Gk} = 0.05$. Generally, the feature of all the curves is less "oscillatory", comparing to the previous case, and is very close to the situation presented in Fig. 3.26. Again, Fig. 3.29.c) and 3.29.d) represent only the case when $|\Omega_{max}^{15}| = 7$ [rad/s$^2$]

The next two figures, Fig. 3.30. and Fig. 3.31. correspond to the case of the initial angular displacement at ankle joint, $\Delta q^4(0) = 0.1$ [rad], and ZMP displacement is compensated by the trunk. In both figures, the cases for $\Omega_{max}^4$ having all three values (3, 5 and 7) coincide with each other. Also, the global feedback with respect to ZMP, acting out of region $\Delta x_{NPER} = \pm3$ [cm], has been adopted.

In the first case, given in Fig. 3.30. we have that $k_{ZMP}^{Gk} = 0.1$. The behaviour of the trunk, ankle and hip is shown in Fig. 3.30.a),b), and of control input to the actuators and of torque deviations in Fig. 3.30.c),d). It is interesting to note that the trunk at the begining of the gait (in fact, while ZMP is in the $\Delta x_{NPER}$ zone) tracks its nominal perfectly well and then, starts to diverge rapidly. Obviously, additio-

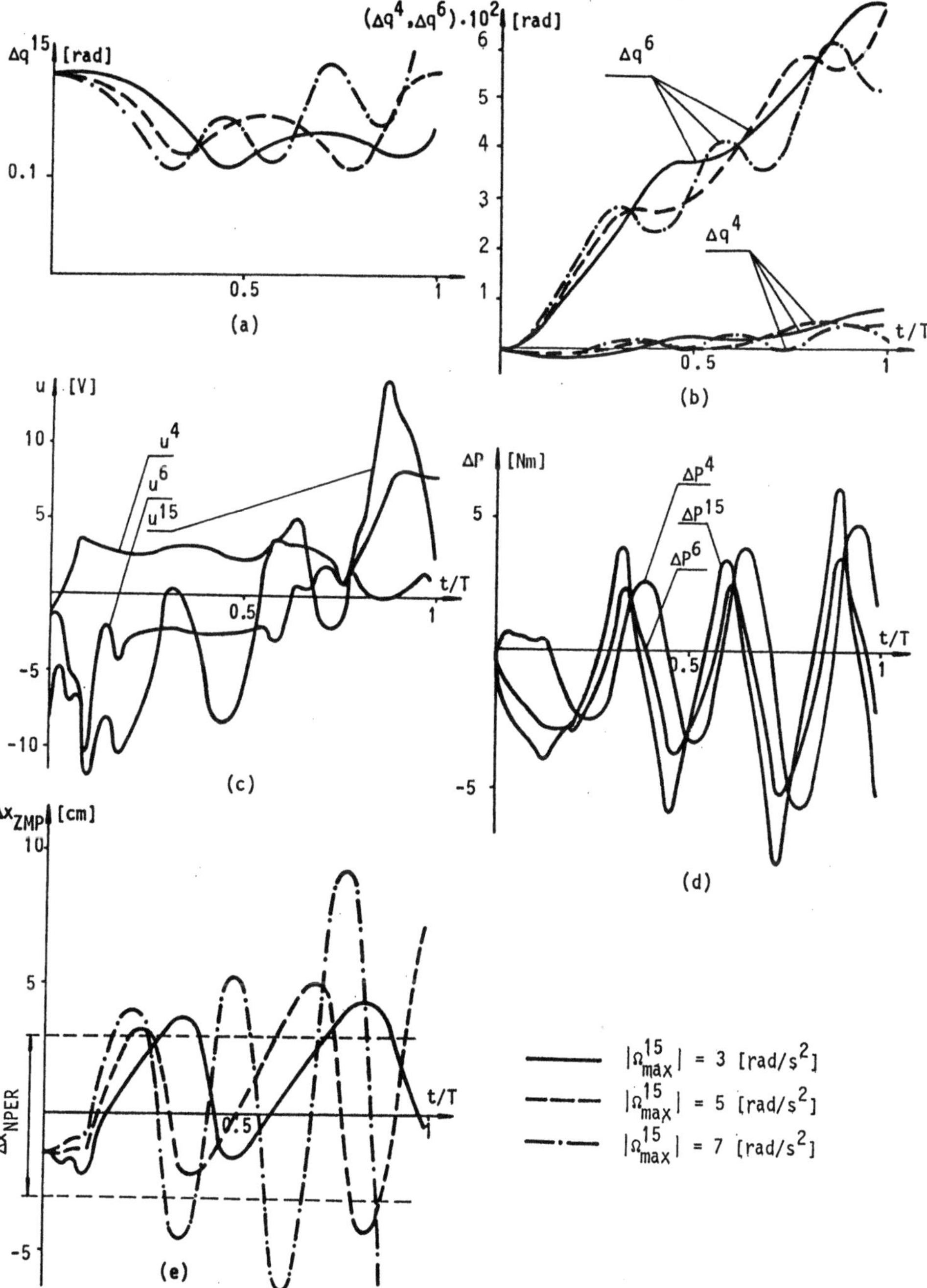



Fig. 3.28. Walk simulation, ZMP deviation compensated by trunk
$k_{ZMP}^{Gk} = 0.1$, $\Delta q^{15}(0) = 0.2$ [rad]

266

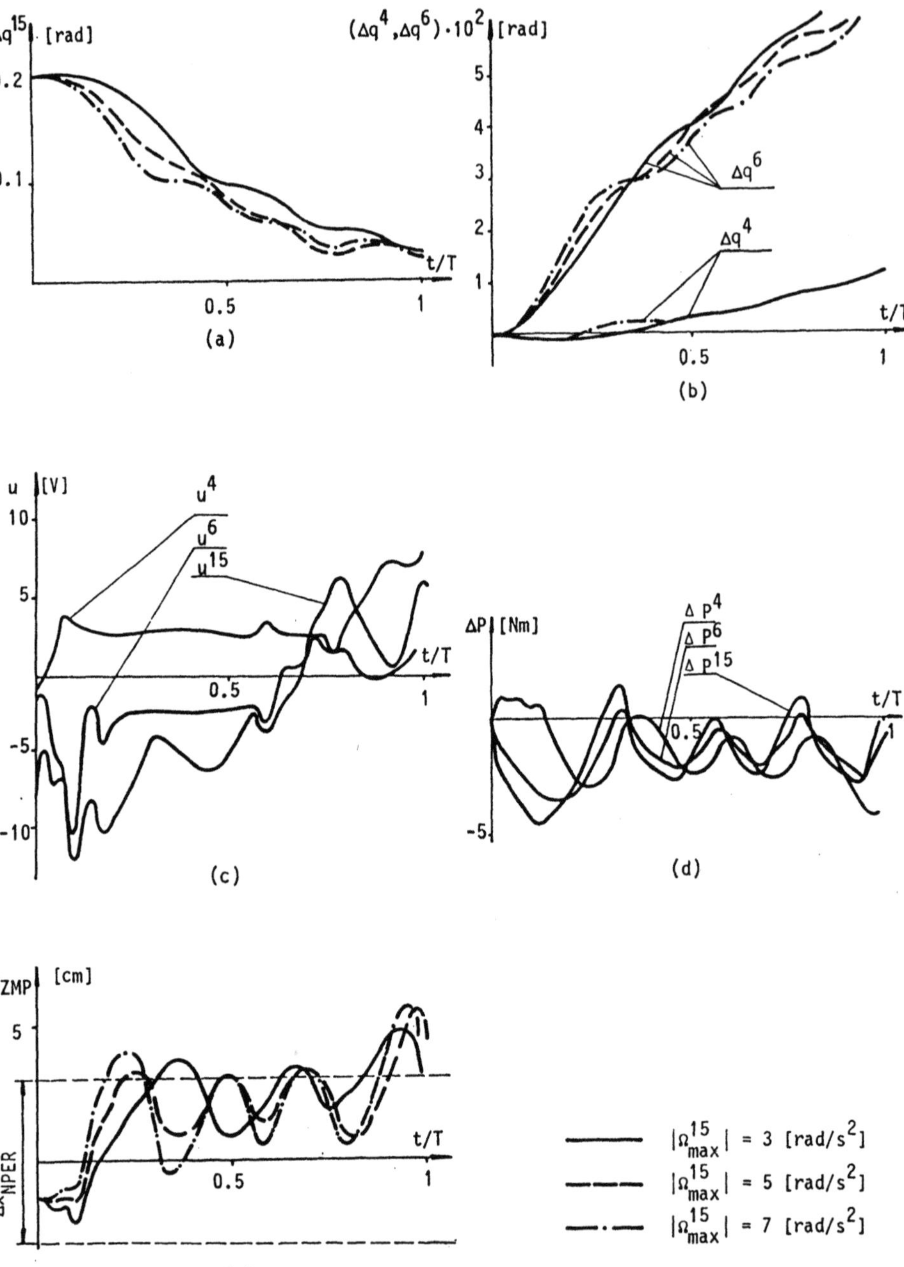

Fig. 3.29. Walk simulation, ZMP deviation compensated by trunk $k_{ZMP}^{Gk}$ = 0.05, $\Delta q^{15}(0)$ = 0.2 [rad]

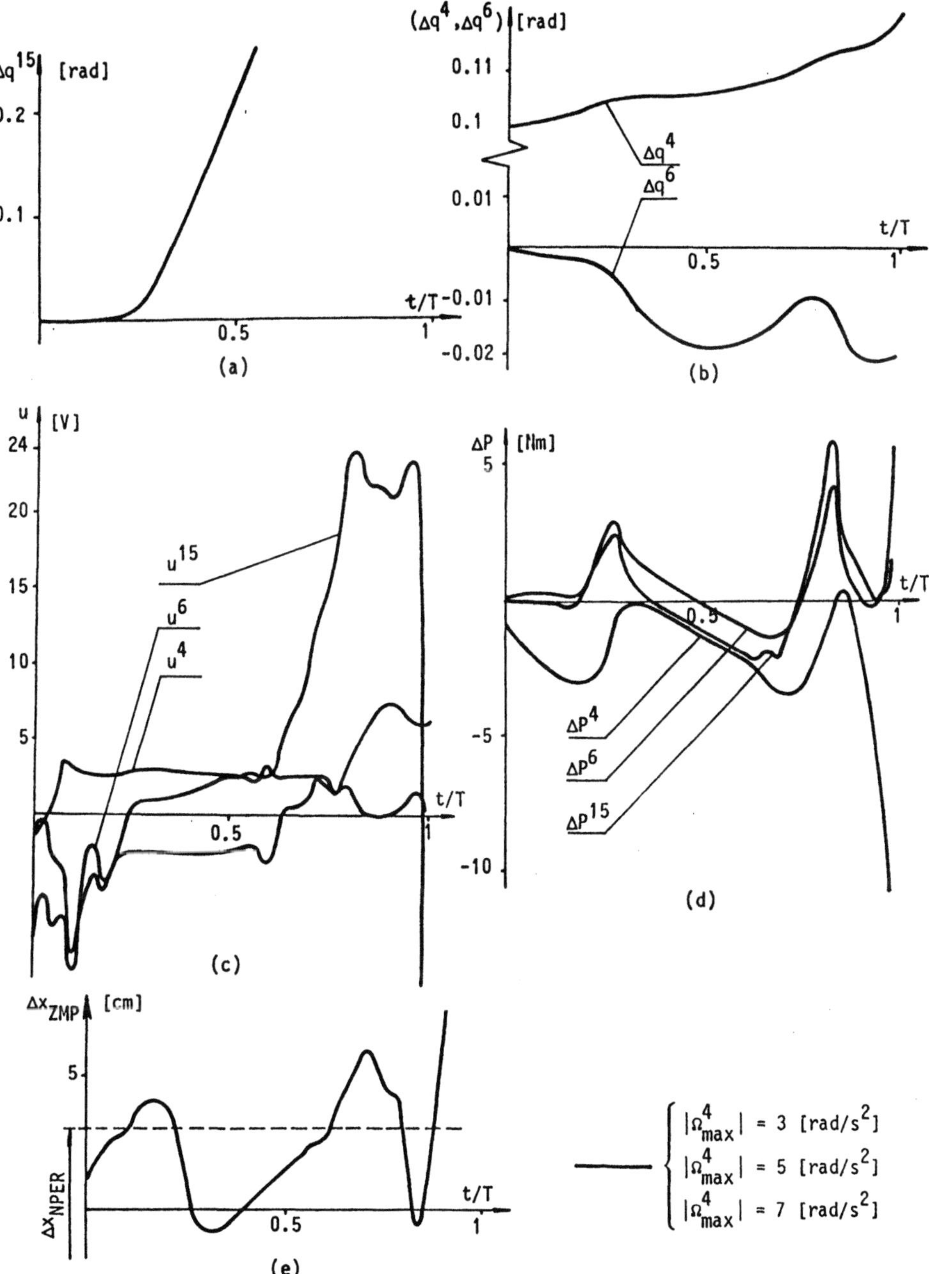



Fig. 3.30. Walk simulation, ZMP deviation compensated by trunk, $k_{ZMP}^{Gk} = 0.1$, $\Delta q^4(0) = +0.1$ [rad]

nal task of monitoring ZMP position spoiled to the extreme the quality of tracking of the joint's nominal angle. The ankle, to which initial displacement is assigned, does not diverge significantly, while the hip in spite of the fact that neither initial displacement, nor any additional requirement has been assigned to it shows an oscillatory behaviour. Let us note that at a certain moment the trunk actuator achieved saturation. The ZMP behaviour shown in Fig. 3.30.e) is quite unacceptable because the ZMP would go out of the surface covered by the sole, what means that the mechanism collapsed by rotation about the foot edge.

Fig. 3.31. gives the simulation results for $k_{ZMP}^{Gk} = 0.05$. All the diagrams are in their appearance similar quite to those shown above, but with the extremly high values diminished. The hip angular trajectory has lost its oscillatory character, the control voltage for the considered actuators and corresponding joints torque deviations are smaller than in Fig. 3.30, but the ZMP excursion is again very large, and very close to the foot edge. (The largest excursion of ZMP is 17.5 [cm]).

The next example illustrates the situation when ZMP displacement is compensated by the hip joint (Fig. 3.32). The initial disturbance of $\Delta q^{15}(0) = 0.2$ [rad], is assigned to the trunk. In the simulation, for $|\Omega_{max}^{15}| = 5$ [rad/s$^2$] and for $|\Omega_{max}^{15}| = 7$ [rad/s$^2$] is adopted $k_{ZMP}^{Gk} = 0.1$ while for $\Omega_{max}^{15} = 3$ [rad/s$^2$], $k_{ZMP}^{Gk} = 0.05$. For the larger values of $k_{ZMP}^{Gk}$ it is not possible to achieve a stable gait. The same set of diagrams is given as before. As in previous case, the global feedback with respect to ZMP position is not activated in the region around its nominal value; $\Delta x_{NPER} = \pm 3$ [cm]. Now, when the trunk has no additional control task, except the tracking of its own nominal joint trajectory, its convergence is very fast. Also, the ankle joint tracks its nominal almost perfectly well, while the hip joint does not converge for any value of $\Omega_{max}^{15}$. The additional requirement of ZMP position maintenance influences the behaviour of hip joint to such an extent to hinder tracking of the nominal joint trajectory. In contrast to this, the ZMP displacement is kept under control and the ZMP behaviour can be considered as moderately good. Also, the diagrams of control inputs and driving torques deviations from the nominals are given only for $|\Omega_{max}^{15}| = 7$ [rad/s$^2$]. Again, the ZMP, as an overall characteristic indicator of mechanism's motion, exibits a behaviour similar to that in previous cases.

The further set of examples illustrates the compensation of ZMP displacement by the ankle joint.

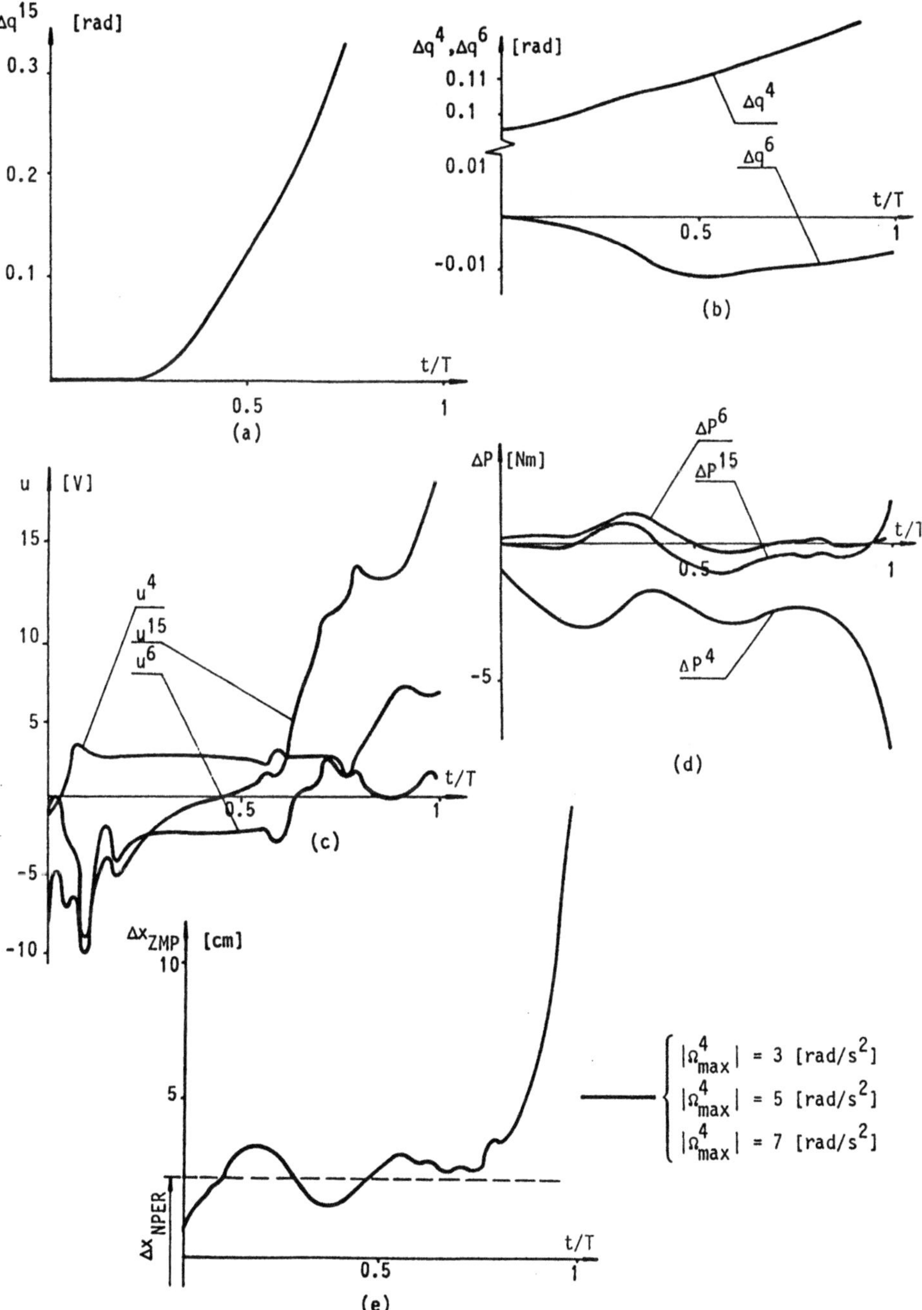



Fig. 3.31. Walk simulation, ZMP deviation compensated by trunk, $k_{ZMP}^{Gk} = 0.05$, $\Delta q^4(0) = +0.1$ [rad]

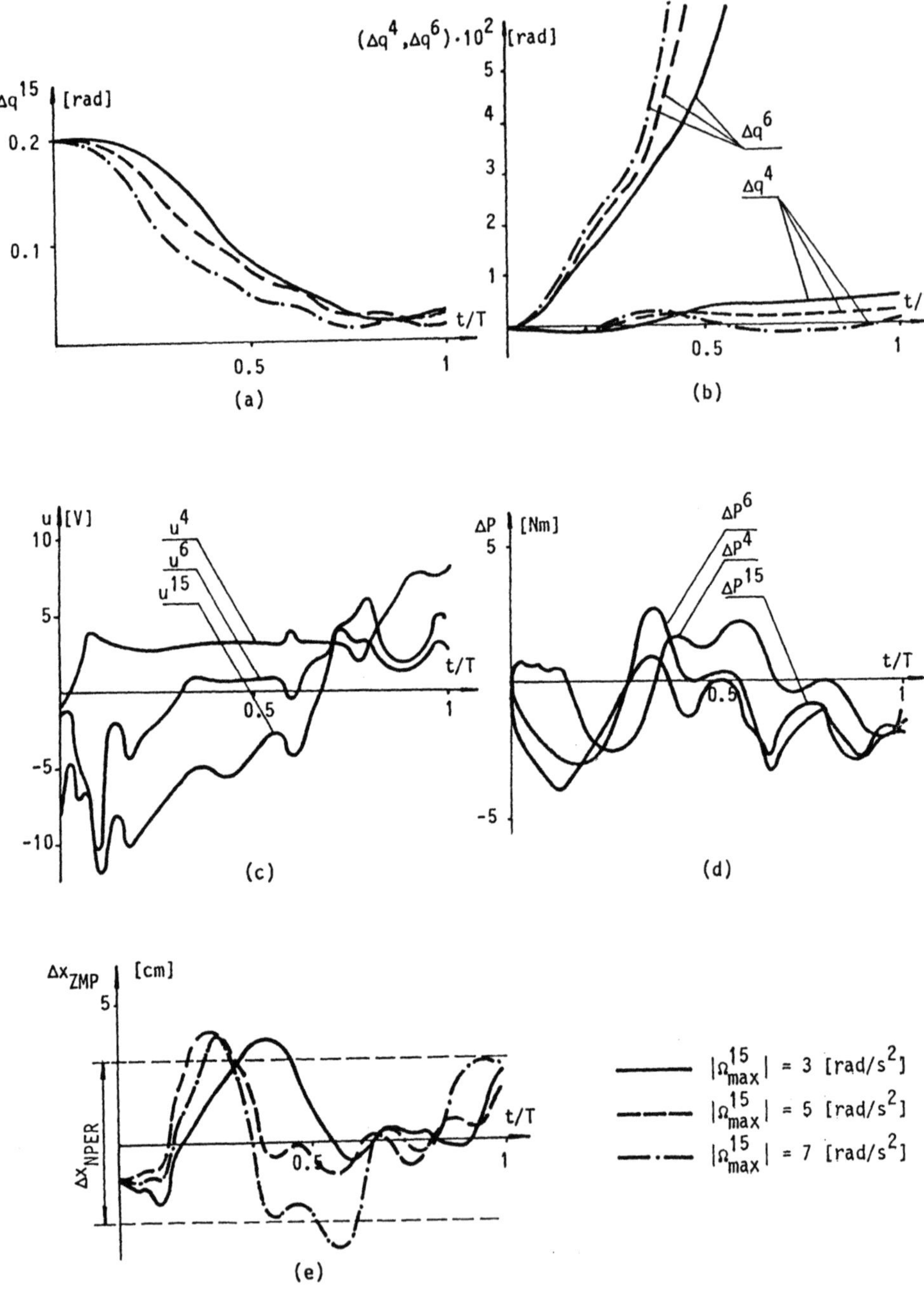

Fig. 3.32. Walk simulation, ZMP deviation compensated by hip joint, $\Delta q^{15}(0) = +0.2$ [rad]

Two types of initial disturbances are adopted: the trunk angular deviation by $\Delta q^{15}(0) = 0.2$ [rad], and ankle angular deviation by $\Delta q^4(0) = 0.1$ [rad]. All three values for $\Omega^i_{max}$ (i=4, 15) are applied. As before, at each joint, the complete feedback defined by (3.4.9) has been applied, and then additional feedback, to maintain ZMP position, introduced to the ankle. In contrast to the previous cases, the global feedback with respect to ZMP is permanently acting. Two quantities for $k^{G4}_{ZMP}$ are adopted: 0.5 and 1.

Consider the case when the initial disturbance is assigned to the trunk, and $k^{Gk}_{ZMP} = 0.5$. Fig. 3.33. presents the simulation results in the same manner as before. The trunk converges to its nominal in a similar way as in previous example, but the behaviour of the ankle and hip joints is different. Both of them converge in a similar way for all three values of $\Omega^{15}_{max}$. At the end of the simulated gait, the deviations are almost the same. From the diagrams for $u^i$ and $\Delta P^i$ (Fig. 3.33.c) and d)), it can be seen that the control is close to the nominal, as well as the torque deviations is close to zero. It is interesting to note that the smallest torque deviations occur at the ankle joint which additionally controls the ZMP motion.

The most interesting is Fig. 3.33.e), illustrating the ZMP behaviour. It can be noticed that ZMP excursions from the nominals are smallest up to now. Note that the scale for $\Delta x_{ZMP}$ is enlarged approximately three times comparing to previous cases, and it will be used for all subsequentent examples. But, the behaviour of other joints, especially of the ankle, is not spoiled much by such very "strong" keeping ZMP position under control.

Next example, Fig. 3.34, considers the same gait as in Fig. 3.33, but for $k^{Gk}_{ZMP} = 1$. After initial oscillations, the behaviour of ZMP is even better, and for $|\Omega^{15}_{max}| = 3$ [rad/s$^2$] it almost coincides with the nominal value.

For an easier comparison of ZMP behaviour with respect to the global feedback gain value, $k^{Gk}_{ZMP}$, in Fig. 3.35. are given the diagrams for the same $\Omega^{15}_{max}$ and $k^{Gk}_{ZMP} = 0.5$ and $k^{Gk}_{ZMP} = 1$.

As can be seen, the larger the $k^{Gk}_{ZMP}$ values are, the smaller are the

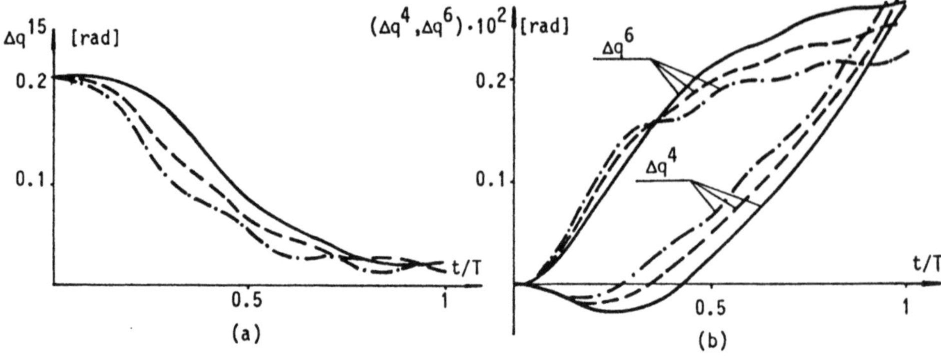

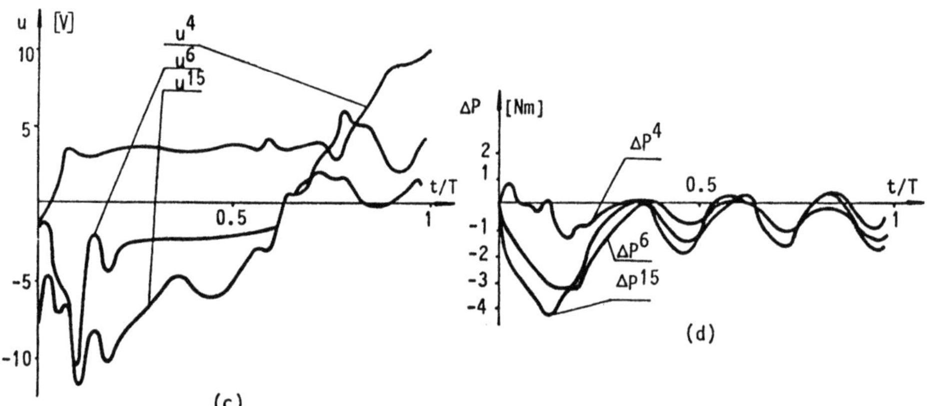

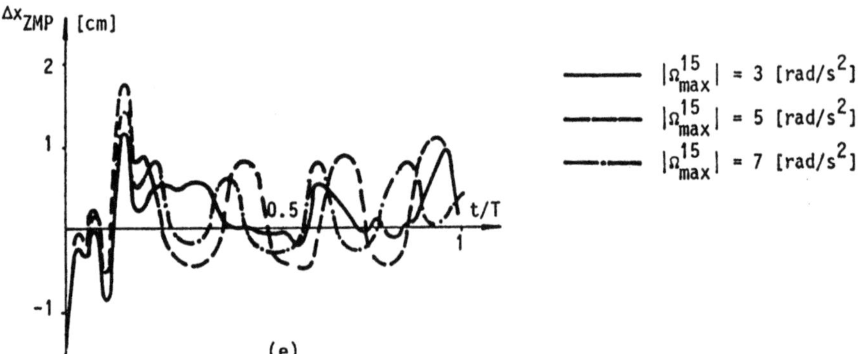

Fig. 3.33. Walk simulation, ZMP deviation compensated by ankle joint, $k_{ZMP}^{Gk} = 0.5$, $\Delta q^{15} = 0.2$ [rad]

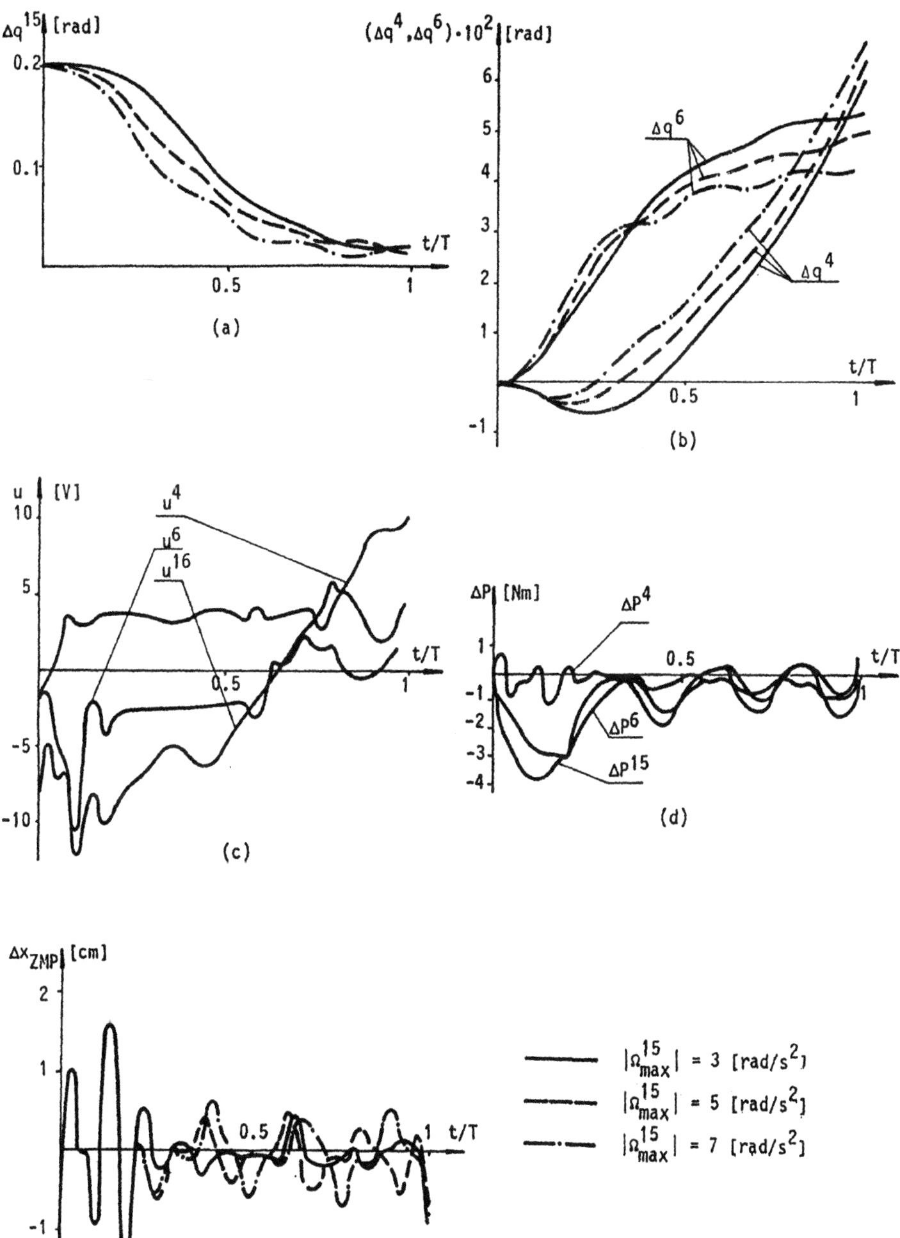

Fig. 3.34. Walk simulation, ZMP deviation compensated by ankle joint, $k_{ZMP}^{Gk} = 1$, $\Delta q^{15} = 0.2$ [rad]

274

ZMP deviations, with only somewhat larger oscillations at the beginning of the gait.

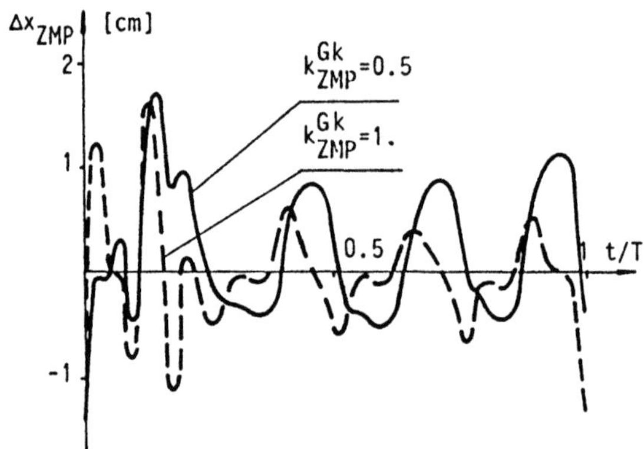


Fig. 3.35. Comparison of the ZMP behaviour for
various global feedback gains

The last set of simulation results illustrates the situation when both the initial angular deviation and the task of ZMP displacement compensation are assigned to the ankle joint. Fig. 3.36. presents the case for $k_{ZMP}^{Gk} = 0.5$, and Fig. 3.37. for $k_{ZMP}^{Gk} = 1$.

In Fig. 3.36.a), all three angle deviations, for the ankle, hip and trunk, are presented. All deviations are enhanced in time. For the hip and trunk they remain within a very narrow region, while for the ankle, this deviation is much larger. The joint torque deviations show a permanent, but a very slow increase without oscilations. A very interesting feature exibits the diagram representing ZMP displacement which is of the similar character as the diagram of joint torque deviations. All three cases for $\Omega_{max}^{4}$ coincide so that they are presented by the same line, as well as in Fig. 3.36.a). The ZMP position motion is very smooth, and for large period of step duration is almost constant. Quite similar are the results shown in Fig. 3.37. illustrating the case when $k_{ZMP}^{Gk} = 1$. Let us focus on the diagram of ZMP behaviour, Fig. 3.37.d). Now, the situation is even better than in previous example. For a significant period of gait duration the ZMP practically does not deviate from the nominal position.

In all previous examples, the mechanism motion was described and analyzed by consideration of the biped system's internal synergy, i.e. of joint coordinates and driving torque deviations, as well as the ZMP

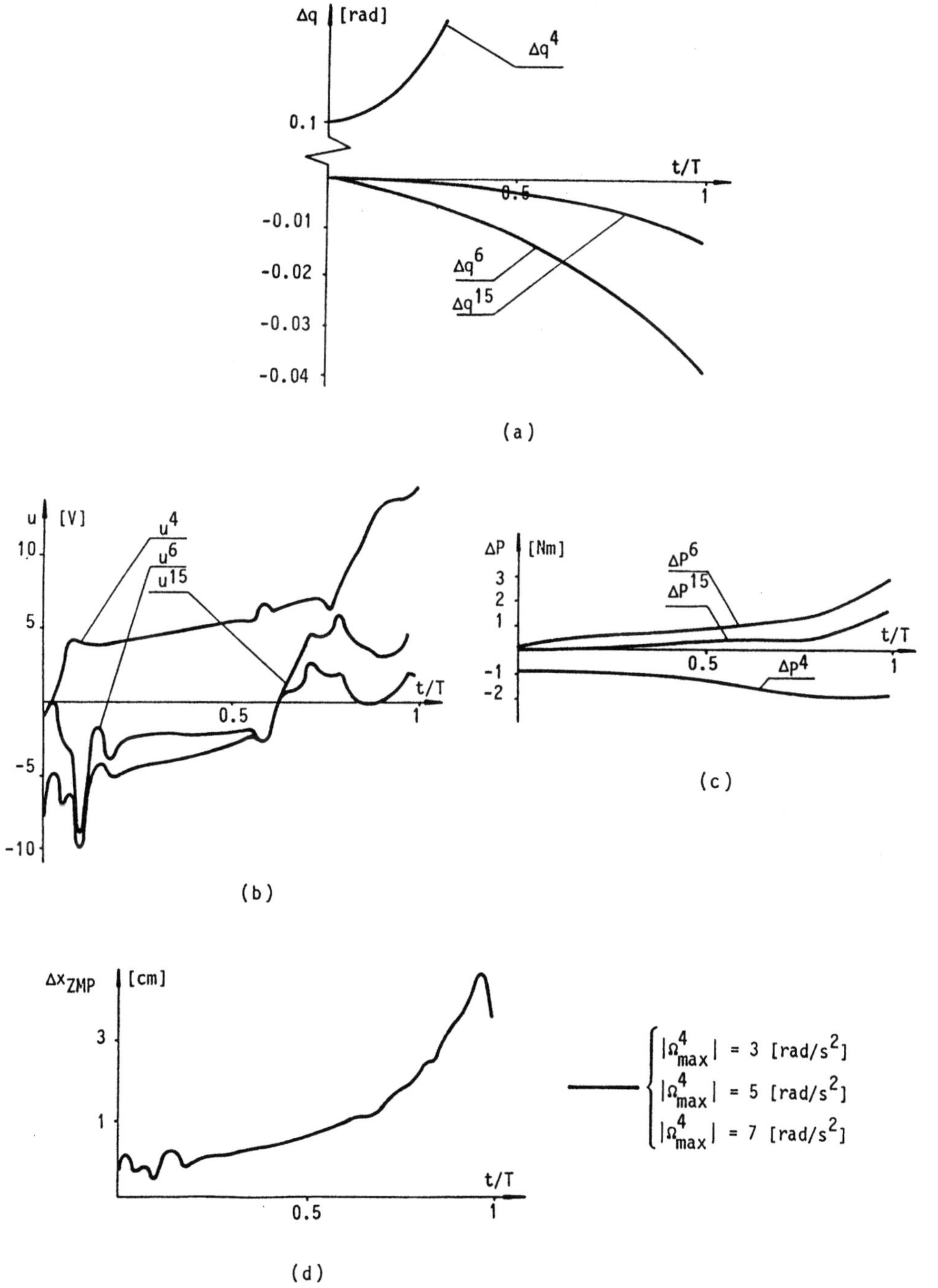

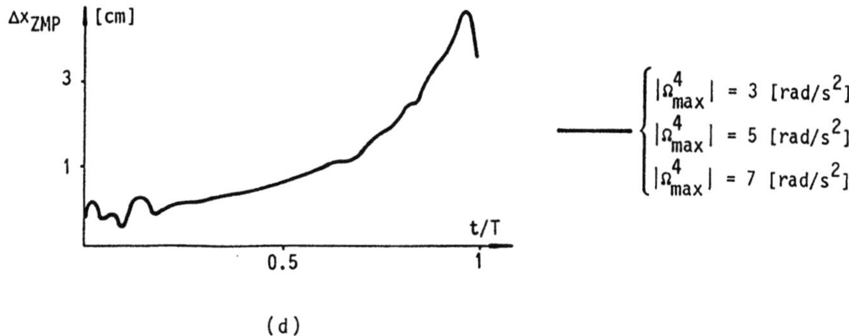

Fig. 3.36. Walk simulation, ZMP deviation compensated by ankle joint, $k_{ZMP}^{Gk} = 0.5$, $\Delta q^4(0) = 0.1$ [rad]

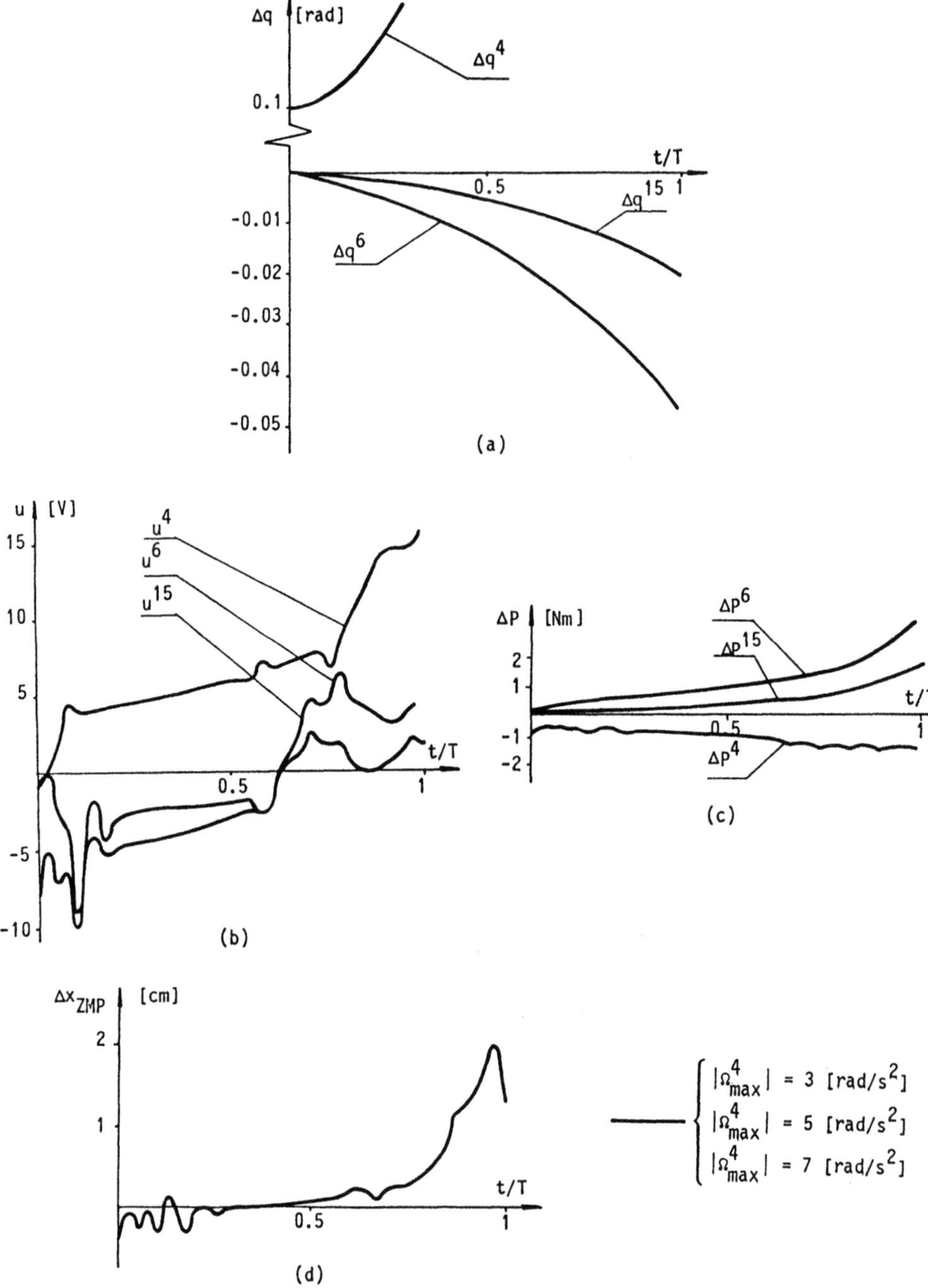



Fig. 3.37. Walk simulation, ZMP deviation compensated by ankle joint, $k_{ZMP}^{Gk} = 1.$, $\Delta q^4(0) = 0.1$ [rad]

behaviour. These information are indispensable for observing the motion of a particular joint and of the mechanism's behaviour when the overall stability is concerned, but do not enable an easy monitoring of the mechanism's motion with respect to the external (Cartesian) coordinate frame.

The quality of "external motion" is primarily judged by its anthropomorphic shape, i.e. by the resemblance of the motion to the human pattern. As in the nominal control synthesis, at the level of perturbed regimes repeatability conditions have to be fulfilled as best as possible. Any deviation from the nominal value at the end of a step is an initial disturbance for the subsequent one. To reduce these deviations, it is essential for a continuous walk. Obviously, the minimization all deviations brings the system closer to the nominal motion, which satisfies all requirements, including the anthropomorphicity. But, if the angular deviation exceeds an allowable range, even at a single joint, it may cause an unacceptable mechanism posture, and some additional corrections may be needed.

To illustrate the mechanism "external" motion the four examples of stick diagrams are presented. In two of them, initial disturbance is assigned to the trunk joint, $\Delta q^{15}(0) = 0.2$ [rad] (Fig. 3.38. and Fig. 3.39), and in the two others to the ankle joint $\Delta q^4(0) = 0.1$ [rad] (Fig. 3.40. and Fig. 3.41).

Consider first the cases when the initial disturbance is assigned to the trunk. In both figures, the nominal trunk position is shown by the dashed line. The stick diagram in Fig. 3.38. corresponds to the simulation presented in Fig. 3.33. for $\Omega_{max}^{15} = 7$ [rad/s$^2$]. The compensation of ZMP displacement is performed by the ankle joint. The stick diagram in Fig. 3.39. illustrates the case previously presented in Fig. 3.32, also for $\Omega_{max}^{15} = 7$ [rad/s$^2$]. The ZMP displacement is maintained by the hip. From this example it can be concluded that the hip compensating motion causes some irregularities in the legs motion. At the end of the considered period of time, the mechanism is not ready for supporting foot switching. The gait has to be continued in single support phase till legs disposition is appropriate to start the double support phase.

The next two stick diagrams (Figs. 3.40. and 3.41) illustrate the mechanism behaviour for the initial disturbance assigned to the ankle joint. Fig. 3.40. shows the case when the ZMP deviations are compensated by the ankle joint, and Fig. 3.41. shows compensation by trunk joint.

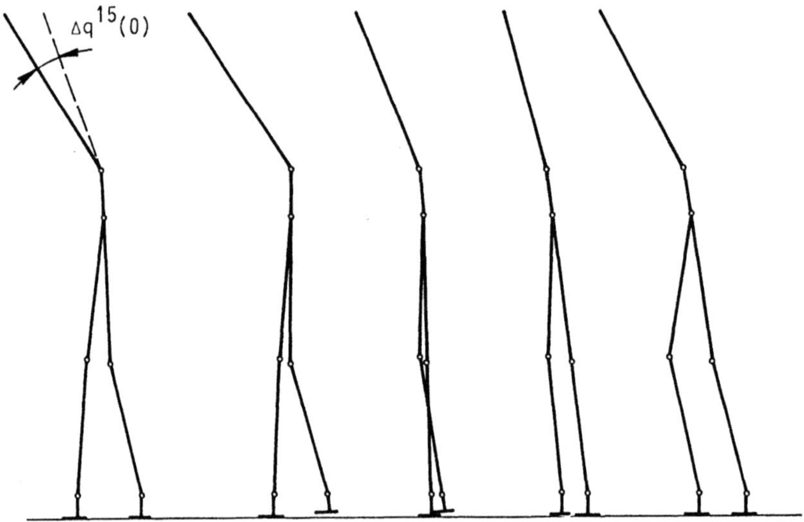


Fig. 3.38. Stick diagram of the walk simulation presented at Fig. 3.34. for $|\Omega_{max}^{15}|=7$ [rad/s]. ZMP displacement compensated by ankle, $\Delta q^{15}(0) = 0.2$ [rad]

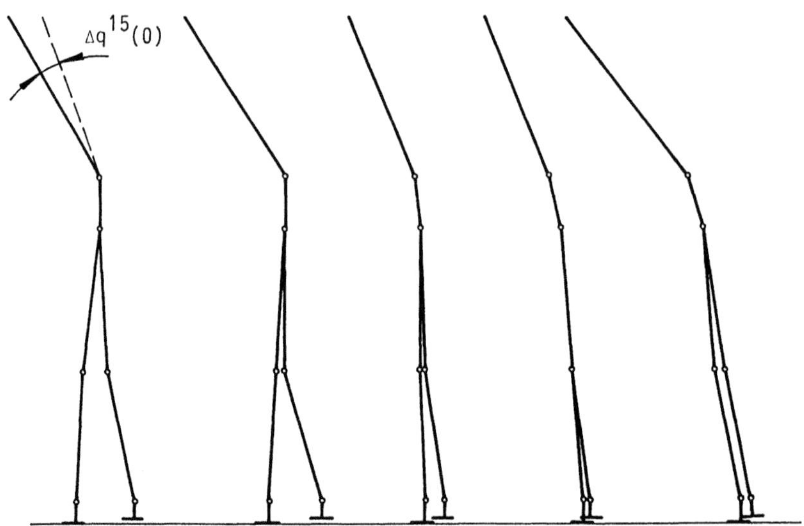


Fig. 3.39. Stick diagram of the walk simulation presented at Fig. 3.32. for $|\Omega_{max}^{15}|=7$ [rad/s$^2$]. ZMP displacement compensated by hip, $\Delta q^{15}(0) = 0.2$ [rad]

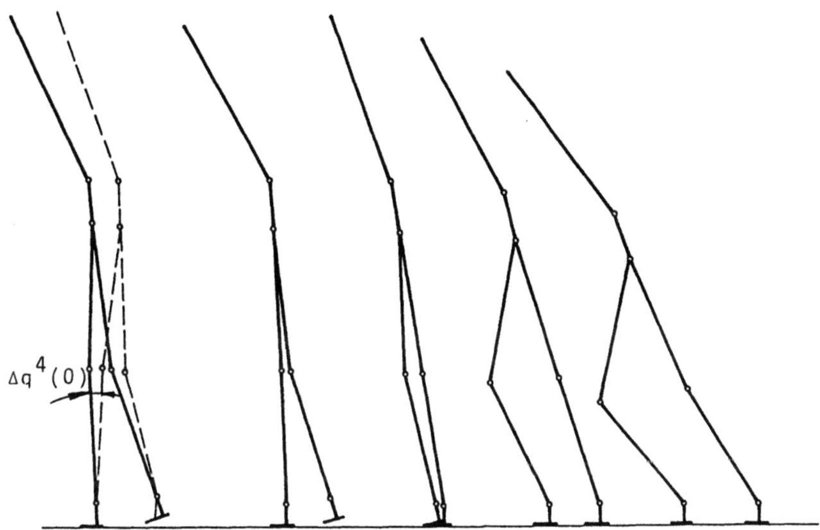



Fig. 3.40. Stick diagram of the walk simulations presented at Fig. 3.37. for $|\Omega^4_{max}|$=7 [rad/s$^2$]. ZMP displacement compensated by ankle, $\Delta q^4(0)$=0.1 [rad]

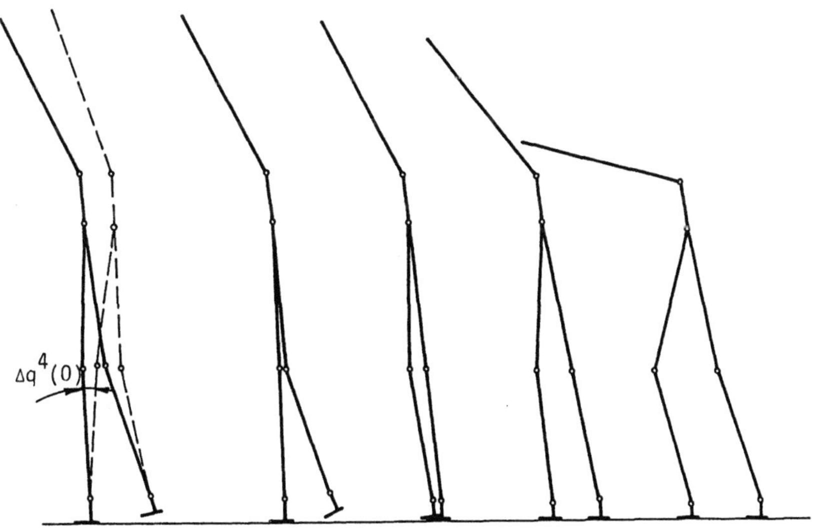



Fig. 3.41. Stick diagram of walk simulation presented at Fig. 3.30. for $|\Omega^4_{max}|$=7 [rad/s$^2$]. ZMP displacement compensated by trunk, $\Delta q^4(0)$ = 0.1 [rad]

Fig. 3.40. illustrates the simulation already presented in Fig. 3.37, while stick diagram in Fig. 3.41. corresponds to simulation presented in Fig. 3.30. In both cases, the dashed line shows the nominal mechanism position at the initial moment.

In both cases, the relatively large deviations from the anthropomorphic pattern are observed. The entire mechanism in Fig. 3.40. is inclined forward to a large extent, in spite of an excelent ZMP behaviour (Fig. 3.37). In Fig. 3.41, an extremely large trunk inclination, occurred at the end of the step, which spoiled the gait anthropomorphicity. In both cases, the terminal mechanism positions are not acceptable as the initial positions for the next step. Such large deviations from the nominal at the ankle joint are caused by strong initial disturbance. The angular deviations of the ankle joint cause a displacement of the entire mechanism's part above the ankle. The motion of the ankle joint, required by compensation of ZMP deviations, results in such nonanthropomorphic behaviour of the biped system.

These examples show that when the perturbed motion of the biped system around the nominal (desired) gait is considered, some serious problems may arise in regard to the mechanism movement in respect to the external coordinate frame. If the mechanism posture exceeds the admissible range (i.e. if the disturbance acting upon the system is too strong) the control strategy has to be changed, to prevent the mechanism motion from becoming nonanthropomorphic. This means, when some large deviations of the joint angles from the nominal trajectories appear, the system may completely change the strategy, i.e., the biped can step out in order to enlarge the stability margin by the double-support posture [49]. The control proposed in the previous section is valid for small perturbations, while for the large perturbations the system cannot track the desired nominal trajectory and this has to be changed. After the change of nominal trajectory, when the system is stabilized and its collapsing is avoided, the system may be driven back to the original nominal trajectory (and the nominal desired gait may be continued). However, here we shall not deal with the system's accomodation to large perturbations.

## 3.5. Stability Analysis

Up to now we have considered synthesis of control for biped systems
for tracking of desired nominal trajectories (which have been computed
to ensure stable nominal gait). We have proposed several control laws
which enable tracking of the desired trajectories and maintenance of
the system equilibrium during this tracking. In the previous section
we have shown, by simulation, the efficiency of these control laws.
However, in order to prove the validity of the control law proposed,
we have to analyze the stability of the entire mechanical system when
the selected control is applied. The locomotion systems are highly
nonlinear and therefore their stability analysis requires application
of complex methods for stability analysis of large-scale nonlinear
systems.

In this section we shall analyze stability of the locomotion systems
when various control laws are applied. The purpose of this analysis is
not only to prove the validity of the proposed control laws, but it
also may be used to select control parameters (feedback gains etc.).
In other words, we may use this analysis to establish an iterative algorithm for computer-aided synthesis of control law and control parameters.

To analyze stability of the locomotion mechanisms we use the aggregation-decomposition method via Lyapunov vector functions in bounded regions of state space which originally has been developed for manipulation robots [31]. As it is valid for the mechanisms with all joints
powered, this method cannot be directly applied to the locomotion mechanisms containing unpowered d.o.f. Because of that, we modify the syb-
systems modelling by incorporating the models of unpowered d.o.f. into
the composite subsystems models. In this way, the complete mechanism
is considered in stability analysis.

The analysis of the system with all powered joints is performed in the
following way. The system is considered as a set of subsystems each of
which is associated with one powered joint (joint with a corresponding
actuator). First, the stability of each subsystem is checked (neglecting the coupling) and then, dynamic coupling between the subsystems
is analyzed. The tests for stability of the overall system are established by taking into account all dynamic interconnections between the
subsystems. However, these tests require that all subsystems are stable.

In order to analyze stability of the mechanisms which include unpowered joints, we introduce the so-called "composite" subsystems which
consist of one unpowered and one powered joint. Thus we obtain a subsystem which, if considered as decoupled from the rest of the system,
might be stabilized. Further, the interconnections of the "composite"
subsystem with the rest of subsystems are taken into account, and
tests for stability of the overall mechanism are established.

In the text to follow we shall present in detail the proposed approach
to stability analysis [41].


### 3.5.1. Modelling of composite subsystems

As was already said in the previous section, the mathematical model of
the complete system S consists of two parts: the model of mechanical
structure $S^M$, and the model of actuators $S_a^i$. These models are:

$$S^M: \quad P = H(q) \cdot \ddot{q} + h(q, \dot{q}) \tag{3.3.1}$$

$$S_a^i: \quad \dot{x}_c^i = A_c^i \cdot x_c^i + b_c^i \cdot N(u^i) + f_c^i \cdot P_c^i \tag{3.3.2}$$

The notation is the same as before.

The mechanical structure of n d.o.f. is powered by m (m>n/2) actuators
(i.e. DC motors). Since (n-m) joints are unpowered, the driving torques $P^i$ around the axes of these joints are assumed to be zero, i.e.
the vector of driving torques P has the following form P = ($P_c^1$, $P_c^2$,...
...,$P_c^m$, 0,...,0)$^T$. The model of the mechanical part (3.3.1) and the
models of actuators (3.3.2) represent the complete dynamic model of
the system S (active spatial mechanism having unpowered joints). In
this models, no approximation has been made (save that we assume the
links are rigid bodies).

In order to apply the method for stability analysis we shall rearrange
this model in another way. Namely, we should write this model as a set
of subsystems that can be stabilized if they are considered free from
coupling with the rest of the system.

The model of the $\ell$-th unpowered joint follows from (3.3.1):

$$-H_{\ell\ell}\ddot{q}_N^{\ell} = \sum_{\substack{j=1 \\ j \neq \ell}}^{n} H_{\ell j}\ddot{q}^{j} + h_{\ell}(q, \dot{q}) \qquad (3.5.1)$$

where $q_N^{\ell} \in R^1$ is the angle of the $\ell$-th unpowered joint, $H_{\ell j}$ are the members of matrix $H(q)$, and $h_{\ell}$ is the member of vector $h$. The subscript $N$ denotes unpowered d.o.f. However, instead of this model, let us describe the motion of the system around the axis of the $\ell$-th unpowered joint as the motion of an inverted pendulum. In fact, the motion of locomotion mechanisms rotating around the foot edge is the most similar to the motion of a massive inverted pendulum.

The equation of inverted pendulum motion in a plane is [2]

$$\ddot{q}_N^{\ell} = \frac{M}{I_o + M\rho^2} g \sin q_N^{\ell} + \frac{1}{I_o + M\rho^2} P_N^{\ell} \qquad (3.5.2)$$

where $M$ and $I_o$ are the mass and inertia moment of the pendulum (in our case the pendulum corresponds to the whole system), $\rho$ is the distance from the supporting point to the pendulum mass centre, $P_N^{\ell} \in R^1$ is the resultant generalized force acting on the pendulum.

If we suppose that the angle $q_N^{\ell}$ is small we can introduce the approximation $(\sin q_N^{\ell}) \approx q_N^{\ell}$. If the term multiplying $q_N^{\ell}$ is denoted by $C_o^{\ell*}$, and the term multiplying $P_N^{\ell}$ by $\bar{f}_N^{\ell}$, and $x_N^{\ell} = [q_N^{\ell}, \dot{q}_N^{\ell}]^T$ is adopted as state vector, then (3.5.2) can be written in the matrix form

$$\begin{bmatrix} \dot{q}_N^{\ell} \\ \ddot{q}_N^{\ell} \end{bmatrix} = \begin{bmatrix} 0 & 1 \\ C_o^{\ell*} & 0 \end{bmatrix} \begin{bmatrix} q_N^{\ell} \\ \dot{q}_N^{\ell} \end{bmatrix} + \begin{bmatrix} 0 \\ \bar{f}_N^{\ell} \end{bmatrix} P_N^{\ell} \qquad (3.5.3)$$

what in compact form is

$$\dot{x}_N^{\ell} = A_N^{\ell} \cdot x_N^{\ell} + f_N^{\ell} \cdot P_N^{\ell}$$

Since we want the models (3.5.3) and (3.5.1) to coincide we, shall define the force $P_N^{\ell}$ as

$$P_N^{\ell} = \frac{-H_{\ell\ell}^{-1}}{\bar{f}_N^{\ell}} [\sum_{\substack{j=1 \\ j \neq \ell}}^{n} H_{\ell j}\ddot{q}^{j} + h_{\ell}(q, \dot{q})] - \frac{C_o^{\ell*}}{\bar{f}_N^{\ell}} q_N^{\ell} \qquad (3.5.4)$$

In this way we ensure that (3.5.3) is an exact model of the system

284

motion about the axis of unpowered joint. Since this joint has no actuator, it cannot be stabilized directly if it is considered as decoupled from the other joints. Let us now form the "composite" subsystem containing one unpowered d.o.f. (3.5.3) and the k-th powered d.o.f. (3.3.2). If we suppose the mechanical structure is powered by DC motors, the composite subsystem model can be arranged in a diagonal form of the models of unpowered and powered d.o.f.

$$
\begin{bmatrix} [\dot{x}_N^\ell] \\ [\dot{x}_C^k] \end{bmatrix} = \begin{bmatrix} [A_N^\ell] & 0 \\ 0 & [A_C^k] \end{bmatrix} \begin{bmatrix} [x_N^\ell] \\ [x_C^k] \end{bmatrix} + \begin{bmatrix} [f_N^\ell] & 0 \\ 0 & [f_C^k] \end{bmatrix} \begin{bmatrix} P_N^\ell \\ P_C^k \end{bmatrix} + \begin{bmatrix} 0 \\ [b_C^k] \end{bmatrix} N(u^k)
$$

$$(3.5.5)$$

The subscript N corresponds to the unpowered and the subscript c to the powered d.o.f. Here, $x_N^\ell \in R^{n_N^\ell}$ and $x_C^k \in R^{n_C^k}$ are the state vectors of the $\ell$-th unpowered $x_N^\ell = (q_N^\ell, \dot{q}_N^\ell)$ and k-th powered d.o.f. $x_C^k = (q_C^k, \dot{q}_C^k, i_R^k)$; $i_R^k$ is the rotor current of the corresponding DC motor, whereas $n_N^\ell = 2$ and $n_C^k = 3$ are their orders. $A_N^\ell \in R^{n_N^\ell \times n_N^\ell}$ and $A_C^k \in R^{n_C^k \times n_C^k}$, $f_N^\ell \in R^{n_N^\ell}$ and $f_C^k \in R^{n_C^k}$, $P_N^\ell \in R^1$ and $P_C^k \in R^1$ are the system matrices, force distribution vectors, and generalized forces of unpowered and powered d.o.f., respectively. $A_N^\ell$ and $f_N^\ell$ are defined by (3.5.3). Taking into account the form of actuator matrix $A_C^k$, and the form of model of unpowered d.o.f. (3.5.3), expression (3.5.5) can be written in an explicit form as

$$
\begin{bmatrix} [\dot{q}_N^\ell] \\ [\ddot{q}_N^\ell] \\ [\dot{q}_C^k] \\ [\ddot{q}_C^k] \\ [\dot{i}_R^k] \end{bmatrix} = \begin{bmatrix} 0 & 1 & 0 & 0 & 0 \\ C_o^* & 0 & 0 & 0 & 0 \\ 0 & 0 & 0 & 1 & 0 \\ 0 & 0 & 0 & a_{2,2}^k & a_{2,3}^k \\ 0 & 0 & 0 & a_{3,2}^k & a_{3,3}^k \end{bmatrix} \begin{bmatrix} q_N^\ell \\ \dot{q}_N^\ell \\ q_C^k \\ \dot{q}_C^k \\ i_R^k \end{bmatrix}
$$
$$
+ \begin{bmatrix} 0 & 0 \\ \bar{f}_N^\ell & 0 \\ 0 & 0 \\ 0 & \bar{f}_C^k \\ 0 & 0 \end{bmatrix} \begin{bmatrix} P_N^\ell \\ P_C^k \end{bmatrix} + \begin{bmatrix} 0 \\ 0 \\ 0 \\ 0 \\ \bar{b}_C^k \end{bmatrix} N(u^k)
$$

$$(3.5.6)$$

where $a_{i,j}^k$ are the elements of matrix $A_C^k$, or, in a compact form as

$$\dot{x}_z^k = A_z^k \cdot x_z^k + f_z^k \cdot P_z^k + b_z^k \cdot N(u^k) \qquad \forall k \in J$$

where $x_z^k \in R^{n_z^k}$ is the state vector of the composite subsystem, $A_z^k \in R^{n_z^k \times n_z^k}$, $f_z^k \in R^{n_z^k \times 2}$, $b_z^k \in R^{n_z^k}$ are the subsystem matrix, the matrix of force distribution and the vector of control distribution, respectively. Thus, $n_z^k = n_N^\ell + n_C^k$, $x_z^k = (x_N^{\ell T}, x_C^{kT})^T$ and $P_z^k = (P_N^\ell, P_C^k)^T$. Obviously, (3.5.6) defines only the k-th composite subsystem model. Set J is defined as $J = \{j, j = 2m-n+1, \ldots, m\}$. It is assumed that the k-th powered joint is associated with the $\ell$-th unpowered joint.

In the stability analysis, described in the next paragraph it is required that all decoupled subsystems are exponentially stable. If this is not fulfilled, the adopted aggregation-decomposition method does not work. If the subsystems correspond to joints of kinematic chain, their coupling are represented by the moments around the joint axis. In fact, decoupling means an investigation of the subsystem model without the term which corresponds to the generalized force. In the case of a composite subsystem this is the term $(f_z^k \cdot P_z^k)$, i.e. the decoupled composite subsystem can be written as

$$\dot{x}_z^k = A_z^k x_z^k + b_z^k N(u^k) \tag{3.5.7}$$

However, such a decoupled composite subsystem will decompose into two independent models of unpowered and powered d.o.f. having no interaction. The interaction between these d.o.f. is, in fact, the only way to control the motion of the unpowered d.o.f. In order to preserve the integrity of the decoupled composite model, some additional elements should be introduced into the matrix $A_z^i$ in the places which represent the influence of powered d.o.f. on the unpowered d.o.f. and vice versa. Then, the model of the composite subsystem is of the final form

$$
\begin{bmatrix} \begin{bmatrix} \dot{q}_N^\ell \\ \ddot{q}_N^\ell \end{bmatrix} \\ \begin{bmatrix} \dot{q}_C^k \\ \ddot{q}_C^k \end{bmatrix} \\ \begin{bmatrix} \dot{i}_R^k \end{bmatrix} \end{bmatrix}
=
\begin{bmatrix}
\begin{bmatrix} 0 & 1 \\ C_o^{\ell *} & 0 \end{bmatrix} & \begin{bmatrix} D_1^{1k} & D_1^{2k} & D_1^{3k} \end{bmatrix} \\
\begin{bmatrix} 0 & 0 \\ D_2^{1k} & D_2^{2k} \end{bmatrix} & \begin{bmatrix} 0 & 1 & 0 \\ 0 & a_{2,2}^k & a_{2,3}^k \\ 0 & a_{3,2}^k & a_{3,3}^k \end{bmatrix}
\end{bmatrix}
\begin{bmatrix} \begin{bmatrix} q_N^\ell \\ \dot{q}_N^\ell \end{bmatrix} \\ \begin{bmatrix} q_C^k \\ \dot{q}_C^k \end{bmatrix} \\ \begin{bmatrix} i_R^k \end{bmatrix} \end{bmatrix}
+
$$

$$+ \begin{bmatrix} 0 & 0 \\ \bar{f}_N^\ell & 0 \\ 0 & 0 \\ 0 & \bar{f}_c^k \\ 0 & 0 \end{bmatrix} \begin{bmatrix} P_N - \dfrac{D_1^k \cdot x_c^k}{\bar{f}_N^\ell} \\[2mm] P_c^k - \dfrac{D_2^k \cdot x_N^\ell}{\bar{f}_c^k} \end{bmatrix} + \begin{bmatrix} 0 \\ 0 \\ 0 \\ 0 \\ \bar{b}_c^k \end{bmatrix} N(u^k) \qquad (3.5.8)$$

or

$$\dot{x}_z^k = A_z^{k*} x_z^k + f_z^k P_z^{k*} + b_z^k N(u^k)$$

The vector $D_1^k = [D_1^{1k} \ D_1^{2k} \ D_1^{3k}]$ represents the influence of the powered d.o.f. on the unpowered one, whereas $D_2^k = [D_2^{1k} \ D_2^{2k}]$ represents an opposite effect. The elements of vectors $[D_1]$ and $[D_2]$ have to be chosen in such a way that the matrix $A_z^k$ has eigenvalues at the desired location in the left half of complex plane, what guarantees the desired degree of exponential stability of the "free" (decoupled) subsystem. Since the vectors $D_1^k$ and $D_2^k$ are chosen arbitrarily, $D_1^k \cdot [x_k]^T$ and $D_2^k \cdot [x_N^\ell]^T$ will be subtracted from the term $(f_z^k P_z^k)$, i.e. $P_z^{k*} = (P_N^{\ell*}, P_c^{k*})$, $P_N^{\ell*} = P_N^\ell - [(D_1^k \cdot x_c^k)/\bar{f}_N^\ell]$, $P_c^{k*} = P_c^k - [(D_2^k \cdot x_N^\ell)/\bar{f}_c^k]$.

The composite subsystem model formed in this way is suitable for stability investigation and enables a stability analysis of the system having joints without actuators. It should be emphasized that the models of "composite" subsystems (3.5.8) are exact, i.e. they contain no approximations. The model (3.5.8) coincides with the original model of the $\ell$-th unpowered joint (3.5.1) and the model of the k-th powered joint with the actuator (3.3.2) which is driving the k-th joint. We only rearranged the model in order to obtain it in a convenient form.

The mathematical model of the mechanism part which consists of composite subsystems is

$$\dot{x}_z = \hat{A}_z^* x_z + f_z P_z^* + b_z N(u_z) \qquad (3.5.10)$$

where $x_z \in R^{N_z}$ is a state vector; $x_z = (x_z^{(2m-n+1)T}, \ldots, x_z^{mT})^T$. $\hat{A}_z^* \in R^{N_z \times N_z}$, $\hat{A}_z^* = \mathrm{diag}\{A_z^{k*}\}$ is the system matrix, while $b_z = \mathrm{diag}\{b_z^k\}$ and $f_z = \mathrm{diag}\{f_z^k\}$, $b_z \in R^{N_z \times (n-m)}$, $f_z \in R^{N_z \times 2(n-m)}$, are the distribution matrices of control force; $N(u_z) \in R^{n-m}$ and $P_z^* \in R^{2(n-m)}$ are the corresponding control and force (defined by (3.5.8)), $N_z$ is the order of the model formed of composite subsystems,

$$N_z = \sum_{k=2m-n+1}^{m} n_z^k \qquad (3.5.11)$$

Thus, the mathematical model of a complete biped mechanism S with composite subsystems included, can be obtained by uniting the model of composite subsystems (3.5.10), (including (n-m) unpowered and (n-m) powered d.o.f.) and the rest (2m-n) powered d.o.f.

$$S: \quad \dot{x} = Ax + FP + BN(u) \qquad (3.5.12)$$

where $x \in R^N$, $x = (x_c^{1T}, \ldots, x_c^{(n-m)T}, x_z^T)^T$ is the system state vector $P = (P_c^1, P_c^2, \ldots, P_c^{2m-n}, P_z^{2m-n+1*T}, \ldots, P_z^{m*T})^T$. Matrices $A \in R^{N \times N}$, $B \in R^{N \times m}$ and $F \in R^{N \times n}$ are of the form

$$A = \begin{bmatrix} \hat{A}_c & 0 \\ \hline 0 & \hat{A}_z \end{bmatrix}, \qquad B = \begin{bmatrix} \hat{b}_c & 0 \\ \hline 0 & b_z \end{bmatrix}, \qquad F = \begin{bmatrix} \hat{f}_c & 0 \\ \hline 0 & f_z \end{bmatrix}$$

$$N = N_z + \sum_{i=1}^{2m-n} n_i$$

$\hat{A}_c = \text{diag}[A_c^i]$, $\hat{b}_c = \text{diag}[b_c^i]$ and $\hat{f}_c = \text{diag}[f_c^i]$, $\forall i \in I_2$, $I_2 = \{i, i=1, \ldots \ldots, 2m-n\}$. In this way, a mathematical model of the complete system is formed which will be used in the stability analysis.

The complete system S (3.5.12) is obviously composed of m subsystems: (2m-n) subsystems correspond to the powered joints modelled as in (3.3.2), and (n-m) composite subsystems modelled as in (3.5.10). In fact, all the subsystems (powered and composite) can be written in the same form:

$$\dot{x}^i = A^i x^i + b^i N(u^i) + f^i \hat{P}_i(x), \qquad \forall i \in I_1 \qquad (3.5.13)$$

where $x^i$ stands for $x_c^i$ if $i = 1, 2, \ldots, 2m-n$, and for $x_z^i$ if $i = 2m-n+1, \ldots, m$. The same holds for $A^i$, $b^i$, $f^i$ and $u^i$, while $\hat{P}_i$ stands for $P_c^i$ if $i = 1, 2, \ldots, 2m-n$, and for $P_z^{i*}$ if $i = 2m-n+1, \ldots, m$. The order of subsystems (3.5.13) are denoted by $n_i$ (though, it might be either $n_i$, or $n_z^i$, depending on i).

Thus we obtain a model of the system S in the form which is convenient for application of the chosen method of stability analysis. The model

288

(3.5.13) is complete and it coincides with the original models (3.3.1) and (3.3.2).


## 3.5.2. Stability analysis

In regard to the task imposed on locomotion systems, which is in detail described in Paragraph 3.4.1, the most suitable for stability analysis seems to be the definition of practical stability [31]. Accordingly, the system is considered to be practically stable if $\forall x(0) \in X^I$ implies $x(t) \in X^F$, $\forall t \in T_s$, where $T_s = \{t: t \in (\tau_s, \tau)\}$, and $x(t) \in X^t(t)$, $\forall t \in T$ where $X^I \subseteq X^t(0)$ and $X^F \subseteq X^t(t)$, $\forall t \in T_s$. As already explained, the two-stage approach to control synthesis has been adopted.

At the first level, the level of nominal regimes, such system trajectories and corresponding control are synthesized to ensure the system's motion (in the absence of disturbances) satisfies all conditions and requirements imposed by the specific nature of biped locomotion. If, however, a disturbance occurred, the system state would deviate from its nominal, and, the practical stability about nominal trajectory is to be considered.

Now, we are going to present the aggregation-decomposition method for stability analysis via Lyapunov's function in bounded regions of the state space. This method has been already used for robots having all joints powered [31]. In the previous paragraph we proposed such mathematical modelling of composite subsystems, including also unpowered d.o.f., which enables the application of this method to biped locomotion systems.

Let us consider the overall system model S defined as in (3.5.13), which can be considered as a set of m subsystems $S^i$ (either of the composite or powered joints) which are coupled through the term $(f^i \cdot \hat{P}_i)$,

$$\dot{x}^i = A^i x^i + b^i N(u^i) + f^i \hat{P}_i(x), \qquad \forall i \in I_1$$

Let us assume the nominal trajectory of the state vector $x^o(t)$, $\forall t \in T$ is given in such a way that it satisfies $x^o(0) \in X^I$, $x^o(t) \in X^F$, $\forall t \in T_s$ and $x(t) \in X^t(t)$, $\forall t \in T$. Further, let us assume the nominal trajectory $x^o(t)$ has been selected in such a way that we can find a nominal (programmed) control $u^o(t)$, which is a function of time, and which satisfies

$$\dot{x}^{Oi} = A^i x^{Oi} + b^i N(u^{Oi}) + f^i \hat{P}^O_i(x^O), \quad \forall i \in I_1, \ \forall t \in T \qquad (3.5.14)$$

where $x^O(t) = (x^{O1T}(t), x^{O2T}(t), \ldots, x^{OmT}(t))^T$, $u^O = (u^{O1}, u^{O2}, \ldots, u^{Om})^T$. Here, $\hat{P}^O_i(x^O)$ denotes nominal values of $\hat{P}_i(x)$. Because the subsystems (3.5.13) include the composite subsystems, this means that the nominal trajectory $x^O(t)$ satisfies the composite subsystems. In other words, the nominal trajectory $x^O(t)$ must satisfy equilibrium of the system under the nominal conditions. If the biped system is considered, the nominal trajectory must ensure equilibrium of the biped dynamics during the gait (i.e. during the transfer of the system state from the region $X^I$ to region $X^F$). The synthesis of such nominal trajectories has been already considered in detail in Chapter 2 [22]. Here we assume that the nominal trajectory $x^O(t)$ and the corresponding nominal control $u^O(t)$, satisfying (3.5.14), can be determined.

However, due to the perturbation actions upon the system, a deviation of the system state from its nominal trajectory must appear. The model of deviation from the nominal trajectory can be written (according to (3.5.13) and (3.5.14)) as

$$\Delta\dot{x}^i = A^i \Delta x^i + b^i N(t, \Delta u^i) + f^i \Delta\hat{P}_i(t, \Delta x, x^O(t)), \quad \forall i \in I_1 \qquad (3.5.15)$$

where $\Delta x^i = x^i - x^{Oi}(t)$, $\Delta u^i = u^i - u^{Oi}(t)$, $\Delta\hat{P}_i = \hat{P}_i - \hat{P}^O_i(x^O)$. Now, the problem is to stabilize the model of deviation (3.5.15) from the nominal trajectory $x^O(t)$, i.e. we have to synthesize the control $\Delta u^i$ such that the model of deviation from $x^O(t)$ (3.5.15) is stabilized. The aim is to ensure practical stability of the system around the nominal trajectory $x^O(t)$, such that for each $\Delta x(0) \in X^I - x^O(0)$ it is fulfilled $\Delta x(t) \in X^F - x^O(t)$, $\forall t \in T_s$, and $\Delta x(t) \in X^t(t) - x^O(t)$, $\forall t \in T$.

Let us synthesize a decentralized control. To do this let consider an approximate model of deviation in its decoupled form (i.e. the model in which the coupling terms between subsystems ($f^i \Delta\hat{P}_i$) are neglected):

$$\Delta\dot{x}^i = A^i \Delta x^i + b^i N(t, \Delta u^i), \quad \forall i \in I_1 \qquad (3.5.16)$$

The decoupled model of system (3.5.16) represents a set of decoupled linear subsystems which can be stabilized by simple linear feedback control

$$\Delta u^i = -k^{LT}_i \Delta x^i, \quad \forall i \in I_1 \qquad (3.5.17)$$

290

where $k_i^L \in R^{n_i}$, is the vector of local feedback gains. We select feedback gains such that the subsystem

$$\Delta \dot{x}^i = (A^i - b^i k_i^{LT}) \Delta x^i = \tilde{A}^i \Delta x^i, \qquad \forall i \in I_1 \qquad (3.5.18)$$

(where $\tilde{A}^i$ is a closed-loop subsystem matrix) is exponentially stable. In (3.5.18) we have neglected the amplitude saturation upon the input $N(t, \Delta u^i)$. If this nonlinearity is taken into account, it can be shown that subsystem (3.5.18) is exponentially stabilized in the finite region $X_i$ in the state space, with a desired stability degree $\Pi_i$. If decoupled subsystems (3.5.18) are considered, it is obvious that this model will be exponentially stable in the region

$$X = X_1 \times X_2 \times \cdots \times X_n. \qquad (3.5.19)$$

Now we shall analyze stability of the complete system (3.5.15) if a decentralized control (3.5.17) is applied. We are going to use the method of analysis of asymptotic stability of complex systems in finite regions, which has been proposed by Weissenberger [47].

Let us express the subsystems characteristics by the Lyapunov functions (positive definite functions of $\Delta x^i$). The Lyapunov functions and their derivatives along solutions for decoupled subsystems have to satisfy

$$\Pi_{i1} ||\Delta x^i|| < v_i(\Delta x^i) < \Pi_{i2} ||\Delta x^i|| \qquad (3.5.20)$$

$$-\Pi_{i3} ||\Delta x^i|| < \dot{v}_i(\Delta x^i) \Big|_{\text{along solution of (3.5.18)}} < -\Pi_{i4} ||\Delta x^i|| \qquad (3.5.21)$$

for $\forall i \in I_1$, $\Pi_{ik} > 0$ are real numbers, for $k = 1, 2, 3, 4$, $v_i > 0$, $v_i : R^{n_i} \to R^1$. The analysis concerning the stability on finite regions using aggregation - decomposition method can be conservative. Therefore, it was shown [43] that functions $v_i$ should be chosen in such a way to be the best estimates of the degree of exponential stability $\Pi_i$ of the decoupled subsystems. Thus [43], we should select such Lyapunov function $v_i$ which satisfies:

$$\dot{v}_i(\Delta x^i) = (\text{grad } v^i)^T \Delta x^i \leq -\Pi_{i4} \Pi_{i2}^{-1} v_i \leq -\Pi_i v_i, \qquad \forall i \in I_1 \qquad (3.5.22)$$

where $\dot{v}_i$ is taken along the trajectory of decoupled subsystem (3.5.18).

Let us select the Lyapunov function in the form

$$v_i = (\Delta x^{iT} H^i \Delta x^i)^{1/2}, \qquad \forall i \in I_1 \tag{3.5.23}$$

where matrix $H^i \in R^{n_i \times n_i}$ (symmetric and positive definite) can be derived as the solution of the Lyapunov matrix equation [44]

$$\tilde{A}^{iT} H^i + H^i \tilde{A}^i = -G^i \tag{3.5.24}$$

where $G^i \in R^{n_i \times n_i}$ is an arbitrarily defined, symmetric and positive definite matrix. If we select $G^i$ to be equal to $\pi_i H^i$, then the selected Lyapunov function (3.5.23) obviously satisfies (3.5.22). Since the control signal is of limited amplitude, the condition (3.5.22) can be satisfied only in the finite region of initial conditions $x_i(0) \in X_i$, i.e. the decoupled system is asymptotically stable in the region X, defined by (3.5.19). The region $X_i$ can be estimated via Lyapunov functions by regions $\tilde{X}_i$ with an adequate choice of constants $v_{io}$

$$\tilde{X}_i = \{\Delta x^i: v_i(\Delta x^i) < v_{io} \text{ and } \Delta x^i \in X_i\}, \quad \forall i \in I_1, \quad \tilde{X}_i \subseteq X_i \tag{3.5.25}$$

where $v_{io} > 0$ are positive numbers. Then, the region

$$\tilde{X}(0) = \tilde{X}_1 \times \tilde{X}_2 \times \cdots \times \tilde{X}_m, \qquad \tilde{X}(0) \subseteq R^N \tag{3.5.26}$$

is the best estimate of the region of asymptotic stability X of the set of decoupled subsystems (3.5.18). However, in (3.5.18) we have neglected the coupling terms $(f^i \cdot \Delta \hat{P}_i)$.

Now it should be investigated how the coupling influences the stability of the overall system S. Since $\lim_{\Delta x \to 0} \Delta \hat{P}_i \to 0$ the coupling influence can be estimated by the numbers $\xi_{ij}$ ($\xi_{ij} \geq 0$ for $i \neq j$) which satisfy [45, 46]

$$(\text{grad } v_i)^T f^i \Delta \hat{P}_i(t, \Delta x) \leq \sum_{j=1}^{m} \xi_{ij} v_j, \quad \forall i \in I_1, \quad \forall t \in T, \quad \forall \Delta x \in \tilde{X} - x^o(t) \tag{3.5.27}$$

A sufficient condition that the whole system S is asymptotically stable in the region $\tilde{X}(0)$ is [47]

$$G v_o < 0 \tag{3.5.28}$$

where $v_o$ is the $m \times 1$ vector, and $v_o = (v_{1o}, \ldots, v_{mo})^T$, $v_o \in R^m$ and the elements of the $m \times m$ matrix G are defined as

$$G_{ij} = -\Pi_i \delta_{ij} + \xi_{ij}. \tag{3.5.29}$$

Here, $\delta_{ij}$ is the Kronecker symbol.

It is necessary to point out that (3.5.28) is only a sufficient, but not necessary condition. If this condition is not fulfilled, we cannot say anything about the system stability. If (3.5.28) is fulfilled, then $\tilde{X}(0)$ is an estimate of the region of the overall system stability. Then, it is possible to estimate the region $\tilde{X}(t)$ which contains the system state during the tracking of the nominal trajectory $x^o(t)$ by

$$\max_{i \in I_1} (v_i(\Delta x^i(t))/v_{io}) < \max_{i \in I_1} (v_i(\Delta x^i(0))/v_{io}) \exp(-\beta t) \tag{3.5.30}$$

where $\beta > 0$ can be computed from

$$\beta = \min_{i \in I_1} (-v_{io}^{-1} \sum_{j=1}^{m} G_{ij} v_{jo}) = \min_{i \in I_1} (\beta_i) \tag{3.5.31}$$

where $\beta_i = -v_{io}^{-1} \sum_{j=1}^{m} G_{ij} v_{jo}.$

Inequality (3.5.30) is an estimation of shrinkage of the region $\tilde{X}(t)$ which contains a solution of system S. Now the practical stability of the system can be checked. If

$$x^I \subseteq \tilde{X}(0) \text{ and } \tilde{X}(t) \subseteq x^t(t), \; \forall t \in T, \; \tilde{X}(t) \subseteq x^F(t), \; \forall t \in T_s \tag{3.5.32}$$

is satisfied, then it can be stated that the system S is practically stable around the nominal $x^o(t)$.

It should be mentioned that the practical stability can also be analyzed in a direct way [48].

If the local linear feedback controllers defined by (3.5.17), are not sufficient to stabilize the system, and additional control input should be introduced. The task of this control is to take care of the overall system characteristics and behaviour. The feedbacks have to minimize the destabilizing influence of the coupling and to prevent rotation of the entire system about the foot edge. As we have explained in Section 3.4. we may introduce the global control in the form ($\Delta \hat{P}_i^*$ instead $\Delta P_c^{i*}$ is used)

$$\Delta u_i^G = k_{i4}^G \Delta \hat{P}_i^* + k_{i5}^G \tag{3.5.33}$$

where $k^G_{14}$ and $k^G_{15}$ are the scalar gains which are defined in (3.4.10). Here, $\Delta\hat{P}^*_i$ represents a value which corresponds to the coupling $\Delta\hat{P}_i$. Two ways of obtaining $\Delta\hat{P}^*_i$ have been suggested in Paragraph 3.4.2: a) The driving torques around joint axes are directly measured by force transducers at the joints, and the force feedback is established as the global control (i.e. $\Delta\hat{P}^*_i = \Delta\hat{P}_i$), b) The driving torques $\Delta\hat{P}^*_i(t)$ are computed on-line and the global control is obtained via the on-line computation of the mechanism dynamics. These two ways for obtaining $\Delta\hat{P}^*_i$ have been discussed in [31, 48]. In both cases, $\Delta\hat{P}^*_i$ corresponding to both the powered joints and composite subsystems can be obtained. Namely, by measuring forces at the contact point between the sole of the supporting leg and the ground, we get information on the effects of coupling upon the unpowered joint $\Delta P^\ell_N$. Therefore, we can establish a global control from the unpowered joint to the one of the powered joints (i.e. to its actuator) and by this to compensate for the effects of coupling upon the unpowered joint, as we have explained in Paragraph 3.4.3. In fact, the role of this global control is to stabilize the entire composite subsystem. Therefore, the global feedback loop is introduced between the unpowered joint and that powered joint with which the composite subsystem has been formed. The second way to obtain information on $\Delta P^\ell_N$ (coupling upon the unpowered joint) is by the on-line computation of these moments, using either the complete or some approximate model of the system dynamics. In these ways, global control $\Delta u^G_i$ might be applied at powered joints and in composite subsystems to reduce the destabilizing effects of couplings between the subsystems. If a global control is introduced, the stability analysis can be performed as described above. However, the numbers $\xi^*_{ij}$ estimating coupling are now defined to satisfy the following inequalities (instead of (3.5.27))

$$(\text{grad } v_i)^T f^i \Delta\hat{P}_i + (\text{grad } v_i)^T b^i \Delta u^G_i \leq \sum_{j=1}^{m} \xi^*_{ij} v_j, \quad \forall i \in I_1, \quad \forall t \in T,$$

$$\forall \Delta x \in \tilde{X} - x^0(t) \quad (3.5.34)$$

The next step is to check conditions (3.5.28), i.e. to test whether the system with applied local and global control is asymptotically stable in the region $\tilde{X}(0)$. In doing this, the numbers $\xi_{ij}$ in (3.5.29) have to be replaced by numbers $\xi^*_{ij}$. The role of the global control is to compensate for the effects of coupling upon the global system stability. Therefore, if the global control is properly selected, then the numbers $\xi^*_{ij}$ have to satisfy

$$\xi^*_{ij} \leq \xi_{ij}, \quad \forall i,j \in I_1$$

Therefore, the fulfillment of stability test has to be easier if the global control is introduced, than if only the local control is applied.


### 3.5.3. Stability analysis of a specific biped structure

The proposed approach to stability analysis of the mechanisms containing unpowered d.o.f. will be applied to a biped. The contact of a foot and the ground surface is characterized by the presence of several d.o.f. (one, two, or three, depending on the adopted model complexity) and none of them is powered. Their influence on stability of the system as a whole is of paramount importance. Namely, it is quite possible that the system realizes exactly the desired trajectories of all powered joints, and at same time, falls down by rotating about the foot edge. Obviously, the control system has to hinder this situation and prevent the system from collapsing.

The nominal trajectories are synthesized using prescribed synergy method. The legs trajectories are prescribed in advance (adopted from the human pattern) while trajectories of the upper part of the body (trunk with fixed arms) are calculated to ensure the desired position of ZMP. For such nominal trajectories $x^o(t)$, the nominal control $u^o(t)$ is computed which has to ensure their exact tracking if there are no perturbations acting upon the system. This synthesis of nominal trajectories has been presented in detail in Chapter 2.

At the level of perturbed regime, such control is derived to force the biped system state vector to its nominal value and, at the same time, to prevent the system from collapsing because of the presence of unpowered d.o.f. [39]. Thus, as already explained in Section 3.4. the control input to the i-th actuator consists of two parts

$$u^i = u^{oi} + \Delta u^i \qquad (3.5.35)$$

where $u^{oi}$ is the nominal control input to the i-th actuator while $\Delta u^i$ is the corrective input to the same actuator, synthesized at the level of perturbed regimes. The corresponding feedback gains are given by (3.4.10).

The control law (3.4.9) holds for the subsystems $i = (1,2,...,2m-n)$, and a similar control is derived for the composite subsystems, taking into account that $\Delta\hat{P}_i^*$ for composite subsystems are the vectors of dimension $(2\times 1)$.

In (3.4.10), the part depending only on local states of the i-th joint corresponds to the local and the rest to the global control. The global control is introduced in the form of feedback with respect to both the driving torques $\Delta\hat{P}_i^*$, and the bang-bang part $k_{i5}^G$. Here, $\Delta\hat{P}_i^*$ represents the force feedback (i.e. the measured moments about joints). An additional feedback respecting ZMP position, defined by (3.4.18), is also available as explained in Paragraph 3.4.3.

The scheme of adopted mechanical structure of the locomotion mechanism is given in Fig. 3.42. and corresponding parameters are listed in Table 3.4. Again, the mechanism has twenty d.o.f., but only eight of them are powered and involved in the gait. As before, there is one active d.o.f. per each leg joint, and two of them for trunk motion. The rest of them are fixed. In the examples below each powered joint is modelled as one subsystem; the composite subsystem comprises the models of one powered and one unpowered joint. Thus, the inactive d.o.f. are not included into subsystems modelling. To make the examples of stability analysis easier to follow, a redrown scheme of the same mechanism is presented in Fig. 3.43, with only those d.o.f. which will be included in the stability analysis. All the joints represented by the unit rotational axes $\vec{e}_i'$ $(i=1,\ldots,9)$, and the corresponding links are re-enumerated, and this notation will hold for the rest of this section. Let us note that the link representing the upper body comprises the trunk and both hands. Redistribution of subsystems models used in stability analysis is schematically shown in Fig. 3.43. We shall investigate the overall system stability in the sagittal plane only, so that there is only one unpowered d.o.f. Thus, the mechanism which we are to consider here has nine d.o.f. (n=9), eight of them (m=8) being powered. The elements of matrices of the actuator models and their distribution per joints are given in Table 3.5.

Table 3.5. Actuator parameters

| Actuator \ Term | $a_{2,2}$ | $a_{2,3}$ | $a_{3,2}$ | $a_{3,3}$ | $b_3$ | $f_2$ | used at joint |
|---|---|---|---|---|---|---|---|
| $M_1$ | -3.0 | 0.13 | $-10^5$ | -450 | 2000 | $-7\cdot10^{-4}$ | 2, 3, 6, 7 |
| $M_2$ | -1.928 | 3.03 | -6800 | -264 | 400 | -0.179 | 4, 5, 8, 9 |

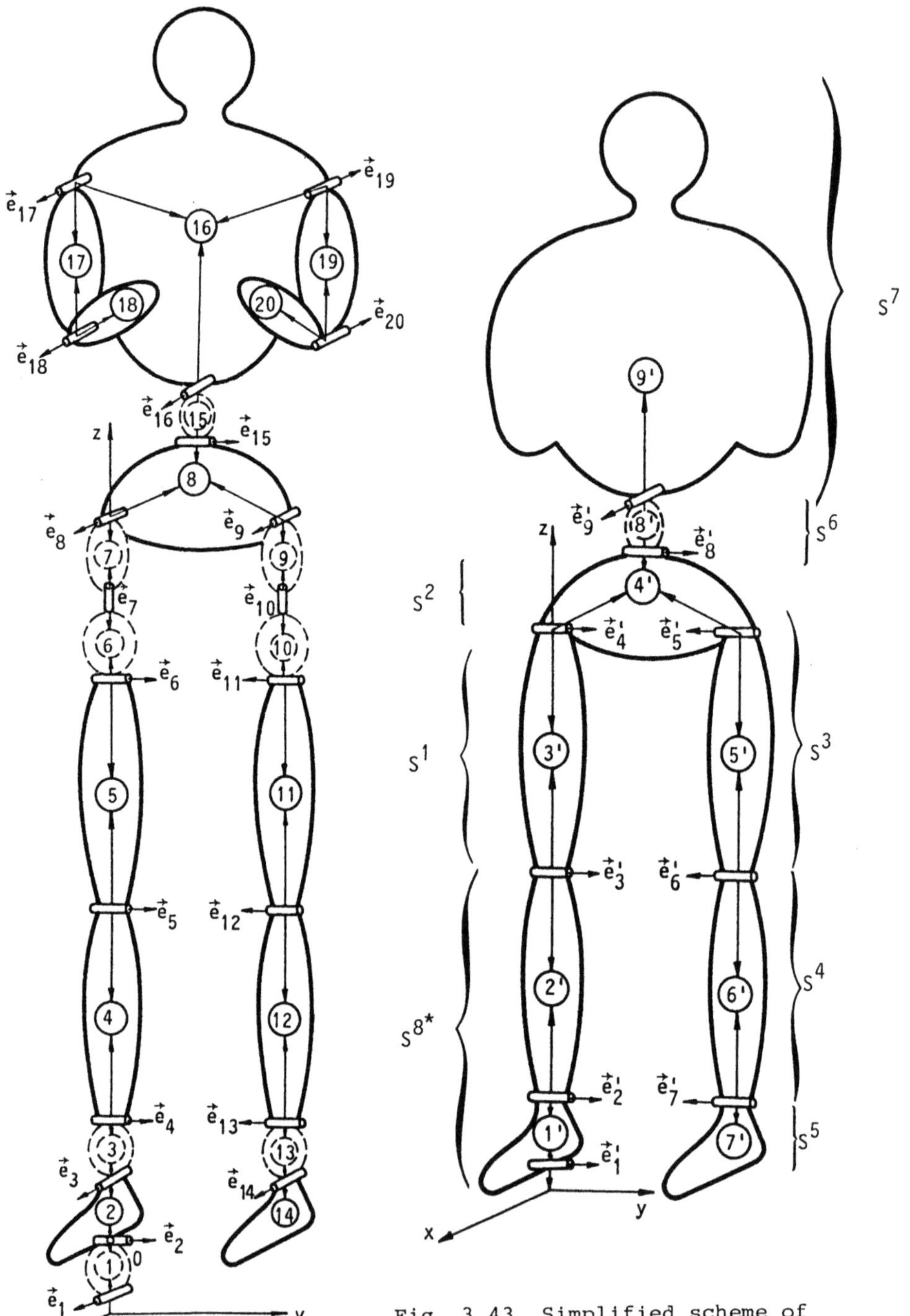

Fig. 3.42. Scheme of complete
mechanical biped
structure

Fig. 3.43. Simplified scheme of
mechanical biped structure
with disposition of modelled subsystems

Table 3.4. Kinematic and dynamic parameters of the mechanism (Fig. 3.42)

| Link | Mass kg | Moment of inertia $[kgm^2]$ | | | Distance of the axes centres of joints from the link centre [m] | Joint unit axis |
| | | $J_x$ | $J_y$ | $J_z$ | | |
| --- | --- | --- | --- | --- | --- | --- |
| 1 | 2 | 3 | 4 | 5 | 6 | 7 |
| 1 | 0.0 | 0.0 | 0.0 | 0.0 | $\vec{r}_{1,1} = (0,\,0,\,0.0001)^T$; $\quad\vec{r}_{1,2} = (0,\,0,\,-0.0001)^T$ | $\vec{e}_1 = (1,\,0,\,0)^T$ |
| 2 | 1.53 | 0.00006 | 0.00055 | 0.00045 | $\vec{r}_{2,2} = (0,\,0,\,0.030)^T$; $\quad\vec{r}_{2,3} = (0,\,0,\,-0.070)^T$ | $\vec{e}_2 = (0,\,1,\,0)^T$ |
| 3 | 0.0 | 0.0 | 0.0 | 0.0 | $\vec{r}_{3,3} = (0,\,0,\,0.0001)^T$; $\quad\vec{r}_{3,4} = (0,\,0,\,-0.0001)^T$ | $\vec{e}_3 = (1,\,0,\,0)^T$ |
| 4 | 3.21 | 0.00393 | 0.00393 | 0.00038 | $\vec{r}_{4,4} = (0,\,0,\,0.200)^T$; $\quad\vec{r}_{4,5} = (0,\,0,\,-0.186)^T$ | $\vec{e}_4 = (0,\,1,\,0)^T$ |
| 5 | 8.41 | 0.01120 | 0.01200 | 0.00300 | $\vec{r}_{5,5} = (-0.03,\,0,\,0.3)^T$ $\quad\vec{r}_{5,6} = (-0.03,\,0,\,-0.13)^T$ | $\vec{e}_5 = (0,\,1,\,0)^T$ |
| 6 | 0.0 | 0.0 | 0.0 | 0.0 | $\vec{r}_{6,6} = (0,\,0,\,0.0001)^T$; $\quad\vec{r}_{6,7} = (0,\,0,\,-0.0001)^T$ | $\vec{e}_6 = (0,\,1,\,0)^T$ |
| 7 | 0.0 | 0.0 | 0.0 | 0.0 | $\vec{r}_{7,7} = (0,\,0,\,0.0001)^T$; $\quad\vec{r}_{7,8} = (0,\,0,\,-0.0001)^T$ | $\vec{e}_7 = (1,\,0,\,0)^T$ |
| 8 | 6.96 | 0.00700 | 0.00565 | 0.00625 | $\vec{r}_{8,8} = (-0.03,\,0.09,\,0.065)^T$; $\vec{r}_{8,9} = (-0.03,\,-0.09,\,0.065)^T$; $\vec{r}_{8,15} = (0.05,\,0,\,-0.015)^T$ | $\vec{e}_8 = (1,\,0,\,0)^T$ |
| 9 | 0.0 | 0.0 | 0.0 | 0.0 | $\vec{r}_{9,9} = (0,\,0,\,-0.0001)^T$; $\quad\vec{r}_{9,10} = (0,\,0,\,0.0001)^T$ | $\vec{e}_9 = (1,\,0,\,0)^T$ |
| 10 | 0.0 | 0.0 | 0.0 | 0.0 | $\vec{r}_{10,10} = (0,\,0,\,-0.0001)^T$; $\quad\vec{r}_{10,11} = (0,\,0,\,0.0001)^T$ | $\vec{e}_{10} = (0,\,0,\,1)^T$ |
| 11 | 8.41 | 0.01120 | 0.01200 | 0.00300 | $\vec{r}_{11,11} = (-0.03,\,0,\,-0.13)^T$; $\quad\vec{r}_{11,12} = (-0.03,\,0,\,0.3)^T$ | $\vec{e}_{11} = (0,\,-1,\,0)^T$ |

Table 3.4. Continued

| 1 | 2 | 3 | 4 | 5 | 6 | 7 |
|---|---|---|---|---|---|---|
| 12 | 3.21 | 0.00393 | 0.00393 | 0.00038 | $\vec{r}_{12,12} = (0, 0, -0.186)^T$; $\quad \vec{r}_{12,13} = (0, 0, 0.200)^T$ | $\vec{e}_{12} = (0, -1, 0)^T$ |
| 13 | 0.0 | 0.0 | 0.0 | 0.0 | $\vec{r}_{13,13} = (0, 0, -0.0001)^T$; $\quad \vec{r}_{13,14} = (0, 0, 0.0001)^T$ | $\vec{e}_{13} = (0, -1, 0)^T$ |
| 14 | 1.53 | 0.00006 | 0.00055 | 0.00045 | $\vec{r}_{14,14} = (0, 0, -0.070)^T$ | $\vec{e}_{14} = (1, 0, 0)^T$ |
| 15 | 0.0 | 0.0 | 0.0 | 0.0 | $\vec{r}_{15,15} = (0, 0, 0.0001)^T$; $\quad \vec{r}_{15,16} = (0, 0, -0.0001)^T$ | $\vec{e}_{15} = (0, 1, 0)^T$ |
| 16 | 30.85 | 0.15140 | 0.13700 | 0.02830 | $\vec{r}_{16,16} = (0.035, 0, 0.4)^T$; $\quad \vec{r}_{16,17} = (-0.005, 0.2, 0)^T$; <br> $\vec{r}_{16,19} = (-0.005, -0.2, 0)^T$ | $\vec{e}_{16} = (1, 0, 0)^T$ |
| 17 | 2.07 | 0.00200 | 0.00200 | 0.00022 | $\vec{r}_{17,17} = (0, 0, -0.0001)^T$; $\quad \vec{r}_{17,18} = (0, 0, 0.0001)^T$ | $\vec{e}_{17} = (1, 0, 0)^T$ |
| 18 | 1.14 | 0.00250 | 0.00425 | 0.00014 | $\vec{r}_{18,18} = (0, 0, -0.132)^T$ | $\vec{e}_{18} = (1, 0, 0)^T$ |
| 19 | 2.07 | 0.00200 | 0.00200 | 0.00022 | $\vec{r}_{19,19} = (0, 0, -0.0001)^T$; $\quad \vec{r}_{19,20} = (0, 0, 0.0001)^T$ | $\vec{e}_{19} = (-1, 0, 0)^T$ |
| 20 | 1.14 | 0.00250 | 0.00425 | 0.00014 | $\vec{r}_{20,20} = (0, 0, -0.132)^T$ | $\vec{e}_{20} = (-1, 0, 0)^T$ |

Let us determine the stability region $X_i$ of the decoupled subsystem. Consider first the local control (3.5.9), (3.5.10) which has to stabilize the decoupled subsystem [31]. If we assume that the complete state vector $\Delta x^i$ is measurable, the closed-loop subsystem is given by (3.5.18). It is clear that in the case of a stable subsystem, the poles have to be at the left-hand side of the complex plane. If we denote the modulus of their real part by $|\sigma_p^i|$, the subsystem will be exponentially stable with a stability degree defined as

$$\Pi_i = \min_{p=1,2,3} |\sigma_p^i| \qquad (3.5.36)$$

which can be guaranteed only if the control inputs are within the limits

$$|k_i^{LT}\Delta x^i| < \bar{u}_m^i = u_m^i - \max_{t \in T} |u^{oi}(t)| \qquad (3.5.37)$$

The actuator velocity-torque characteristics limit the values of the state coordinates. According to this characteristics we can write

$$|\bar{k}_i^2 \Delta \dot{q}^i + \bar{k}_i^3 \Delta i_R^i| \leq \bar{k}_m^i \rightarrow |\bar{k}_i^T \Delta x^i| \leq \bar{k}_m^i \qquad (3.5.38)$$

where $\bar{k}_i = (0, \bar{k}_i^2, \bar{k}_i^3)^T$ and $\bar{k}_m^i$ are defined by the motor characteristics. Further, the regions of allowable angle deviations for each d.o.f. are introduced. In this way, the stability regions are constrained, for both the subsystems corresponding to powered d.o.f. and composite subsystems.

We may define a finite region $X_i$ (according to (3.5.19)) in the state space $R_i^n$, in which the subsystem $S^i$ is exponentially stable with a stability degree $\Pi_i$

$$X_i = \{\Delta x^i, \ |k_i^{LT}\Delta x^i| < |\bar{u}_m^i| \wedge |\bar{k}_i^T \Delta x^i| \leq \bar{k}_m^i\} \qquad (3.5.39)$$

Now, we have to investigate the stability of the whole system. For this purpose, the Lyapunov subsystem functions have to be chosen according to (3.5.23) taking into account relation (3.5.22) which has to be satisfied in the region $X_i$. $\tilde{X}_i$ will be an estimate of $X_i$ according to

$$\tilde{X}_i = \{\Delta x^i : v_i(\Delta x^i) \leq v_{io}\}, \qquad \forall i \in I_1$$

To investigate the asymptotic stability of the overall system, the values $\xi_{ik}^*$, which estimate the subsystem coupling, have to be determined according to (3.5.34). This expression for the composite subsystem is of the form

$$(\text{grad}v_i)^T [f_z^i \Delta P_z^{i*}(t, \Delta x) - b_z^i (k_{i4}^G \cdot \Delta \hat{P}_i^* + k_{i5}^G)] \leq \sum_{k=1}^{m} \xi_{ik}^* v_k, \quad i=2m-n+1,\ldots,m$$

where global control by both $\Delta P_N^\ell$ and $\Delta P_C^i$ is introduced. If for $\xi_{ij}^*$ thus defined, the condition (3.5.28) is satisfied, it can be claimed that region $\tilde{X}$, defined by (3.5.26) is an estimate of the region of the overall system stability.

Now, we have to select the composite subsystem, i.e. to choose to which of powered joints we shall associate the model of unpowered d.o.f. This selection is not unique. In the particular mechanism considered, there is only one composite subsystem. Two selections of composite subsystems will be considered. In the first case we will form the composite subsystem model of the models of unpowered d.o.f., and ankle joint. First, let us denote the models of powered subsystems: the subsystem model $S^1$ corresponds to the model of joint 3 powered by the actuator, $S^2$ corresponds to joint 4, $S^3$ to joint 5, $S^4$ to joint 6, $S^5$ to joint 7, $S^6$ to joint 8, $S^7$ to joint 9. As the last model of powered subsystem, $S^8$, is adopted the model of powered subsystem which will be included into the composite subsystem. Thus, to $S^8$ corresponds the model of joint 2 with the model of the corresponding actuator. According to this, the model of composite subsystem will be denoted as $S^{8*}$ and it will comprise the models of unpowered subsystem and $S^8$.

The composite subsystem matrices $A_z^{8*}$ and $f_z^8$ and vector $b_z^8$ are defined as

$$A_z^{8*} = \begin{bmatrix} 0 & 1 & 0 & 0 & 0 \\ 0.882 & 0 & -80 & -10 & 0 \\ 0 & 0 & 0 & 1 & 0 \\ 300 & 100 & 0 & -3 & 0.13 \\ 0 & 0 & 0 & -100000 & -450 \end{bmatrix}$$

$$b_z^8 = \begin{bmatrix} 0 \\ 0 \\ 0 \\ 0 \\ 2000 \end{bmatrix}, \quad f_z^8 = \begin{bmatrix} 0 & 0 \\ 0.01387 & 0 \\ 0 & 0 \\ 0 & 0.0007 \\ 0 & 0 \end{bmatrix}$$

Obviously, the vectors $D_1^8$ and $D_2^8$ from (3.5.8) are $D_1^8 = [-80, -10, 0]$ and $D_2^8 = [300, 100]$. The Lyapunov functions of all subsystems are selected in the form of (3.5.23). The matrices $H^i$ are selected to satisfy (3.5.24) and they are obtained as

$$H^8 = \begin{bmatrix} 100422.00 & 33545.10 & 8660.65 & -98.68 & 0.02 \\ 33545.10 & 11425.23 & -3273.52 & -335.46 & -0.07 \\ 8660.65 & -3273.52 & 183099.94 & 8965.84 & 1.95 \\ -98.68 & -335.46 & 8965.84 & 1097.767 & 0.086 \\ 0.02 & -0.07 & 1.95 & 0.086 & 0.0009 \end{bmatrix}$$

The Lyapunov matrices corresponding to the models of powered subsystems $S^i$ are

$$H^i = \begin{bmatrix} 62777.42 & 2291.71 & 0.56 \\ 2291.71 & 161.15 & 0.02 \\ 0.56 & 0.02 & 0.00011 \end{bmatrix} \qquad i = 1,4,5$$

$$H^i = \begin{bmatrix} 32912.207 & 408.621 & 4.535 \\ 408.621 & 6.482 & 0.065 \\ 4.535 & 0.065 & 0.00092 \end{bmatrix} \qquad i = 2,3,6,7$$

The regions of joints angle deviations (the superscripts correspond to the subsystems model numbers), in which stability is investigated are (in radians)

$$\Delta q^1 = \pm 0.044, \ \Delta q^2 = \pm 0.0422, \ \Delta q^3 = \pm 0.0126, \ \Delta q^4 = \pm 0.03,$$

$$\Delta q^5 = \pm 0.03, \ \Delta q^6 = \pm 0.099, \ \Delta q^7 = \pm 0.01, \ \Delta q_z^1 = \pm 0.01, \ \Delta q_z^2 = \pm 0.077$$

where $\Delta q_z^1$ and $\Delta q_z^2$ correspond to the joints comprising the composite subsystem i.e. the unpowered and powered d.o.f. (ankle joint with axis of rotation $\vec{e}_2$).

Constants $v_{io}$ which define estimates of stability regions $\tilde{X}_i$ are computed to be

$$v_{10} = 0.65, \qquad v_{20} = 1.0399, \qquad v_{30} = 0.4256, \qquad v_{40} = 0.1545,$$

$$v_{50} = 0.75166, \qquad v_{60} = 0.7395, \qquad v_{70} = 0.1814, \qquad v_{80} = 0.3291$$

302

Here, constant $v_{80}$ corresponds to the composite subsystem.

The results of stability analysis are presented in Table 3.6. Three types of control law are investigated:

a) the complete feedback structure defined by (3.4.9) plus the global control with respect to ZMP displacement defined by (3.4.18) (Fig. 3.20),

b) the local control is introduced ($k_{i1}^L$, $k_{i2}^L$ and $k_{i3}^L$ from (3.4.9)) plus global control with respect to ZMP position (3.5.18),

c) only local control from (3.4.9) is introduced.

To save space and make comparison easier, all three cases are presented together. The first row corresponds to case a), the second to b), and the third one to c).

The first and the second set of three rows correspond to the knee and the hip of the supporting leg, the third, fourth and the fifth to the hip, knee and ankle of the leg in swing phase, while the sixth and seventh set correspond to the trunk motion in the frontal and sagittal plane, respectively. Finally, the last set of three rows corresponds to the composite subsystem.

Each diagonal element in matrix G, defined by (3.5.29), represents the stability degree of the decoupled subsystem, while the rest of each row corresponds to the destabilizing influence of other subsystems on it. If this influence is not destabilizing, in this place should be put zero. The first requirement to be satisfied in the procedure of stability analysis is that the decoupled subsystems are stable, which means that the diagonal elements in G are negative. This is fulfilled in all three cases considered. Then, to derive a final conclusion about the system stability the product $G \cdot v_o$ has to be observed. If this product is negative, the stability under the given conditions is proved. As can be seen from the vector $G \cdot v_o$ for the case (a), the stability for all subsystems is proved. In the case (b) (when the global control at each joint is omitted, but the feedback with respect to the ZMP position still holds) the stability of the composite subsystem is not proved. In case (c) (only local controllers are applied) the stability of the system cannot be proved, which is quite understandable. Then, vector $\eta$ representing shrinkage of bounds of the regions $\check{X}(t)$ is also presented. (Here, $\eta$ denotes $\eta = (\beta_1, \beta_2, \ldots, \beta_m)^T$ and $\beta_i$ are defined by (3.5.31)).

Table 3.6. Results of stability analysis (composite system consisting of ankle joint and unpowered joint)

$$
G = \begin{bmatrix}
-2402.97 & 0 & 0 & 0 & 0 & 0 & 0 & 0 \\
-2360.45 & 0 & 0 & 0 & 0 & 0.195 & 0 & 0 \\
-2360.45 & 0 & 0 & 0 & 0 & 0.195 & 0 & 0 \\
0 & -120.65 & 0 & 0 & 0 & 0 & 0 & 0 \\
0 & -120.43 & 0 & 0 & 0 & 0 & 0 & 0 \\
0 & -120.43 & 0 & 0 & 0 & 0 & 0 & 0 \\
0 & 0 & -76.15 & 0 & 0 & 0 & 0 & 0 \\
0 & 0 & -73.88 & 0 & 0 & 0 & 0 & 0 \\
0 & 0 & -73.88 & 0 & 0 & 0 & 0 & 0 \\
0 & 0 & 0 & -5114.02 & 0 & 0 & 0 & 0 \\
0 & 0 & 0 & -4761.15 & 0 & 0 & 0 & 0 \\
0 & 0 & 0 & -4761.15 & 0 & 0 & 0 & 0 \\
0 & 0 & 0 & 0 & -789.58 & 0 & 0 & 0 \\
0 & 0 & 0 & 0 & -764.16 & 0 & 0 & 0 \\
0 & 0 & 0 & 0 & -764.16 & 0 & 0 & 0 \\
0 & 0 & 0 & 0 & 0 & -67.56 & 0 & 0 \\
0 & 0 & 0 & 0 & 0 & -66.40 & 0 & 0 \\
0 & 0 & 0 & 0 & 0 & -66.40 & 0 & 0 \\
0 & 0 & 0 & 0 & 0 & 0 & -32.84 & 0 \\
0.323 & 0.1 & 0 & 0 & 0 & 0.309 & -28.86 & 0.114 \\
0.323 & 0.1 & 0 & 0 & 0 & 0.309 & -28.86 & 0.114 \\
13.04 & 11.62 & 0 & 0 & 0 & 6.14 & 0 & -218.33 \\
33.537 & 24.44 & 2.745 & 0 & 0 & 23.78 & 0 & -177.82 \\
37.065 & 26.65 & 8.138 & 8.872 & 1.278 & 26.89 & 11.38 & -170.84
\end{bmatrix}
$$

$$
G \cdot v_0 = \begin{bmatrix}
-1563.62 & -125.46 & -32.41 & -790.28 & -593.50 & -49.96 & -5.96 & -46.74 \\
-1535.81 & -125.24 & -31.44 & -735.75 & -574.39 & -49.10 & -4.66 & +7.48 \\
-1535.81 & -125.24 & -31.44 & -735.75 & -574.39 & -49.10 & -4.66 & +23.35
\end{bmatrix}^T
$$

$$
\eta = \begin{bmatrix}
2402.97 & 120.65 & 76.15 & 5114.01 & 789.58 & 67.56 & 32.84 & 142.03 \\
2360.23 & 120.43 & 73.88 & 4761.14 & 764.16 & 66.40 & 25.66 & - \\
2360.23 & 120.43 & 73.88 & 4761.14 & 764.16 & 66.40 & 25.66 & -
\end{bmatrix}^T
$$

In the previous example, the composite subsystem model consisted of a model of unpowered joint and the model of ankle joint subsystem, both belonging to the supporting leg. However, as we have said, the selection of powered joint which will be included into the composite subsystem, is not unique. We shall present results for the case when the composite subsystem comprises the models of unpowered subsystem (foot with respect to ground) and the model of powered hip joint subsystem.

Both joints belong to the supporting leg. Now, the relation between the subsystem ordinal number and the corresponding joint is somewhat changed. Subsystem $S^1$ corresponds to the model of joint 2 powered by corresponding actuator, $S^2$ corresponds to joint 3, $S^3$ to joint 5, $S^4$ to joint 6, $S^5$ to joint 7, $S^6$ to joint 8, $S^7$ to joint 9. Model of the composite subsystem $S^{8*}$ consists of the models of unpowered d.o.f. (represented by $\vec{e}_1$) and the model of joint 4 with its actuator. In Fig. 3.44. is shown the redistribution of the subsystems for this choice of composite subsystem.

Thus, as in the previous example, the matrices $A_z^{8*}$ and $f_z^8$ and vector $b_z^8$ are

$$A_z^{8*} = \begin{bmatrix} 0 & 1 & 0 & 0 & 0 \\ 0.882 & 0 & -80 & -10 & 0 \\ 0 & 0 & 0 & 1 & 0 \\ 300 & 300 & 0 & -1.93 & 3.036 \\ 0 & 0 & 0 & -6800 & -264 \end{bmatrix}$$

$$f_z^8 = \begin{bmatrix} 0 & 0 \\ 0.01387 & 0 \\ 0 & 0 \\ 0 & -0.179 \\ 0 & 0 \end{bmatrix}, \quad b_z^8 = \begin{bmatrix} 0 \\ 0 \\ 0 \\ 0 \\ 400 \end{bmatrix}$$

where the vectors $D_1^8$ and $D_2^8$ are now: $D_1^8 = [-80 \ -10 \ 0]$ and $D_2^8 = [300 \ 300]$. The Lyapunov matrix for the composite subsystem is

$$H^8 = \begin{bmatrix} 167.23 & 166.89 & 79.88 & -0.49 & -0.0013 \\ 166.98 & 167.82 & 82.52 & -0.56 & -0.0017 \\ 79.88 & 82.52 & 666.43 & 43.64 & -0.0454 \\ -0.49 & -0.56 & 43.65 & 5.23 & 0.0029 \\ -0.0013 & -0.0017 & -0.045 & 0.0029 & 0.0010 \end{bmatrix}$$

while the matrices for the powered subsystems $S^i$ (which are re-enumerated with respect to previous example) are

$$H^i = \begin{bmatrix} 62777.42 & 2291.71 & 0.56 \\ 2291.71 & 161.15 & 0.02 \\ 0.56 & 0.02 & 0.00011 \end{bmatrix} \quad i = 1,2,4,5$$

$$H^i = \begin{bmatrix} 32912.207 & 408.621 & 4.535 \\ 408.621 & 6.48238 & 0.06515 \\ 4.535 & 0.06515 & 0.00092 \end{bmatrix} \qquad i = 3,6,7$$

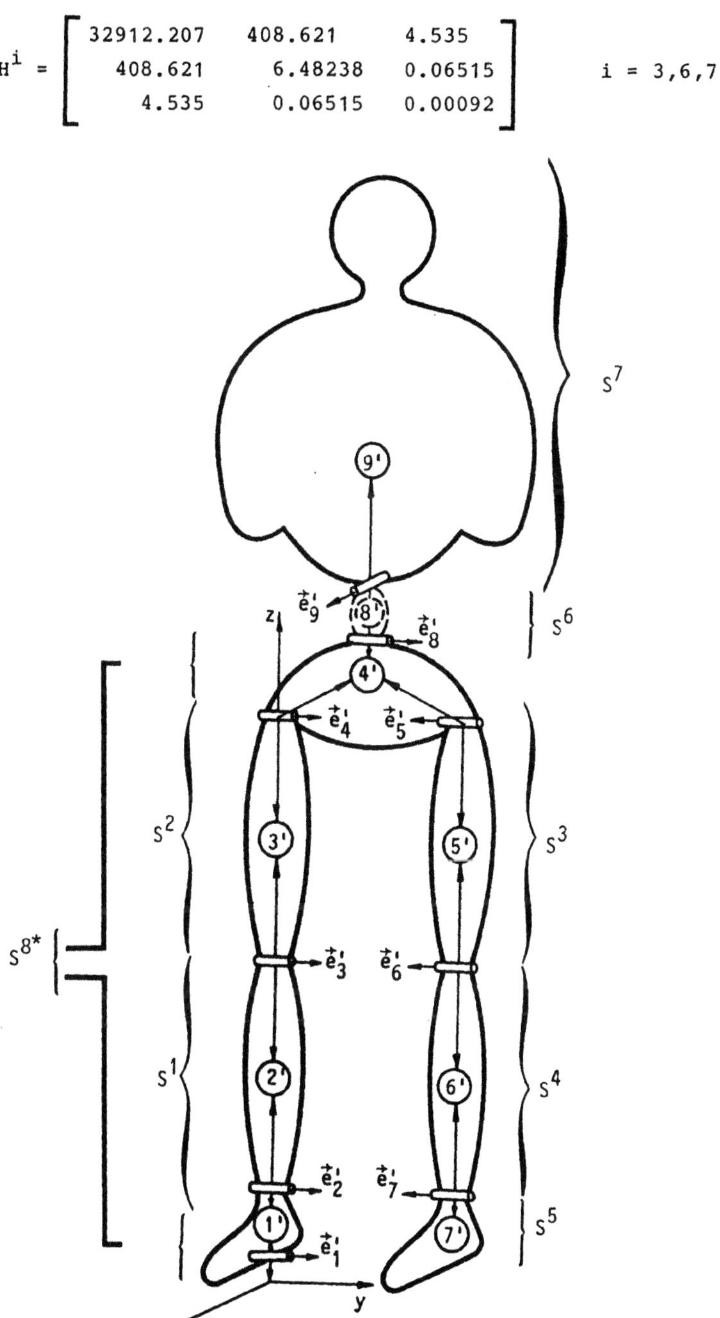


Fig. 3.44. Simplified scheme of mechanical biped structure with distribution of the subsystems

306

The regions of joint angle deviations are the same as in the previous example, while the constants $v_{io}$ are

$$v_{10} = 2.055, \quad v_{20} = 0.65, \quad v_{30} = 0.4256, \quad v_{40} = 0.1545$$

$$v_{50} = 0.7517, \quad v_{60} = 0.7395, \quad v_{70} = 0.1814, \quad v_{80} = 0.3291$$

Three cases of control structure applied are the same as in the first example: a) the complete feedback defined by (3.4.9) plus global control with respect to ZMP defined by (3.4.18) (Fig. 3.20), b) the local control $(k_{i1}^L, k_{i2}^L, k_{i3}^L)$ from (3.4.9) plus global feedback (3.4.18), and c) only local control from (3.4.9). The control which has to maintain ZMP position is applied to the powered joint which is included in the composite subsystem $S^8$, i.e. to the hip.

The results are presented in Table 3.7. Again, three matrices $G \epsilon R^{8 \times 8}$, the vectors $(G \cdot v_o) \epsilon R^8$ and $\eta \epsilon R^8$ are given in the same manner as in the previous example. The first set of rows corresponds to the ankle (now not included into the composite subsystem) and second to the knee of the leg in the supporting phase. The third set of rows and the subsequent ones correspond to the same subsystems as in previous example. Note that the hip of supporting leg is omitted because it is included into the composite subsystem. Thus, the composite subsystem is now composed of the models of unpowered d.o.f. and hip joint of the supporting leg.

The obtained stability tests might be used to compare various control laws and various global feedback structures. As can be seen from Tables 3.6. and 3.7., the introduced global control reduces the effects of coupling upon both the powered joints and composite subsystems. It can be also seen that the global control aimed at compensating of the deviations of ZMP position from its nominal values (i.e. (3.4.18)) increases significiantly the stability degree of the composite subsystems (in both considered cases of composite subsystems). This is quite expectable since this global control has to ensure stabilizing effects of the powered joint upon the unpowered joint, both being "within" the composite subsystem.

It is evident that these stability tests might be used to compare various global feedback structures. As already mentioned, the problem is to select the powered joint to which the global control (3.4.18) has to be assigned, i.e. which joint has to compensate for the ZMP deviations and maintain the entire system stability. By comparison of stability tests it might be concluded which compensation gives better stability results. However, in doing this we must bear in mind that the proposed stability test might be conservative: if the stability test is not fulfilled, that does not mean that the actual system is unstable. Therefore, the comparison of various control schemes must be done carefully. We should always iteratively search for the "best" selection of the subsystems Lyapunov functions for the specific control law and specific global feedback structure.

Table 3.7. Results of stability analysis (composite system consists of the hip joint and unpowered joint)

$$
G = \begin{bmatrix}
-53.92 & 0 & 0 & 0 & 0 & 0 & 0 & 0 \\
-50.91 & 0 & 0 & 0 & 0 & 0 & 0 & 0 \\
-50.91 & 0 & 0 & 0 & 0 & 0 & 0 & 0 \\
0 & -2404.16 & 0 & 0 & 0 & 0 & 0 & 0 \\
0 & -2360.11 & 0 & 0 & 0 & 0.172 & 0.003 & 0 \\
0 & -2360.11 & 0 & 0 & 0 & 0 & 0 & 0 \\
6.021 & 0 & -76.15 & 0 & 0 & 0 & 0 & 0 \\
6.491 & 0 & -73.88 & 0 & 0 & 0 & 0 & 0 \\
6.491 & 0 & -73.88 & 0 & 0 & 0 & 0 & 0 \\
0 & 0 & 0 & -5114.02 & 0 & 0 & 0 & 0 \\
0 & 0 & 0 & -4761.15 & 0 & 0 & 0 & 0 \\
0 & 0 & 0 & -4761.15 & 0 & 0 & 0 & 0 \\
0 & 0 & 0 & 0 & -789.58 & 0 & 0 & 0 \\
0 & 0 & 0 & 0 & -764.16 & 0 & 0 & 0 \\
0 & 0 & 0 & 0 & -764.16 & 0 & 0 & 0 \\
6.351 & 0 & 0 & 0 & 0 & -67.56 & 0 & 0 \\
6.767 & 0 & 0 & 0 & 0 & -66.40 & 0 & 0 \\
6.767 & 0 & 0 & 0 & 0 & -66.40 & 0 & 0 \\
0 & 0.89 & 0 & 0 & 0 & 0 & -32.84 & 0 \\
0.045 & 1.547 & 0 & 0 & 0 & 0.198 & -28.86 & 0.109 \\
0.045 & 1.547 & 0 & 0 & 0 & 0 & -28.86 & 0.109 \\
1.236 & 16.534 & 0 & 16.375 & 7.77 & 11.14 & 70.56 & -218.31 \\
1.544 & 17.505 & 0 & 20.465 & 6.93 & 11.99 & 67.1 & -216.53 \\
3.242 & 22.87 & 3.8 & 43.052 & 2.29 & 16.72 & 47.86 & -206.7
\end{bmatrix}
$$

$$
G \cdot v_0 = \begin{bmatrix}
-110.81 & -1564.4 & -20.03 & -790.28 & -593.5 & -36.91 & -5.38 & -34.82 \\
-104.63 & -1535.6 & -18.1 & -735.75 & -574.4 & -35.2 & -3.95 & -32.93 \\
-104.63 & -1535.6 & -18.1 & -735.75 & -574.4 & -35.2 & -3.95 & -20.84
\end{bmatrix}^T
$$

$$
\eta = \begin{bmatrix}
53.92 & 2404.16 & 47.08 & 5114.01 & 789.6 & 49.9 & 29.64 & 98.04 \\
50.91 & 2359.92 & 42.54 & 4761.14 & 764.16 & 47.6 & 21.78 & 92.7 \\
50.91 & 2359.92 & 42.54 & 4761.14 & 764.16 & 47.6 & 21.78 & 58.67
\end{bmatrix}^T
$$

It is obvious that the determination of numbers $\xi_{ij}$ which estimate the interconnections between the subsystems according to (3.5.27), or (3.5.34), requires application of digital computer. It is necessary to search in the finite region $\tilde{X}$ for the minimal numbers $\xi_{ij}$ or $\xi_{ij}^*$ which satisfy (3.5.27), or (3.5.34). This means that computation of the coupling $\Delta\hat{P}_i(t, \Delta x)$ (i.e. the driving torques at the mechanism's joints) has to be repeated many times, and therefore, the stability analysis may be time consuming, even if powerful current computers are applied. Nevertheless, the presented stability analysis may be used as an efficient tool for selection of the appropriate control law, since it guarantes the stability of the entire system (assuming the model of the system is sufficiently reliable). An algorithm can be established which will iteratively search for the most appropriate control law and feedback structure, by analyzing the stability of the locomotion system with various control laws. Using this analysis, we may determine to which joint it is necessary to apply global control in order to compensate for the coupling from the rest of the system [48], and at which joints the global control might be ommitted. It is obvious that the global control need not to be applied at all joints. Similarly, if we want to apply global control by on-line computation of the coupling between the subsystems, then on the basis of the above stability analysis we may draw conclusions about an approximative dynamic model that should be applied. Namely, the aim is to select the most simple control law which still meets all the imposed requirements within a specific control task. Therefore, we should try to find the most simple approximative dynamic model which could be used for the on-line computation of coupling between the subsystems, but which will ensure practical stability of the entire system [48].

It should be mentioned that the presented method for stability analysis may be used to analyze the stability of the biped when some other control laws are implemented. However, since the method is based upon the aggregation-decomposition approach, it is the most appropriate for the decentralized control structures which we have considered. (In fact, all three control laws which we have analyzed, have a decentralized structure with addition of the necessary global feedback loops).

## 3.6. Conclusion

In this chapter we have considered the control of the biped locomotion system. The various control algorithms have been reviewed and several new control schemes have been proposed. The simulation of the complete biped dynamics with the proposed control laws has been presented. An algorithm for stability analysis of the entire locomotion system has also been presented, which can be used for the synthesis of the most appropriate control law for each specific locomotion system. This is the first attempt to synthesize the control of the biped system starting from the complete nonlinear dynamic model of the system, and to analyze stability of the entire biped system.

On the basis of these considerations we can conclude that the synthesis of the control of a biped system is extremely complex. Bipeds are very complex and highly nonlinear dynamic systems. To ensure their equilibrium during the gait, it is necessary to take into account the entire dynamics of the system. Therefore, a dynamic control must be applied in order to stabilize the biped system around desired trajectories, i.e. around the nominal trajectories. However, due to the high nonlinearity of the biped system, the control which has to take into account dynamics of the system, would be extremely complex. The dynamic control laws would have to include computation of the biped dynamics. This means that the control computer would have to compute dynamic moments around the biped joints at each 5-10 [ms] in order to achieve sampling rate compatible with the system dynamics. Even current powerful microprocessors are not capable to compute the dynamic moments fast enough. Therefore, various approximate methods have to be used to stabilize the biped system during the gait.

A special problem in controlling of biped system is to maintain the system dynamic equilibrium during the walk. The presence of unpowered d.o.f. between the legs soles and the ground makes this problem extremely complex, as these d.o.f. are "responsible" for the system equilibrium, and they cannot be controlled directly. As we have explained in Chapter 2, the dynamic equilibrium of the system requires the ZMP be kept within the certain region at the sole of the supporting leg. To achieve this, the motion of the biped joints must satisfy the equations of dynamic equilibrium of the system (i.e. the condition that the resultant of all dynamic forces of the system act at certain point (ZMP) within the given region at the sole). If the task of maintaining equilibrium is assigned to the trunk joints, then at each sampling interval, for the actual positions, velocities, and accelerations of the legs joints, we may compute the necessary motion of the trunk to keep the system at equilibrium. However, such computation of the compensating movements of the trunk can hardly be implemented on-line. Therefore, we have proposed another approximate approach. We compute off--line desired nominal trajectories of all biped joints. The trajectories of the legs joints are selected to ensure a desired type of the biped gait: these trajectories may be obtained by recording the motion of the human legs during the regular walk. The trajectories of the trunk are computed off-line from the equations of dynamic equilibrium in order to ensure that ZMP is kept at the desired position(s) during the walk. The computed trajectories are stored in the control computer memory, and, in on-line control of biped, the corresponding values are taken from the memory, and realized.

The nominal trajectories thus obtained ensure that the system is in equilibrium under ideal conditions: if the model of the system is perfect, if the trajectories are perfectly realized and if no perturbation is acting upon the system. Since these conditions are never fulfilled, additional control is necessary which will ensure tracking of these nominal trajectories and preserve the system equilibrium when disturbances are present. Again, at the level of perturbed regime, the problem of equilibrium maintenance arises. The control which has to ensure the system stability must take into account the dynamics of the system. Because of its complexity and nonlinearity, we have proposed some approximate control schemes. We have started from the simple decentralized controller most frequently used in the control of robotic systems. This controller assumes that each joint is controlled independently from the rest of the system. The feedback structure and implementation of such controller is very simple. However, having in mind that the equilibrium of the entire system involves all inertial and other dynamic forces of the system, we have modified the simple PD regulators around the joints in order to ensure that the accelerations of the joints, during tracking of the desired nominal trajectories, do not exceed certain limits. The synthesis of such local controllers has been presented in Paragraph 3.4.2.

However, such decentralized controller cannot stabilize the entire system because of the strong dynamic coupling between the joints, and the necessity to indirectly control unpowered d.o.f. and to preserve the system equilibrium in the perturbed regimes. Therefore, additional feedback loops have to be applied. We have proposed the application of global control in two forms: by force feedback and by on-line computation of dynamic coupling between the joints. By measuring moments at the joints (by means of force transducers) we get direct information on coupling between the joints. Thus, we may easely establish the global control which should compensate for the effects of coupling between the joints. Similarly, by measuring the forces between the sole of the supporting leg and the ground we may establish the global control which would maintain the system equilibrium. An approximate control law for the compensation of the ZMP deviations from the nominal position(s), based on the finformation from force transducers at the sole, has been proposed (Paragraph 3.4.3).

These global control laws require detailed elaboration in order to precisely define the feedback structure and feedback gains. As explained in Section 3.4, the selection of the joint which has to ensure the compensation of the ZMP deviations from its nominal position(s) is not unique and it has to be done in the scope of the control synthesis. The simulation results presented in Section 3.4.4. can be used to determine the most appropriate control structure and global feedback gains. However, precise determination of all the relevant parameters of the system and selection of feedback structure can be achieved by stability analysis of the entire nonlinear model of the system, as presented in Section 3.5. This stability analysis gives an insight into the system performance with various control laws and, therefore, it enables the synthesis of the most suitable control law.

The simulation results, as well as the results of stability analysis, show that the proposed control law can be successfully applied to control the biped system. However, further elaboration is needed to precisely define the control law which would cover different situations which might arise during the system walk along the regular or irregular terrain. (For example, we have not considered perturbed regimes when the system climb the staircases etc.). The simulation results show that the system exibits various performances depending on the perturbation acting upon it: the performance depends on the joint at which the initial perturbation appears, and, it depends on how large is the initial error, etc. Some further studies might show that it is preferable to compensate ZMP deviations by two or more powered joints, and not only by one, as we have considered. In this, the problem of

distribution of global feedback loops from the unpowered joints to powered joints will arise. It has been shown that if we want to ensure
the anthropomorphic behaviour of the system for large perturbations we
should change the control strategy; these problems, however, have not
been considered in this book.

We have considered only the global control which include force feedback (both for compensation of coupling between the joints and for compensation of the ZMP deviations from its nominal position). We have not
considered in detail the possibility of implementing global control by
on-line computation of dynamic forces, since the necessary rate of this
computation cannot be achieved by current microprocessors. However, it
is quite possible that in the near future, the implementation of array
processors, or transputers, or some new microprocessor architectures,
will make possible the application of such dynamic control law which
will include the on-line (in the scope of 5-10 [ms]) computation of all
relevant dynamic forces, or even of the compensating movements of the
joints to maintain the system equilibrium. However, it should be borne
in mind that the biped locomotion system is very sensitive to parameter variations and model uncertainities, as shown by both the simulation results and stability tests. Therefore, it is questionable whether the dynamic control including the on-line computation of the dynamic system might ensure system robustness. The proposed global force
feedback control is to a great extend robust to both parameter and model uncertainities, since the dynamic forces at the joints and at the
soles are directly measured. However, the force feedback suffers from
some drawbacks regarding the effects of elastic modes in the system.
Therefore, in the future work, a combination of force feedback and on-
-line computation of dynamic forces might lead to better results and
enable high performances of the biped controller.


## References


[1] Bernstain N.A., "On the Motion Synthesis", (in Russian), Medgiz,
    Moscow, 1947.

[2] Chow C.K., Jacobson D.H., "Postural Stability of Human Locomotion",
    Mathematical Biosciences, Vol. 15, pp. 93-107, 1972.

[3] Hemami H., Katbab A., "Constrained Inverted Pendulum Model of Eva-
    luting Upright Postural Stability", Journal of Dynamic Systems,
    Measurement and Control, Vol. 104, December 1982.

[4] Bavarian B., Wyman F., Hemami H., "Control of Constrained Planar Simple Inverted Pendulum", Int. Journal of Control, Vol. 37, No. 4, 1983.

[5] Hemami H., Wall C., Block F., Golliday G.Jr, "Single Inverted Pendulum Biped Experiments", Journal of Interdis. Modelling & Simulation, Vol. 2, No. 3, 1979.

[6] Hemami H., Cvetković V., "Postural Stability of Two Biped Models via Lyapunov Second Method", IEEE Transactions on Automatic Control, February 1977.

[7] Hemami H., Wyman B., "Indirect Control of the Forces of Constraint in Dynamic Systems", Journal of Dynamic Systems, Measurement and Control, Vol. 101, December 1979.

[8] Hemami H., Wyman B., "Modelling and Control of Constrained Dynamic Systems with Application to Biped Locomotion in the Frontal Plane", IEEE Trans on Automatic Control, Vol. 24, No. 4, 1979.

[9] Hemami H., "A Feedback on-off Model of Biped Dynamics", Proc. of Int. Conf. on Cybernetics and Society, Denver, Colorado, 1979.

[10] Hemami H., Robinson C.S., Ceranowicz A.Z., "Stability of Planar Biped Models by Simultaneous Pole Assignment and Decoupling", Int. Journal of Systems Sci., Vol. 11, No. 1, 1980.

[11] Goddard R., Hemami H., Weimer F.C., "Biped Side Step in the Frontal Plane", IEEE Trans on Automatic Control, Vol. 28, No. 2, 1983.

[12] Hemami H., Yuan-Fang Z., Hines M., "Initiation of Walk and Tiptoe of a Planar Nine Link Biped", Mathematical Biosciences, Vol. 61, pp. 163-189, 1982.

[13] Hemami H., Hines M., Goddard R., Friedman B., "Biped Sway in the Frontal Plane with Locked Knees", IEEE Trans. on Systems, Man and Cybernetics, Vol. 12, No. 4, 1982.

[14] Hemami H., Yuan-Fang Z., "Dynamic and Control of Motion on the Ground and in the Air with Application to Biped Robots", Jour. of Robotic Systems, Vol. 1, No. 1, 1984.

[15] Yuan-Fang Z., Hemami H., "Impact Effects of Biped Contact With the Environment", IEEE Trans. On Systems, Man and Cybernetics, Vol. 14, No. 3, 1984.

[16] Wongchaisuwat C., Hemami H., Buchner H., "Control of Sliding and Rolling at Natural Joints", Trans. of ASME Journal of Biomechanical Engineering, Vol. 106, November, 1984.

[17] Wongchainsuwat C., Hemami H., Hines M., "Control Exerted by Ligaments", Journal of Biomechanics, Vol. 18, No. 7, 1984.

[18] Hemami H., Ben-Ren C., "Stability Analysis and Input Design of a Two Link Planar Biped", The International Journal of Robotics Research, Vol. 3, No. 2, 1984.

[19] Hill J.C., "A Dynamic Model of the Human Postural Control System", Final Report 1970. School of Engineering Oakland, University Rochester, Michigan.

[20] Gubina F., "Stability and Dynamic Control of Certain Types of Biped Locomotion", IV Symp. on External Control of Human Extremities, Dubrovnik, 1972.

[21] Gubina F., McGhee B., Hemami H., "On the Dynamic Stability of Biped Locomotion", IEEE Trans. on Biomedical Engineering, Vol. 21, March 1974.

[22] Vukobratović M., Legged Locomotion Robots and Anthropomorphic Mechanisms, Monograph, Mihailo Pupin Institute, Belgrade, 1975.

[23] Vukobratović M., Stokić D., "Postural Stability of Anthropomorphic Systems", Mathematical Biosciences, Vol. 25, No. 3/4, 1975.

[24] Vukobratović M., Stokić D., "Significance of Force-Feedback in Controlling Artificial Locomotion - Manipulation Systems", IEEE Trans. on Biomedical Engineering, Vol. 27, No. 12, 1982.

[25] Beletskii V.V., Biped Walk, (in Russian), Nauka, Moscow, 1984.

[26] Vukobratović M., Frank A., Juričić D., "On the Stability of Biped Locomotion", IEEE Trans on Biomedical Engineering, Vol. 17, No. 1, 1970.

[27] Formal'skii A.M., Locomotion of Anthropomorphic Mechanisms, (in Russian), Nauka, Moscow, 1982.

[28] Miura H., Shimoyama I., "Dynamic Walk of Biped", The Inter. Journal of Robotic Research, Vol. 3, No. 2, 1984.

[29] Furusho J., Masubuchi M., "Control of Dynamical Biped Locomotion Systems for Steady Walking", The ASME, Journal of Dynamic Systems, Measurement and Control, Vol. 108, June 1986.

[30] Takamishi A., Egusa Y., Tochizawa M., Tokeaya T., Kato I., "Realization of Dynamic Biped Walking Stabilized with Trunk Motion", Seventh CISM-IFToMM Symposium on Theory and Practice of Robots and Manipulators, Udine, 1988.

[31] Vukobratović M., Stokić D., Control of Manipulation Robots, Monograph, Springer-Verlag, 1982.

[32] Bernstein N.A., Physiology of Motions and Activity Physiology, (in Russian), Publ. Comp. "MEDECINA", Moscow, 1966.

[33] Okhotsimskii D.E., et al., "Control of Integral Locomotion Robots", (in Russian), Proc. of VI IFAC Symp. on Automatic Control in Space, Erevan, USSR 1974.

[34] Medvedov B.S., Leskov A.G., Yuschenko A.S., Systems of Manipulation Robots Control, (in Russian), Series "Scientific Fundamentals of Robotics", Edited by E.P. Popov, "Nauka", Moscow, 1978.

[35] Vukobratović K.M., "How to Control the Artificial Anthropomorphic Systems", IEEE Trans. on Systems, Man and Cybernetics, SMC-3, No. 5, 1973.

[36] Vukobratović K.M., Stokić M.D., Gluhajić V.N., Hristić S.D., "One Method of Control for Large-Scale Humanoid Systems", Mathematical Biosciences, Vol. 36, No. 3/4, 1977.

[37] Vukobratović K.M., Stokić M.D., "Simplified Control Procedure of Strongly Coupled Complex Nonlinear Mechanical Systems", (in Russian), Avtomatika and Telemekhanika, No. 11, 1978.

[38] Vukobratović K.M., Stokić M.D., "Contribution to the Decoupled Control of Large-Scale Mechanical Systems", Automatica, Vol. 16, No. 1, 1980.

[39] Borovac B., Vukobratović M., Surla D., "An Approach to Biped Control Synthesis", Robotica, Vol. 7, pp. 231-141.

[40] Athans M., Falb P., Optimal Control, (in Russian), Mashinostroenie, Moscow, 1968.

[41] Borovac B., Vukobratović M., Stokić D., "Stability Analysis of Mechanisms Having Unpowered Degrees of Freedom", Robotica, (to appear), 1989.

[42] Vukobratović M., Stokić D., "Dynamic Control of Unstable Locomotion Robots", Mathematical Biosciences, Vol. 24, pp. 129-157, 1975.

[43] Šiljak D.D., "Multilevel Stabilization of Large-Scale Systems: A Spinning Flexible Spacecraft", Automatica, Vol. 12, pp. 309-320. 1976.

[44] Šiljak D.D., Large-Scale Dynamic Systems: Stability and Structure, North-Holland, 1978.

[45] Morari M., Stephanopulos G., Aris R., "Finite Stability Regions for Large Scale Systems with Stable and Unstable Systems", Inter. Journal of Control, Vol. 26, No. 5, 1977.

[46] Bitoris G., Comments of "Finite Stability Regions for Large Scale Systems with Stable and Unstable Systems", Inter. Journal of Control, Vol. 27, No. 6, 1978.

[47] Weissenberger S., "Stability Regions of Large-Scale Systems", Automatica, Vol. 9, pp. 653-663, 1973.

[48] Vukobratović M., Stokić D., Kirćanski N., Non-Adaptive and Adaptive Control of Manipulation Robots, Monograph, Springer-Veralg, 1985.

[49] Stokić D., Vukobratović M., "Dynamic Control of Biped Posture", Mathematical Biosciences, Vol. 44, No. 2, 1979.

# Chapter 4:
## Realization of Anthropomorphic Mechanisms

## 4.1. Introduction

Locomotion activity, and especially the human gait, belongs to a class of highly automated motion. Bernstein was the first who noticed this fact [1]. It is known that man has at the disposal for his complete sceletal (locomotion-manipulation) activity several hundreds of muscles which form over three hundred and fifty equivalent d.o.f. In view of such a high number of biological actuators through which man exercises his motor activity the imitation of this activity seems to be a hopelessly dificult task. To grasp the essence of the gait mechanism control and of other skeletal activity is also a task of extreme complexity if one has in mind the detailed insight into the multilevel structure of the extremely complex and perfect control of the human gait, i.e. of human locomotion and manipulation activity.

Nevertheless, in spite of these facts, the attempts have been made to represent the complicated motions of the upper and lower human extremities by an appropriate mathematical apparatus, and to encompass the control system of these complex motion by the corresponding hierarchical structure of control offered by advanced systems theory.

The gait is described by differential equations of motion of the links interconnected by joints, for the imposed repeatability conditions and fulfilment of the conditions of dynamic equilibrium during the motion. As for the concept of control, it has followed the global control scheme of the locomotor act, conceived, on the basis of the intuition and erudition of a brilliant scientist, by Bernstein [1]. In the both segments of research and development (modelling, and control system synthesis) in the domain of the artificial biped locomotion, the success has been made thanking to the fact that the appropriate approximations have been made in copying the human gait and its stabilization, so that an extremely complex task has been reduced to a relatively simple mechanics-control problem whose solution has provided a sound basis for the realization of the anthropomorphic active mechanisms. As has already been shown in the preceding chapters, the multitude of biological actuators and of corresponding equivalent d.o.f. responsible for the realization of the human gait is reduced to about ten mechanical d.o.f. powered by technical actuators. What has been achieved by such great simplification of the human locomotor act? It can be justly said that this simplification yielded the high-fidelity global results in studying the dynamics of the locomotor act, the determination of dynamic reactions of the gait, and assessing the energy balance of the realized gait and of the loads along skeletal links. Through the joints trajectories of lower extremities, which are practically identical to the trajectories of the human skeleton in the process of gait, then, through an appropriate choice of compensation by the ankle joint and/or trunk with the aim of satisfying the conditions of dynamic equilibrium, a global picture of dynamics of the whole system is obtained mimicing well the natural mechanism of the human gait. Besides, with the aim of maintaining the conditions of dynamic equilibrium and attaining stabilization of the nominal (programmed) motion, the procedures from large-scale system theory have been applied in which, apart from the local stabilization of subsystems, the feedbacks based on the dynamic reactions appearing at the points of contact between the foot and the ground were introduced for the first time into robotics, with the aim to preserve stability of the system as a whole, which is the main requirement resulting from the nature of the biped motion. Thus, it has also been shown that stability of the entire locomotion mechanism of man in the gait process cannot be attained without the regulation of dynamic reactions and of the ZMP. Obviously, this also holds for the case of maintaining the posture of the human skeleton. The artificial paralysis of the man's sense of reaction forces demonstrates in the most efficient way his incapability of maintaining his skeletal configuration in a stable state despite of the existence of the complete motor activity of his biological actuators (muscles). Because of that, in the state of genuine paralysis, e.g. of paraplegia type, the realization of the artificial gait and of a stable locomotion act is not possible without the use of force sensors mounted at the sites of contact between the foot and ground, or at some other places of the exoskeletal mechanism whose function is to restore basic locomotor activity of the handicapped man. Therefore, the measurement of dynamic reactions, described in the preceding chapter can replace practically the complete complex and sophisticated mechanism of the global dynamic equilibrium of the human gait and posture.

In a similar way, it is possible to speak of the validity of results concerning the analysis of energy consumption on the basis of the ma-

318

thematical models describing the dynamics of above locomotion (exoskeletal) mechanisms. Irrespective of the great simplification by lowe-
ring number of d.o.f. of the human locomotor mechanism and its reduction to a system of the exoskeletal type, for the same joints trajectories of the human gait and those imposed on the artificial locomotion mechanism and for approximately equal kinematic and dynamic parameters of the "natural" and "artificial" locomotion mechanism, the dynamic reaction calculated on the basis of the mathematical model correspond in both their distribution and intensity to the reactions measured during the human gait. Therefore, it is quite natural that the driving moments at the skeletal joints, especially of the joints of human
extremities and of the "legs" of the artificial skeleton are practically identical. That what is not identical to the quantities of the
human mechanism are the load moments for the calculated angles of the
compensating part of the artificial locomotion mechanism, especially
in the case when the role of the system compensation (dynamic stabilization is assigned exclusively to the trunk. Therefore, because of
the substantial simplification of the artificial system in comparison
to the human locomotion, we mainly lack the information on the distribution and transmition of forces and loads between the joints along
the human skeleton, because of the large number of muscles which have
not been taken into account, i.e. of their pairs and groups which par-
ticipate in forming driving moments about the human joints, and in the
realization of natural compensation of the human locomotion system.

Taking into account all the above statements, and starting from the
mathematical models developed for the dynamics of the biped gait, as
well as the stabilization scheme and complete dynamic equilibrium, the
problem has been tackled of realization of active exoskeleton, instead
of the measuring (passive) ones, intended for restoring basic locomotor activity of severely handicapped people.

In the text to follow the gradual development of active exoskeletons
will be presented, from the partial (without trunk movement), to the
complete ones, enclosing both the possibility of trunk actuation and
stabilization. Also, the latest results in the field of active orthotic systems will be presented based on modular design, for lower and
upper extremities of handicapped persons of dystrophic type in advanced phases of the disease. Beside other realizations of biped systems,
first results of hybrid joint realizations for anthropomorphic systems will
be presented, enclosing the residual muscular motor power of extremities.

## 4.2. Exoskeletal and Orthotic Systems

### 4.2.1. Introduction

Exoskeletons in general, are structures of rigid links, mounted on the body of some living vertebrae and following the main directions and having the main joints of the living orga-

Fig. 4.1. N.J. Mizen's passive exoskeleton

Fig. 4.2. General Electric's Hardiman

nism's endoskeleton. Most important application of exoskeletons is on man. They were developed in various research centers for two main uses:

1. As measuring structures, enabling experimental study of hyman body motion by recording the time histories of joint angles; first known and fine example of such passive exoskeleton is shown in Fig. 4.1. It was realized and tested by N.J. Mizen at the Cornell Laboratory, USA. This was an attempt to measure the most possible number of degrees of freedom of the human body in motion. This example of passive – measuring exoskeleton possessed 39 degrees of freedom, all being equiped by servo potentiometers [2].

2. As active structures, adding power to the human organism inside the active exoskeleton. The primary intention of these powered structures was, in fact, to enable normal man to perform overload tasks, especially in military applications. One of the rare published results from this area was the so-called hardiman by General Electric, built as prototype and tested in 1968 (Fig. 4.2). It was solved as a master-slave follower system, with electric actuators at the joints. The project was abandoned for lack of interest by the potential users [3].

Fig. 4.3. Measuring device for leg joint (1967)

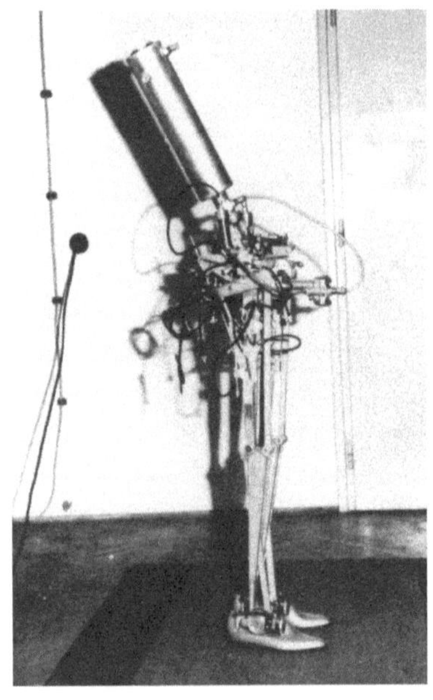

Fig. 4.4. Active model of legs with hydraulic power (1968)

A different application of active exoskeletons was concepted at the Biocybernetic Department of the "Mihailo Pupin" Institute in Beograd, Yugoslavia, as active exoskeletons for rehabilitation of disabled people, with locomotor deficiences of their lower extremities, for various reasons: polio, paraparesis, paralysis, dystrophia, etc.

Experiments and study with exoskeletal and similar structures started in the Biocybernetic Department already in 1967. First a passive--measuring structure was built, imitating the most important degrees of freedom of the human legs (Fig. 4.3), which enabled recording of the time histories of the leg angles during various types of human gait, and after that the first active, hydraulically powered walking system was built and tested (Fig. 4.4), enabling experimental verification of the theoretical results. This system already possessed a dynamic system for maintaining equilibrium during gait, consisting of the hydraulic fluid tank, mounted on top of the system, and performing cyclic motion by kinematic actuation for satisfying the basic conditions of nominal gait upon level horizontal surface [4, 5].

After successful completion of the initial investigations vigorous research and development work with rehabilitation orthotic devices was initiated.

## 4.2.2. <u>Development of active exoskeletons</u>
### <u>for rehabilitation</u>

In 1969, the so-called "kinematic walker" was developed (Fig. 4.5). It consisted of two main actuating pneumatic cylinders, driving via a ki-

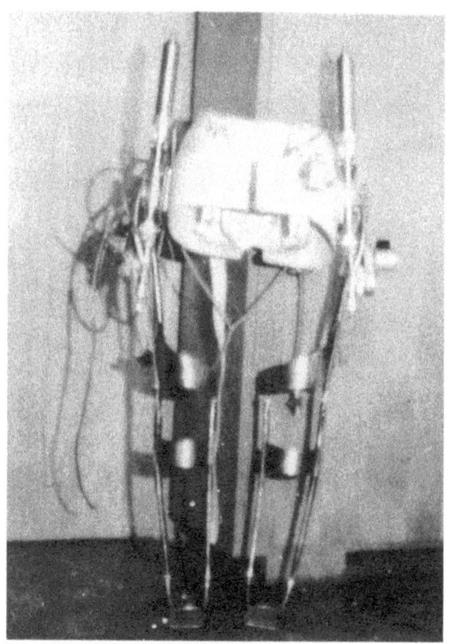

Fig. 4.5. Kinematic walker (1969)

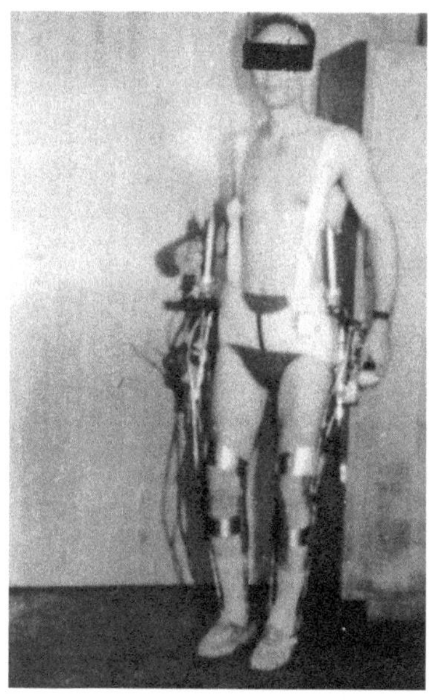

Fig. 4.6. Trials with kinematic walker

nematic linkage (i.e. in a fixed programme) both legs, where the hip and knee joints were driven (active) and the ankle joints passive. The produced gait was of the "sliding foot type". A simple electronic control system provided the necessary signal for triggering the pneumatic control valves, one per leg. Here, motion of the legs could be produced only in the sagittal (longitudinal) plane, which was not sufficient to satisfy even the so-called nominal conditions of dynamic equilibrium during the gait. In any case, extensive trials were conducted with this prototype exoskeleton, with a healthy subject (Fig. 4.6). It was proven, that such a simple structure succeeded in moving a man of medium size and weight who imitated a flaccid state to a sufficient degree, contributing to stability maintenance only. It was also proven, that man can adjust very easily and quickly to a machine, working in a repeating cycle. This experience proved of great importance for further work.

In the course of 1970 the first design of an active exoskeleton with three active degrees per leg, and a supplementary active degree of freedom, for moving the body in the frontal plane (left-right) for sta-

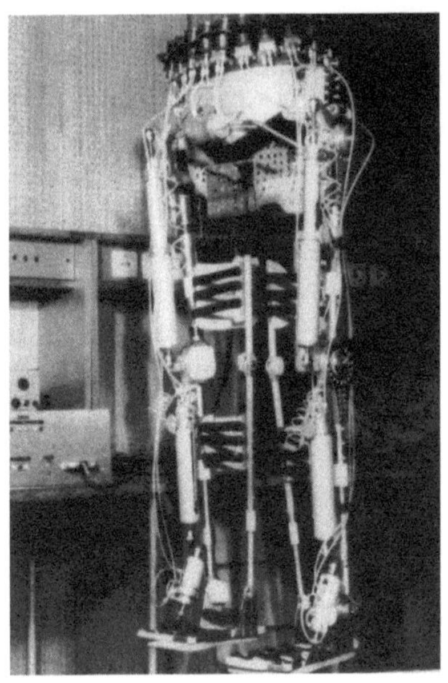

Fig. 4.7. "Partial" active
exoskeleton

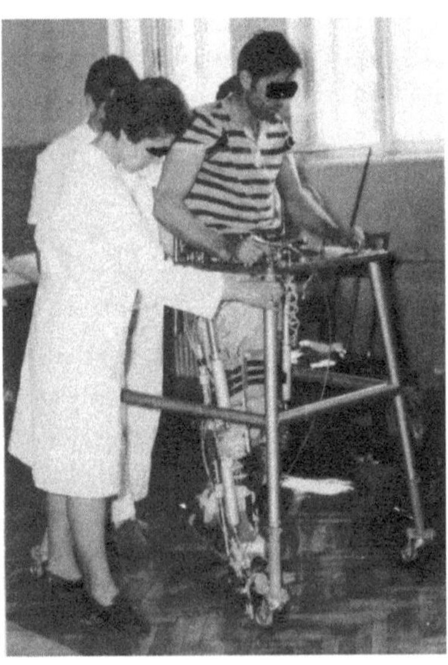

Fig. 4.8. Trial with patient

bility maintenance, was concieved and designed [6-8]. According to certain adopted criteria for evaluation, pneumatic drives were chosen for the first prototype, and also for some later, modified and improved types. Also, a very difficult problem, which had to be solved, was the one of mating the patient's body to the exoskeleton structure, which was burdened by the possibility of development of local wounds (decubitus) at places of higher pressure during prolongued use.

First models of corselets for joining the exoskeleton structure and the patient's body, were made of reinforced phenolic resins, with natural leather padding on the inside and, also padded with natural leather, semi-circular rests for thighs and ankles fastened to the extremities by means of "Velcro" strips. After performing intensive experiments with only one active leg of the exoskeleton, the first model of the so-called "partial" exoskeleton was realized (Fig. 4.7).

It employed 14 solenoid electromagnetic pneumatic valves for the control of seven pneumatic actuators of the exoskeleton, situated in a leather belt around the waist of the pelvic corselet. All pneumatic connections were realized in the form of thin hard nylon tubes formed in helicoids, needed for freedom of joint motion. The control system of the exoskeleton was in the form of an electronic diode function generator connected to the exoskeleton by means of long cable, whereby the compressed air for powering the exoskeleton actuators, was conducted via a plastic hose along the said cable [9 - 12]. Trials with this first "partial" type of exoskeleton (Fig. 4.8) proved, that a full paraplegic can walk with the help of the exoskeleton only when supported by two people or in a rolling aid, as illustrated. Namely, it was not possible to actuate the body trunk of the patient for aiding stability maintenance [13]. For this reason, a new version of the pneumatic active exoskeleton called the "complete" exoskeleton was realized in same (1971) year. It was characterized (Fig. 4.9) by a high corselet, enclosing totally the chest of the patient's body, which enabled controlling the motion of the whole trunk in both the sagittal and frontal plane. In that way, it was possible to realize a nominally much more stable gait enabling thus some of the trial patients to master stable walking with crutches only.

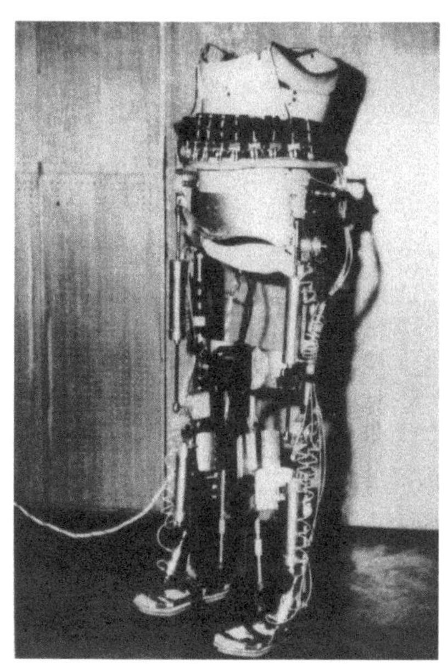

Fig. 4.9. "Complete" exoskeleton

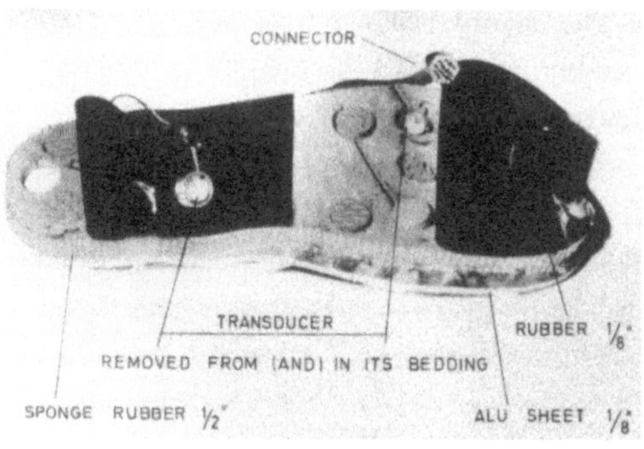


Fig. 4.10. Force feedback

A substantial technical improvement was introduced with the exoskeleton which included force feedback from the soles of the artificial "feet" in the form of three piezo-ceramic force transducers, embedded in the soft sponge rubber on the soles and covered by a layer of medium-hard rubber (Fig. 4.10). The signals of the three force transducers on each foot were processed in the control system and used for correction signals, led to the pneumatic actuators, producing corrective action for keeping the resultant reaction force of the ground to the feet in some prescribed boundaries and ensuring the protection from overturning. This quite sophisticated system in fact did practically work and demonstrated a favourable influence on both

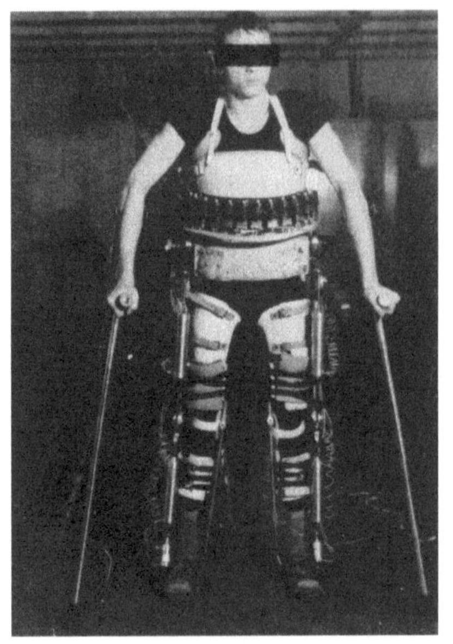

Fig. 4.11. Patient in "complete"
exoskeleton

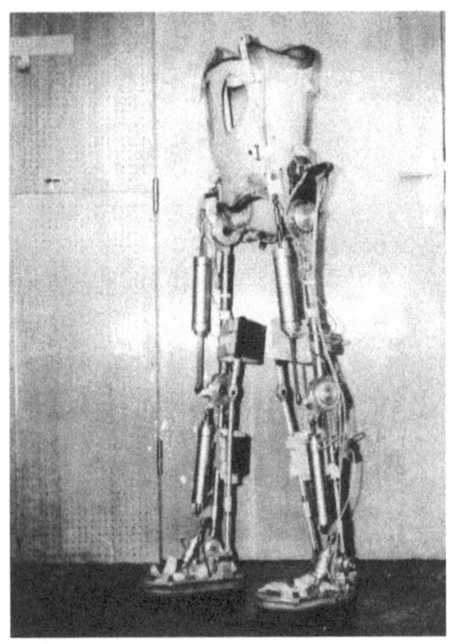

Fig. 4.12. Latest model of active
exoskeleton

the stability of standing (posture) and walking.

Several paraparetics and paraplegics were able to master gait with this type of "complete" exoskeleton, one of them being shown in Fig. 4.11. The patient was able to walk with the aid of the exoskeleton, using crutches, in a walkway with bars and with a light four-legged aluminum support.

After a large series of trials with the "Complete" exoskeleton [13] (over 100 trials at the Orthopaedic Clinic in Belgrade and some other rehabilitation institutions) it was concluded that the main drawback of the existing exoskeleton models was its high weight, mainly due to the large mass of the 14 industrial solenoid valves and steel, used in the construction of the carrying elements of the exoskeleton.

In the meantime, a new series of miniature solenoid valves was released by a major producer and this prompted us to completely redesign the active exoskeleton, using light aluminum alloys and plastics, reducing the initial weight of the "Complete" exoskeleton from 17 kg of the old model to 12 kg of the new one, which presented a very important saving, producing the best active exoskeleton with pneumatic drives of the classic concept, ever built. This exoskeleton model (Fig. 4.12) enabled the patient very agile gait with the light support, passing through doors, turning, etc. (Fig. 4.13).

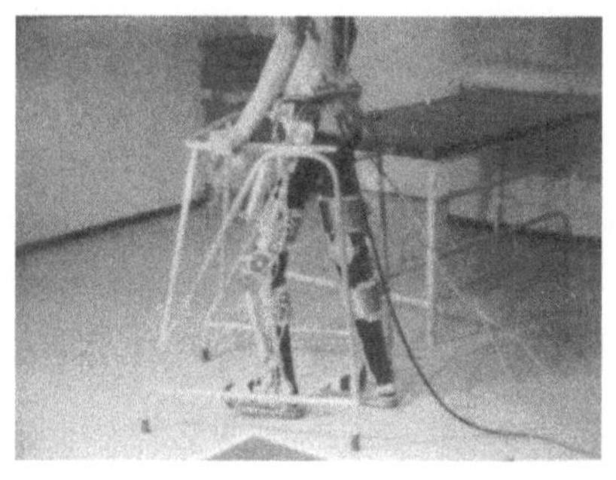

Fig. 4.13. Patient gait with
light support

In that phase of development, during 1973, it was concluded that further progress with pneumatically driven active exoskeletons for rehabilitation, was not practical and promising [12]. The fact, that the motion energy had to be supplied from some source of compressed air, confined this type of walking aids to closed, mostly clinical, environments. Thus, this type of exoskeleton could be used for training and therapeutical purposes in medical institutions only. Concerning weight, some savings could have been achieved by the use of carbon fibre technology, but this was difficult to acquire and very expensive. Otherwise, apart from the exoskeleton the problem of energy source, could not be solved in this way.

Hence, it was decided to switch to development of an active exoskeleton with electromechanical drives [12]. The result of an almost two years' development, was the "Complete electrical" exoskeleton (Fig. 4.14)

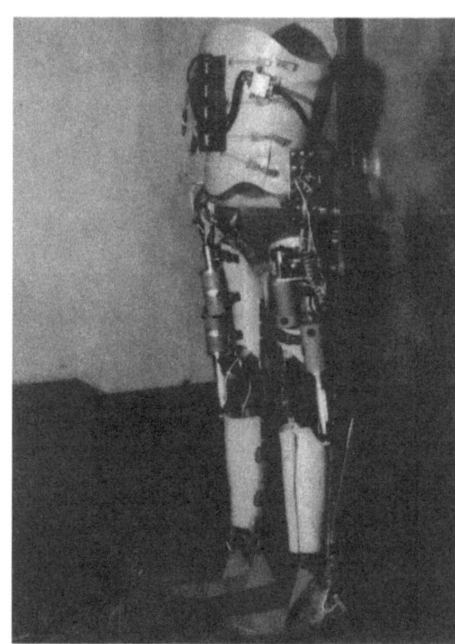

Fig. 4.14. "Complete electrical"
exoskeleton

which proved to be somewhat on the heavy side, weighing about 16 kilograms [12]. There were many rather sophisticated solutions applied in the design of the exoskeleton structure: besides the basic motion of the leg in the sagittal plane, the hip joint performed two kinematically programmed motions of "pelvic twist" and sideways motion, for more stable and realistic gait. Then, the feet had kinematic drives for sideways inclination for stability improvement (the whole foot on the ground during the stance phase) and the knee joints were driven by a specially designed, play-free worm-reducer. However, in 1974, when this exoskeleton type was realized, there were no available yet sufficiently small and light batteries for the power source and the computer techniques had just entered the microprocessor era. For these reasons the power source and control system were outside the exoskeleton, and this greatly limited its applicability.

However, experiments with the servoelectric D.C. drives of the electric exoskeleton  were valuable as they demonstrated about 3-4 times better tracking quality if compared with the pneumatic ones. Moreover, with pneumatic drives it was never possible to achieve smooth motion, because the solenoid valves were triggered by electric pulses in a low frequency range of 10-25 Hz, and the relatively fast response of the pneumatic drives caused the whole system to operate in a "pulsating" form, induced by small-amplitude oscillations around the programmed trajectories. This phenomenon did not discredit operation of the system, but it caused unesthetic impression when looked at; also the pulsating sound of the numerous solenoid valves was irritating.

Hence the appearance of more suitable (lighter and more powerful) electric actuators and new microprocessor control technology enabled to start developing a new generation of orthotic devices for the rehabilitation of disabled persons.


4.2.3. <u>Modular active orthosis - the "active suit"</u>

As continuation of the research and development efforts in the "Mihailo Pupin" Institute in the field of active rehabilitation devices of the exoskeleton type, a new idea evolved in the period of 1974-1978 [14, 15]. This idea was: create an assistive device for producing artificial gait with handicapped persons, the device to be of modular design, as much standardized as possible and to be of semi-soft structure for more comfort of use and less decubital danger. Concerning the type of patients, we had in mind a large  population of severely handicapped, the so-called "dysthrophics", encompassing many types of rather similar deseases, at least as external manifestation is concerned (Duchenne's muscular dystrophy, Becker's muscular dysthrophy, facio-scapulo-humeral dysthrophy, congenital myopathy, spinal muscular athrophy of Kugelberg-Welander type and of adult type, chronic polyomyositis and difficult cases of myasthenia gravis complicated with myasthenic myopathy). All these patients, depending on the type and stage of illness, suffered from more or less expressed difficulties with producing normal gait, even upon flat ground, not to speak of climbing or descending stairs. External demonstration of these diseases was, in short, a

weakness of the neuro-muscular locomotor apparatus. In some cases, main
weakness appeared in the hips only, while in other  in knees, too. Most
of the patients of this group were of the so-called "proximal" type,
which means that the more centrally located joints of the human skeleton with the adjoined muscle groups failed first.

As a results of thorough  studies of the needs of handicapped persons in
the described illness group, a new type of autonomous, modular, active
orthotic device was designed and realized. The "active suit" is an autonomous (self-contained), microcomputer-controlled active skeletal orthosis powered by servoelectric drives, for adding external motion
energy to the hip, or, if needed, to the hip and knee joints of the
handicapped. It has been realized in the form of semi-rigid body cor-

Fig. 4.15. Parts of corselet of
"active suit"

selets (body, thigs and shanks of
legs, 5 in total), manufactured from
strong felt with light alloy stiffeners. These stiffeners serve also
for transmitting lateral driving
forces and reactions to the body
parts (Fig. 4.15). These metal parts
also serve for connecting to the actuator units (in the form of miniature, rather powerfull D.C. servomotors) with special-purpose gear
reducers. For the hip joints, two
such motors were used, giving an actuator of about 100 W output power,
which is sufficient for a medium pace of a medium-weight person upon
level ground, and slow stairs ascent
(Fig. 4.16). In the case of more advanced stages of illness, at each
knee joint, a single-motor actuator
with worm-gear reducer was applied
(Fig. 4.17). Mechanical power of these knee actuators was, of course,
about 50 W. The whole system was controlled by a miniature, chest-mo-
unted microprocessor control system, driving the actuators via transi-
storized servoamplifiers. Main parts of the control system can be seen
in Fig. 4.18. [17]. On top of the control unit, at an easy reach of the
user is the control panel, consisting of several switches, turn-button
commands and signal lamps. By activating the main two-position switch,
the device is energized, shown by lighting of a green LED. Type of

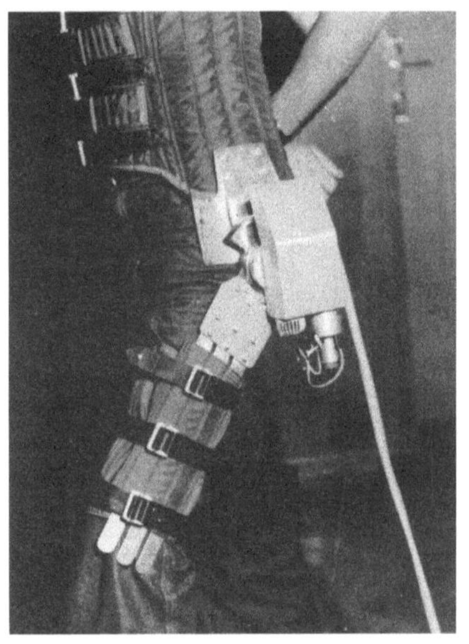

Fig. 4.16. Hip actuator

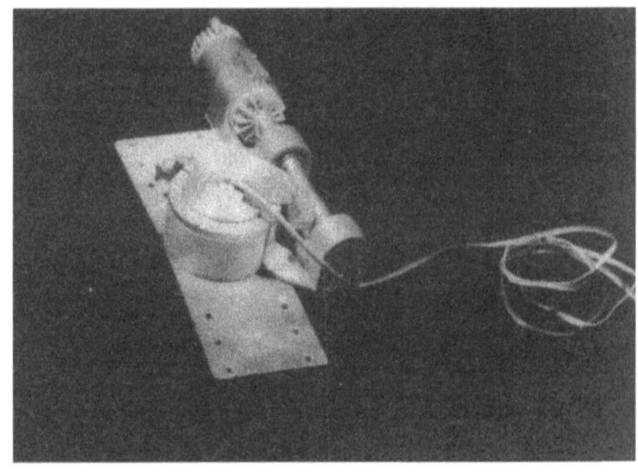

Fig. 4.17. Knee actuator

Fig. 4.18. Control system

gait, i.e. gait upon level ground, upstairs or downstairs is selected by the three-position mode selector switch. By throwing the two-position stop-start switch to START, gait begins, always with the left leg. The user can adjust the gait pace, step stride and turning to left or right, by turning the corresponding button commands. There is also a "battery low" red LED indicator and a RESET button.

The whole system is powered by a nickel-cadmium battery of about 2 kg, situated on the back of the patient, providing a gait autonomy of about 45 minutes uninterrupted gait upon level ground, or climbing 2-3 times the stairs to the third floor. The battery can be fully re-charged during four hours by a specially designed fast charger.

Very good results were obtained during the numerous trials with this type of active orthotic aid. A young dysthrophic with amyothropia spinalis progressiva in the VI stage was chosen for the experiments. He mastered the use of the "active suit" very quickly and was able to use it with no problems. Fig. 4.19. shows a general view of the patient in the "active suit" and Fig. 4.20. gives photograph of the patient climbing stairs with the orthosis. A more advanced type of the "active suit"

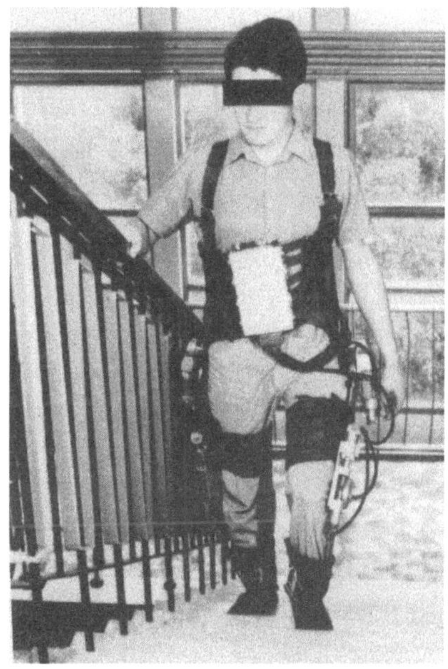

Fig. 4.19. "Active suit" and patient     Fig. 4.20. Climbing stairs

was developed after 1978 when the prototype was tested. It contained an improved control system, lighter battery pack, and at each knee joint one more electric actuator, described before and shown in Fig. 4.17. This new model (Fig. 4.21) was realized in 1980 and it was also tested with success on healthy subjects.

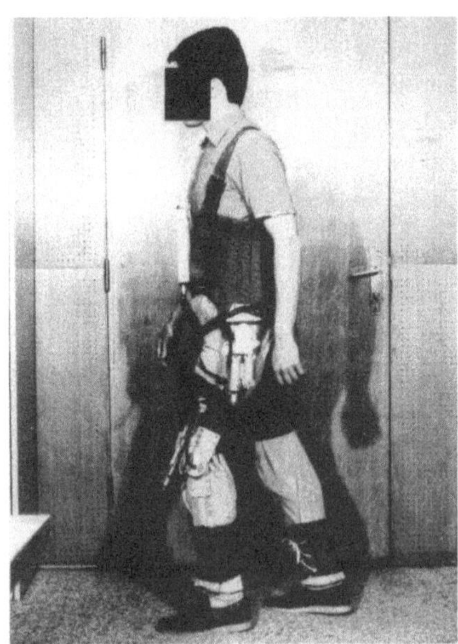

Fig. 4.21. Latest type of "active suit"


4.2.4. <u>Other biped realizations</u>

There have been several attempts to realize working models of artificial bipedal walking devices. Most important centers of this research
and development are in Japan and the USSR.

In the Robotics Laboratory, named after its founder and head prof.
Ichiro Kato, the KATO LABORATORY, at the known technical WASEDA UNIVER-
SITY in Tokyo, a series of walking bipedal robots of the WABOT family
were realized in the course of 1973 - 1986 [16]. Two interesting re-
presentants will be presented here.

The WL-10RD (Fig. 4.22), the <u>W</u>aseda <u>L</u>eg - 10 <u>R</u>efined <u>D</u>ynamic was realized in 1985 and was able to ascend and descend stairs and also ramps
of small inclination, with a pace of 2 to 5 seconds per step. On the
level ground, the system is capable of walking with a gait of 1,3 sec/
/step.

Further refinement of the solution was found in applying the idea of

Fig. 4.22. WABOT WL - 10RD

Fig. 4.23. WABOT WL - 12

keeping the ZMP in some predetermined region inside of foot contour, realized in a similar way as in the hydraulically powered bipedal walking system developed at the "Mihailo Pupin" Institute already in 1968 (Fig. 4.4).

The Japanese solution is shown in Fig. 4.23. It is the WABOT WL - 12 from 1986, realized at a rather high technological level, featuring microcomputer control, modern technology force sensors on the feet and at the joints, and hydraulic actuators from the new generations at the joints. The WL - 12 was capable of walking in a fully stable condition with a 2.6 s/step gait pace, using a newly proposed algorithm, which automatically computed the time trajectory of the body, while arbitrarily prescribing the trajectories of the lower limbs (legs) and the ZMP.

Way back in 1973, at the Tokushima University in Tokushima, Japan, an active exoskeleton for rehabilitation of paraplegics was realized (Fig. 4.24), which had in principle the same kinematic solution of leg drives as the first kinematic walker, built and tested at the "Mihailo Pupin" Institute in 1969 (Fig. 4.5). The essential difference between the two solutions was in the electric drives of the Japanese exoskeleton, realized by means of D.C. servomotors, as compared with the electropneumatic drives of the Yugoslav version.

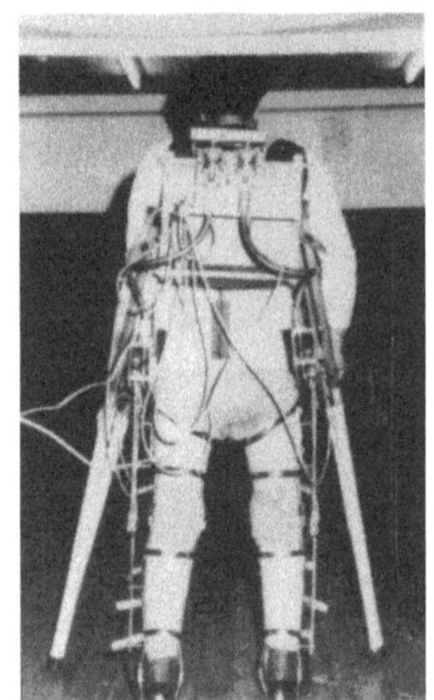

Fig. 4.24. Tokushima active
exoskeleton

At the CITO - Institute in Moscow
(<u>C</u>entral <u>I</u>nstitute for <u>T</u>raumatology
and <u>O</u>rthopaedy) in 1976 it was realized an electrically driven, kinematically programmed, active exoskeleton for the postoperational rehabilitation of orthopaedic character with children. It was reported-
ly successful in the application
and use.

A very interesting project with an
artificial walking mechanism of semi-
-anthropomorphic type was finished
at the Institute for Mechanics of
the Moscow State University Lomono-
sov in 1988 [17]. It  was a walking device consisting of the body and
two telescopic legs (Fig. 4.25). Rotational joints at the knees were

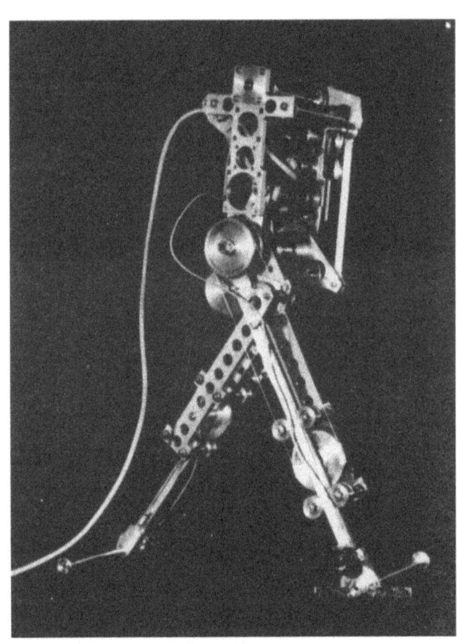

Fig. 4.25. Walking device with
telescopic legs

not present. Each leg was powered
by two D.C. servomotors as actuators: one for driving the rotation
of the hip joint, the other the ex-
tension-retraction of the leg. By
means of computer calculations, the
puls-width control laws were obtained, rendering stable nominal gait
regime of the apparatus. The feedback was realized in the form of a
linear control law, stabilizing the
nominal gait regime. The stability
margin of the essentially nonlinear
system, when stabilized by linear
feedback, was found to be small. It
was found that a much better measuring and actuation accuracy, as well
of the mechanical realization of the
working model, would be desirable.

In Chapter 3 two realizations of the
biped system in the form of their equivalent mechanical schemes were
presented (Figs. 3.12 and 3.13), thus they will not be repeated here.

The new biped version as compared with the one presented in Fig. 3.13 is given in Fig. 4.26 [18]. Kenkyaku-2 is a seven link biped robot with 40 kg weight and 1.1 m height. The stride length is 35-45 cm and the walking period per step is 0.7-1.0 s. Two steel pipes are attached to the feet to maintain the lateral balance, so the robot can realize two dimensional walking.

Fig. 4.26. Kenkyaku-2 robot

A very interesting biped robot of anthropomorphic type BLR-G1 is presented in Ref. [19]. Newest version of this robot is the BLR-G2 which is a nine link biped robot with 25 kg weight and 0.97 m height. The hips, knees and ankles are provided each with a pitch shaft of one degree of freedom. The soles have each a roll shaft of one degree of freedom. BLR-G2 is provided with sole pressure sensors, ankle torque sensors, an inclinometer, an accelerometer and a body-speed sensor. It achieved 3D dynamic walking with kick-action during both single support phase and double support phase. The walking speed is 0.35 m/s with the stride length 0.35 m. Sequence of photos shows one complete stride of the BLR-G2 robot (Fig. 4.27).

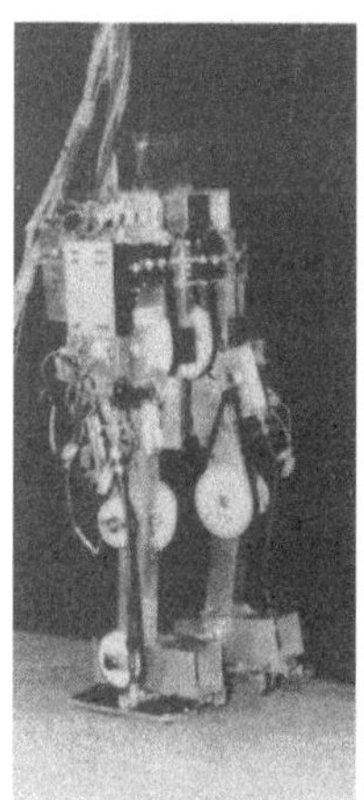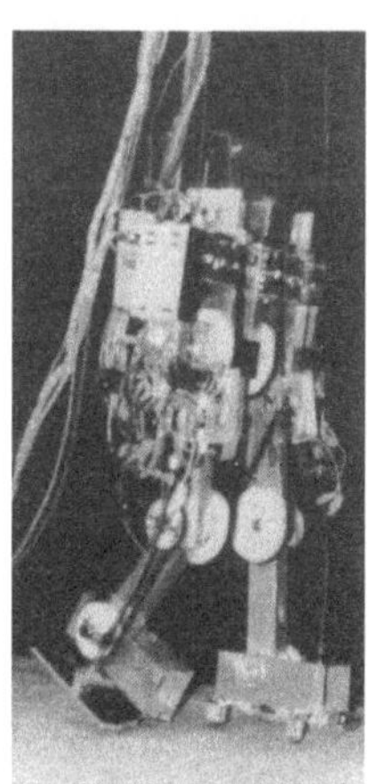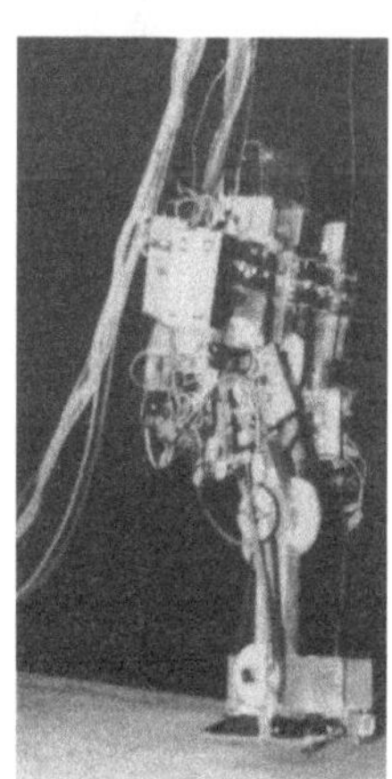

Fig. 4.27. Complete stride of BLR-G2 robot

## 4.3. Hybrid Joint Concept

### 4.3.1. General remarks

Restoring basic skeletal (locomotion and manipulation) activity of man with the aid of exoskeletal externally powered devices is essentially a robotic approach to the rehabilitation of disabled people. Since the subject in question is deprived of motor activity of, for example, lower extremities (paraplegia), the active devices provide the energy necessary for the motion of extremities. However, when there is a certain residual motor activity (in the case of paraparesis of distrophy) it is advisable to make use of it.

Already twenty five years ago, a successful attempt was made and some encouraging initial results were obtained in the domain of functional electric stimulation (FES) of the muscles and muscle groups [20]. The FES assumes the presence of artificial innervation of muscle groups by electrical pulses. The past period has been characterized by an intensive activity in the field of restoring functions of human extremities and organs by means of electrical stimulation with the aim of ensuring their basic functioning. Though the problem of FES goes beyond the scope of this chapter we shall mention some of the most important references concerning the peroneal stimulator and multichannel stimulator [21]. In spite of the indisputable results obtained in the domain of FES the following factors, especially in the case of more complex and prolongued muscle motor activities, limit its application:

- non-repeatable response of stimulated muscles to FES,

- fast fatigue and even pain of muscles during prolongued activity.

These reasons, as well as the possibility of integration of the residual motor activity and external powering of human extremities joints, brought about the idea of the so-called hybrid actuator [22]. In the cases of existence of certain residual motor activity of the muscle system it would be necessary to distinguish two concepts of rehabilitation - the robotic concept and FES. In this way it would be possible to eliminate or substantially reduce the above drawbacks of the FES and the purely robotic approach would be hybridized and advantage taken of its utilization.

Here are presented preliminary results of the research and development of hybrid joint and its control system obtained in the Laboratory for Robotics in the "Mihailo Pupin" Institute [31, 32]. The results obtained up to now are related to the mathematical modelling, simulation, control system synthesis, and the realization of the prototype of hybrid joint, in which the stimulated muscles are supported by external power via an electromechanical orthotic drive.


### 4.3.2. Efferent functional electrical stimulation model


The process of FES provides or improves functional movements of an abnormal neuromuscular system by the application of electrical pulses to the efferent or afferent peripheral neural fibres [21]. These pulses are supplied by either skin electrodes or implantable stimulators, and are controlled by volitional signals by the patient who thus regains, to some extent, voluntary control over his paretic muscle [21].

One can use either efferent or afferent FES to attain functional movements [21]. In efferent stimulation (EFES) the patient activates the stimulator with his volitional control signals (VCS) which excite the efferent nerves or muscle end plates and thus produce direct muscle contraction. According to [23], the force developed by the single muscle contraction is equal to the product of the active state, the length-tension function and the velocity-tension function of the muscle.

According to [24], the active state q can be represented by the following first-order differential equation:

$$\dot{q} + cq = cQA, \qquad 0 \le Q \le 1, \qquad\qquad (4.3.1)$$

where A denotes the maximum tension a single muscle can develop, c is the elasticity constant, $Q = Q(r)$ is the "desired" active state as a function of the input stimulus rate r. In the limiting case $Q = 1$ and the input of the active state remains proportional to the maximum tension. Also, in the relative muscle length interval $1 \le L \le 1.125$ the length-tension function is equal to 1 [24], as well as the velosity-tension ratio $F/F_0$ (where $F = F(x)$ and $F_0 = \max(F)$) is equal to 1 in the maximum limiting case, where x is the muscle shortening, $L = x_0/x$ is the muscle relative length and $x_0$ is the initial muscle length.

The exponential curve q (a solution of equation (4.3.1)) represents the most simple EFES model, global behaviour agreement of which with average empirical force-time curves was obtained both "in vitro" and "in vivo" [25]. But some in-depth analysis of the curve itself suggests the existence of inflexion somewhere arround the start of contraction, while classical mechanics of contraction requires inclusion of the known Hill's force-velocity relation [26] into any macroscopic muscle model. Wilkie's variant of this basic relation [27], describing both contractile and series elastic component behaviour of the shortening muscle can be stated as

$$\dot{x} = \frac{b^*(F_o - F)}{F + a^*} - \frac{d}{dt}(c \cdot x) \qquad (4.3.2)$$

where F is the muscle tension, $a^*$ corresponds to the energy dissipation during the contraction, $b^*$ is the phosphagenic energy transcuding rate [26]. These constants can be put into the dimensionless form by the use of fractions

$$a = a^*/F_o, \qquad b = b^*/\dot{x}_o, \qquad \dot{x}_o = F_o/(a^* b^*) \qquad (4.3.3)$$

The synaptic FM/AM modulation of the EFES signal, playing the role of input transducer to the contraction, can be represented by a function of the mediator (e.g. acetyl-choline, or noradrenaline, or cerotonine) ammount, d, on the actual neuro-muscular connection (end plate) [28]. For the purpose of the hybrid joint simulation, d=1 is adopted as ideal case of neuro-muscular transmission. Finally, the EFES signal transmission through efferent nerves should satisfy the known Hodgkin-Huxley's law of impuls conduction Q [29] taking the Heaviside's excitation form:

$$Q = Q_o(h(t) - h(t-\tau)), \quad h(t-\tau) = \begin{cases} 0, & t-\tau < 0 \\ 1, & t-\tau > 0 \end{cases} . \qquad (4.3.4)$$

Now, the EFES - response mapping of the skeletal muscle can be stated in the form of the force generator F:

$$T^2\ddot{F} + 2aT\dot{F} + cF = bQ, \qquad (4.3.5)$$

where T is the time characteristic of the muscle, and other parameters are as explained above.

As the condition $a^2 > c$ is naturally fulfilled for skeletal muscles, the response has exactly the desired form, completely agreeing with the empirical curves [25], as presented in Fig. 4.28.

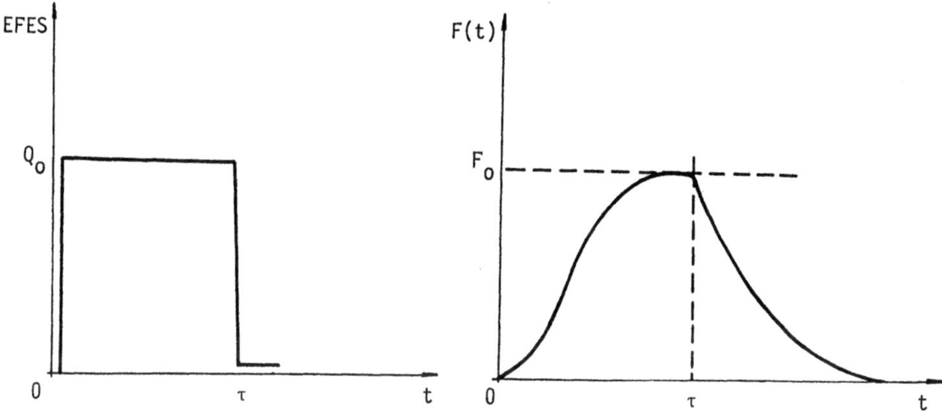

Fig. 4.28. Heaviside's excitation and muscle response

### 4.3.3. The hybrid system model

The EFES orthosis for the pair of mutually antagonistic muscles is combined with an external power drive into a hybrid joint orthosis (Fig. 4.29).

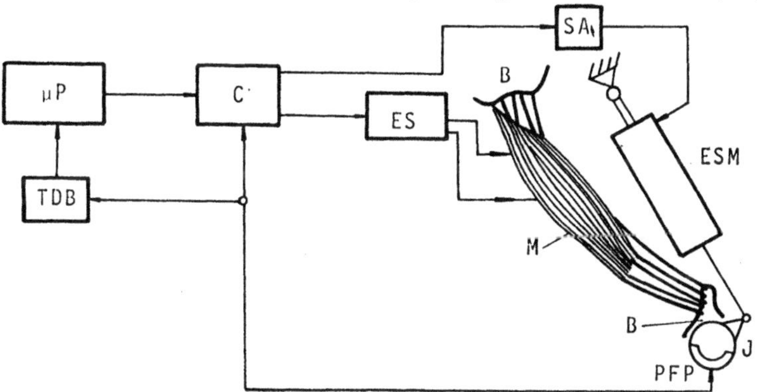

$\mu$P - microprocessor; C - controller; SA - servo-amplifier, ESM - electric servomotor, TDB - time-delay block, ES - electric stimulator; B - bone, M - muscle; J - joint; PFP - position feedback potentiometer

Fig. 4.29. Scheme of hybrid joint orthosis

(a) *External actuator*

A D.C. servo motor is used as an external actuator. Its model is given by the differential equations [30]:

$$L_R \dot{i}_R + r_R i_R + C_E N_v \dot{\theta} = u \qquad J_M N_v \ddot{\theta} + \frac{F_v}{N_m} \dot{\theta} + \frac{M^*}{N_m} = C_M i_R \qquad (4.3.6)$$

where $i_R$ is the rotor current, $\theta$ is the angular position of the output shaft, u is the rotor input voltage, $C_E$ is the electromotor force constant, $C_M$ is the torque constant, $F_v$ is the output shaft viscous friction, $N_v$ is the speed reduction ratio, $N_m$ is the torque multiplier ratio, $M^*$ is the motor external loading torque, $J_R$ is the rotor moment of inertia reduced to output shaft, $J_M$ is the rotor moment of inertia, $J_R = J_M N_v N_m$.

(b) *Equation of motion*

The dynamic equation of rotational joint motion $\ddot{\theta} = f(\theta, \dot{\theta}, i_R, F_k, t)$, representing a unique mapping f is proposed in the linear form:

$$J\ddot{\theta} + F\dot{\theta} + C\theta = \sum_{k=1}^{2} \sigma_k (F_k \cdot r_k - mg\ell\sin\theta) + C_M N_m i_R \qquad (4.3.7)$$

where J is the combined moment of inertia of the actual human body segment and motor output shaft, F is the combined viscous friction of the actual joint tendon and motor output shaft, C is the tendon elasticity coefficient, $\sigma = 1 (k=1)$ for agonistic (i.e. flexor) muscle and $\sigma = -1 (k=1)$ for the antagonistic (extensor) one, $F_k = F_k(t)$ is the actual muscle force (4.5), $r_k = r_k(\theta)$ is the joint lever arm of the actual muscle tendon, $2\ell$ is the actual segment length, m is the mass of the segment.

(c) *The state space model* [31, 32]

The entire mathematical system model, including the equation of motion (4.3.7), two antagonistic muscle force FES generators (4.3.5) and the external actuator (4.3.6) could be put into the canonical form by the substitutions: $x_1 = \theta$, $x_2 = \dot{\theta}$, $x_3 = \dot{F}_1$, $x_4 = F_1$, $x_5 = \dot{F}_2$, $x_6 = F_2$, $x_7 = i_R$, $u_1 = Q_1$, $u_2 = Q_2$, $u_3 = u$. The input signal u is activated for $t > T_m$, where $T_M$ is the time constant of the motor torque application, corresponding to the time necessary to reach 70% of maximum muscle force (the simulation case).

$$\dot{x}_1 = x_2$$

$$\dot{x}_2 = \frac{1}{J}(-Cx_1 - Fx_2 \pm r_1 x_3 \mp r_2 x_5) + \frac{1}{J}(C_M \cdot N_m\, x_7 - mg\ell\sin x_1)$$

$$\dot{x}_3 = x_4$$

$$\dot{x}_4 = \frac{1}{T_1^2}(-c_1 x_3 - 2a_1 T_1 x_4) + \frac{1}{T_1^2}(b_1 u_1) \qquad (4.3.8)$$

$$\dot{x}_5 = x_6$$

$$\dot{x}_6 = \frac{1}{T_2^2}(-C_2 x_5 - 2a_2 T_2 x_6) + \frac{1}{T_2^2}(b_2 u_2)$$

$$\dot{x}_7 = \frac{1}{L_R}(-C_E N_m x_2 - r_R x_7) + \frac{1}{L_R} u_3$$

or in the matrix form:

$$\dot{x} = Ax + Bu + f \tag{4.3.9}$$

where x denotes the state vector, u is the input vector, A is the system matrix, f is the vector of driving torques and loadings distribution and B is the matrix of input distribution:

$$A = \begin{bmatrix} 0 & 1 & 0 & 0 & 0 & 0 & 0 \\ -\dfrac{C}{J} & -\dfrac{F}{J} & \dfrac{r_1}{J} & 0 & -\dfrac{r_2}{J} & 0 & \dfrac{1}{2}C_M N_m \\ 0 & 0 & 0 & 1 & 0 & 0 & 0 \\ 0 & 0 & -\dfrac{C_1}{T_1^2} & -\dfrac{2a_1}{T_1} & 0 & 0 & 0 \\ 0 & 0 & 0 & 0 & 0 & 1 & 0 \\ 0 & 0 & 0 & 0 & -\dfrac{C_2}{T_2^2} & -\dfrac{2a_2}{T_2} & 0 \\ 0 & -\dfrac{C_E N_m}{L_R} & 0 & 0 & 0 & 0 & \dfrac{r_R}{L_R} \end{bmatrix} \tag{4.3.10}$$

$$B = \begin{bmatrix} 0 & 0 & 0 \\ 0 & 0 & 0 \\ 0 & 0 & 0 \\ \dfrac{b_1}{T_1^2} & 0 & 0 \\ 0 & 0 & 0 \\ 0 & \dfrac{b_2}{T_2^2} & 0 \\ 0 & 0 & \dfrac{1}{L_R} \end{bmatrix}, \qquad f = \begin{bmatrix} 0 \\ -\dfrac{1}{J}mg\ell\sin x_1 \\ 0 \\ 0 \\ 0 \\ 0 \\ 0 \end{bmatrix}$$

(d) *Hybrid system control*

The hybrid control system consists of three components:

- external control, which is composed of a nominal (programmed) and a
  local (position and velocity feedback) control [33];
- neural control, which is the result of FES application;
- biochemical control, which is the result of fatigue and pain in muscles, as a consequence of the decline of ATP (Adenosine-Tri-Phospha-
  te) contents in muscle fibres and augmentation of the lactate contents, representing a natural barrier to the quantity and duration
  of the available energy from the muscles innervated by means of FES.


### 4.3.4. Simulation results

Simulation was performed for the elbow joint (total time 2s). In all simula-
tions zero initial conditions for all state variables as well as zero inputs
have been used. Therefore arm was in resting elbow extension at the start of
movement. In all simulation agonistic (flexor) muscle was stimulated by an
input ramp signal of $10t$[V] while the antagonistic (extensor) with
$3t$[V], for $t \leq T_M$, $T_M = 1s$ being the instant of external power application.

In Figs. 4.30 and 4.31 are presented characteristic results of the hybrid joint simulation, where the external actuator inputs were $20h(t-T_m)$
[V] step and $20 \sin(t-T_m)$[V], respectively. It can be seen that the
sine-form input produces much smoother time history of both joint angle and velocity, compared with the response of same joint subjected
to step input of equal amplitude.

Hardware realization of the hybrid joint prototype is given in Fig.
4.32.

Further development of hybrid joints of modular orthotic devices is
both in improving of the fidelity of the muscular system mathematical
models, as well as more appropriate modelling of the joint itself with
greater number of degrees of freedom, being met with some joints of
human extremities.

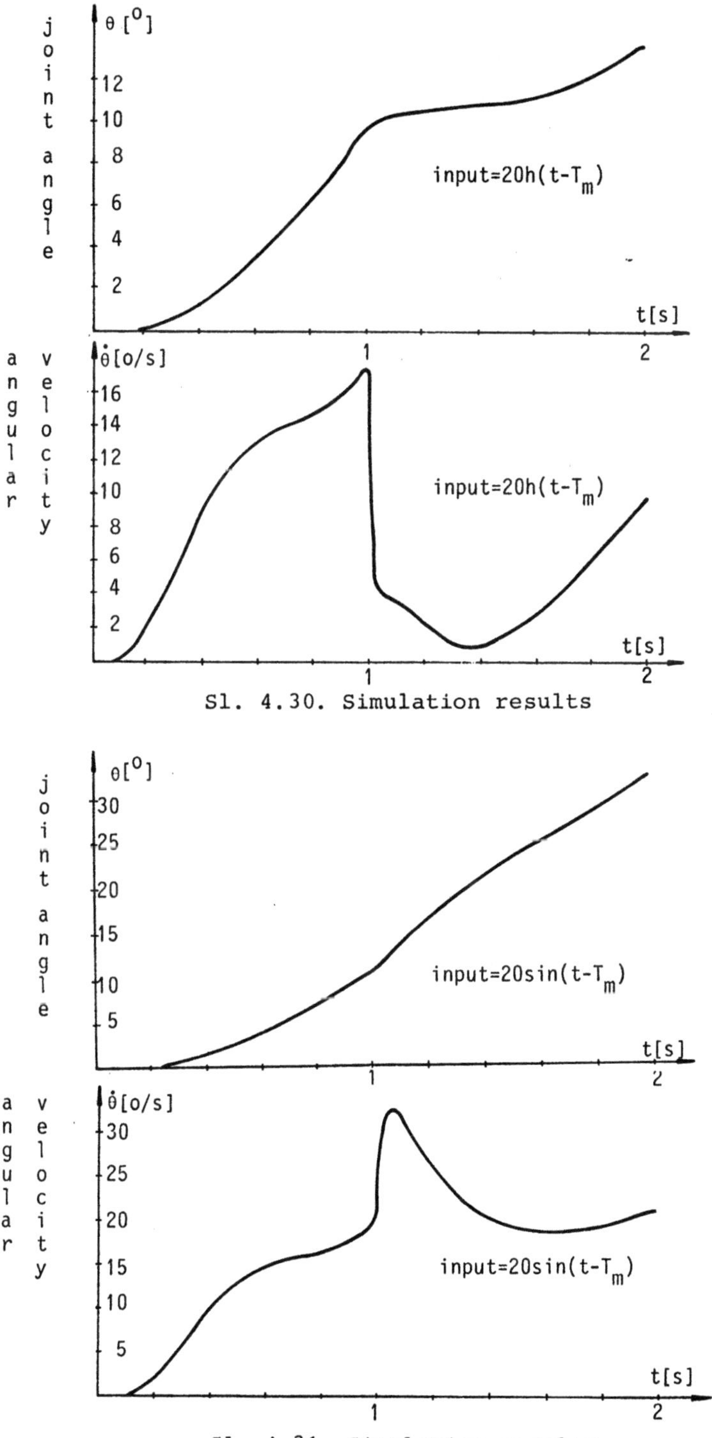



Sl. 4.30. Simulation results



Sl. 4.31. Simulation results

Fig. 4.32. Hardware realization

## 4.4. Legged Robots

Legged robots are outside the tematic scope of this book. Here only
two characteristic representatives of non-anthropomorphic legged robots will be given. The first is the OSU (Ohio State University) hexapod and the second the hopping machine. More detailed information
about them can be found in Ref. [3].

After the pioneering research and laboratory realizations of the four-
-legged machine [34], attention of the researchers has been foccused
during the last ten years on six-legged vehicles [35, 36], as machines
of greater possible practical significance and application.

Research with hexapods is continuing. At the present moment, the lea-
ding is probably the OSU hexapod vehicle (Fig. 4.33). Its basic dedi-
cation is rough-terrain application. The hexapod uses the statically
stable gait. It is 5 m long, 3.3 m high and 1.6 m wide and around
2600 kg weight. It is carrying a driver and is designed for a payload
of 255 kg; the maximum speed is 3.6 m/s.

The coordination of foot placement and leg movement and movement and
the control of body attitude is performed by 13 Intel 86/80 microcom-
puters enabling the driver-operator to attend the more strategic decisions about overall speed and direction. The operator can have continuous simultaneous control of forward speed, lateral speed and rate of
turn, and can also set required values for body height and attitude.
The machine uses vertical gyroscopes, rate gyroscopes and accelerome-

Fig. 4.33. OSU adaptive hexapod
vehicle

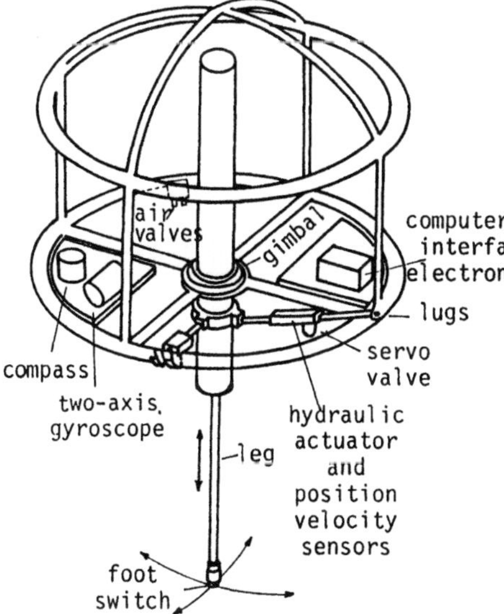


Fig. 4.34. Scheme of 3D hopping
machine

ters to sense body movement and attitude; all legs are fitted with proximity and three-axis force sensors.

The most extreme case of balance in locomotion is leaping on a single leg. The most advanced research is at Carnegie Mellon University (CMU's) Robotics Institute [37]. The hopping machine shown in Fig. 4.34. was designed for experiments on balance in 3D. The main parts are a springy leg and a body, connected by a gimbal--type hip. Actuators control the orientation of the leg with respect to the body and the axial thrust delivered by the leg. Sensors provide state information from the hip, leg and body to a control computer.

The body is made of an aluminium frame, on which are mounted hip actuators, valves, gyroscopes, and computer interface electronics.

The hopper runs at up to 2.2 m/s with strides of up to 0.8 m. When hopping on the spot it keeps its position to within 0.25 m.

# References


[1] Bernstein N., The Coordination and Regulation of Movements, Pergamon Press, 1967.

[2] Mizen N.J., Design and Test of a Full-Scale, Wearable, Exoskeletal Structure, Tech. Rep. Cornell Aeron. Lab., 1964.

[3] Todd D.J., Walking Machines, Kogan Page, London, 1985.

[4] Vukobratović M., Juričić D., "Contribution to the Synthesis of Biped Gait", IEEE Trans. on Bio-Medical Engineering, Vol. 16, No. 1, 1969.

[5]   Hristić D., Vukobratović M., "A New Approach to Rehabilitation
      of Severely Handicaped Persons", Automatika, No. 1, Zagreb, 1971.

[6]   Vukobratović M., Ćirić V., Hristić D., "Contribution to the Study of
      Active Exoskeletons", Proc. 5th IFAC Congress, Paris, 1972.

[7]   Vukobratović M., et al., "Further Development of Active Exoske-
      letons for Rehabilitation", Proc. External Control of Human Ex-
      tremities, Dubrovnik, 1972.

[8]   Vukobratović M., Hristić D., Stojiljković Z., "Development of
      Active Exoskeletons for Rehabilitation", Medical and Biological
      Engineering, Vol. 12, No. 1, 1974.

[9]   Vukobratović M., Ćirić V., Hristić D., "Control of Two-Legged
      Artificial Walking Systems", Proc. IFAC Symp. on Automatic Con-
      trol in Space, Dubrovnik, 1971.

[10]  Vukobratović M., Hristić D., Ćirić V., Zečević M., "Analysis of
      Energy Demand Distribution Within Anthropomorphic Systems",
      Trans. of the ASME, Journal of Dynamic Systems, Measurement and
      Control, Vol. 95, No. 4, 1973.

[11]  Hristić D., Vukobratović M., "Development of Active Aids for Han-
      dicapped", Proc. III International Conf. on Bio-Medical Enginer-
      ing, Sorrento, Italy, 1973.

[12]  Vukobratović M., Legged Locomotion Systems and Anthropomorphic
      Mechanisms, research monograph, Mihailo Pupin Institute, Beograd,
      1975.

[13]  Vukobratović M., Frank A., Juričić D., "On the Stability of Bi-
      ped Locomotion", IEEE Trans. on Bio-Medical Engineering, Vol. 17,
      No. 1, 1970.

[14]  Hristić D., Vukobratović M., "Development of Modular Active Or-
      thosis for Rehabilitation", Orthopadie Technik 4, Duisburg, 1978.

[15]  Vukobratović M., Hristić D., "Modular Active Systems for Rehabi-
      litation of Severely Handicapped", Proc. 7th External Control of
      Human Extremities, Dubrovnik, 1984.

[16]  WASEDA ROBOT, Publ. Kato Laboratory, Waseda University, Tokyo,
      1986.

[17]  Control of Walking Apparatus with Two Actuators Having Puls-
      -Width Time Control, (in Russian), Report, Intitute of Mechanics,
      No. 3605, Moscow, 1988.

[18]  Yamada M., Furusho J., Sano A., "Dynamic Control of Walking Robot
      with Kick-Action", Proc. of 85' International Conference on Ad-
      vanced Robotics, pp. 405-412, 1985, Tokyo.

[19]  Sano A., Furusho J., "3D Steady Walking Using Active Control of
      Body Sway Motion and Foot Pressure", Proc. of USA - Japan Symp.
      on Flexible Automation, pp. 665-672, Minneapolis, 1988.

[20]  Vodovnik L., Functional Electrical Stimulation of Extremities,
      Advances in Electronics and Electron Physics, Vol. 30, 1971, Aca-
      demic Press, Now York.

[21] Vodovnik L., Kralj A., Bajd T., "Modification on Abnormal Motor Control with Functional Electrical Stimulation of Peripheral Nerves", In: Recent Achievements in Restorative Neorology 1 (Eds. Eccles, Dimitrijević), Karger, Basel, 1985.

[22] Vukobratović M., "How to Control Artificial Anthropomorphis Systems", IEEE Trans. on Systems, Man and Cybernetics, Vol. SMC-3, No. 5, 1973.

[23] Hatze H., "A General Myocybernetic Control Model of Skeletal Muscle", Biol. Cybern., Vol. 28, pp. 143-157, 1978.

[24] Zheng Y.F., Hemami H., Stokes B.T., "Muscle Dynamics, Principle and Stability", IEEE Trans. on Biomedical Engineering, Vol. 31, pp. 489-497, 1984.

[25] Viitasalo J.T., Komi P.V., "Force-Time Characteristics and Fiber Composition in Human Leg Extensor Muscles", Eur. J. Appl. Physiol., Vol. 40, pp. 7-15, 1978.

[26] Hill A.V., "Heat of Shortening and Dynamic Constants of Muscle", Proc. Royal Society, B 126, pp. 136-195, London, 1938.

[27] Wilkie D.R., "The Relation Between Force and Velocity in Human Muscle", J. Physiol. 110, pp. 249-280, 1950.

[28] Hoyle G., Castillo J., "Neuromuscular Transmission in Peripatus", J. Exp. Biol. 83, pp. 13-29, 1979.

[29] Hodkin A., Huxley A., "A Quantitative Description of Membrane Current and its Application to Conduction and Excitation in Nerve", J. Physiol. 117, pp. 500-544, London, 1952.

[30] Vukobratović M., Applied Dynamics of Manipulation Robots: Modelling, Analysis and Examples, Springer-Verlag, 1989.

[31] Hristić D., Vukobratović M., Ivančević V., "Hybrid Joint Control System Modelling, Proc. 9th External Control of Human Extremities, Dubrovnik, 1987.

[32] Hristić D., Vukobratović M., Ivančević V., "Hybrid Control System Modelling and Experiments", Proc. World Congr. Orth. Engineering OTTO, Nuremberg, 1988.

[33] Vukobratović M., Stokić D., Applied Control of Manipulation Robots: Analysis, Synthesis and Exercises, Springer-Verlag, 1989.

[34] Frank A.A., McGhee R.B., "Some Consideration Relating to the Design of Autopilots for Legged Vehicles", Journal of Terramechanics, Vol. 6, No. 1, 1969.

[35] Devyanin E.A., Gurfinkel V.S., Kartashev E.V., Lensky V.A., Shneider A.Yu., Shtilman L.G., "A Six-Legged Walking Robot Capable of Terrain Adaptation", Mechanism and Machine Theory, Vol. 18, No. 4, 1983.

[36] Waldron K.J. Vohnout V.J., Pery A., McGhee R.B., "Configuration Design of the Adaptive Suspension Vehicle", International Journal of Robotics Research, Vol. 3, No. 2, 1984.

[37] Raibert M.H., Legged Robots that Balance, MIT Press, Cambridge, Massachusetts, 1985.

# Communications and Control Engineering Series

Editors: A. Fettweis, J. L. Massey, J. W. Modestino, M. Thoma

## Scientific Fundamentals of Robotics

Volume 8
M. Vukobratović

### *Dynamics and Control of Flexible Manipulation Robots*

1990. In prep. ISBN 3-540-17455-9

Volume 6
M. Vukobratović,
V. Potkonjak

### *Applied Dynamics and CAD of Manipulation Robots*

1985. XII, 305 pp. 187 figs. ISBN 3-540-13074-8

Volume 5
M. Vukobratović,
D. Stokić, N. Kirćanski

### *Non-Adaptive and Adaptive Control of Manipulation Robots*

1985. X, 383 pp. 111 figs. ISBN 3-540-13073-X

Volume 4
M. Vukobratović,
N. Kirćanski

### *Real-Time Dynamics of Manipulation Robots*

1985. XII, 239 pp. 43 figs. ISBN 3-540-13072-1

Volume 3
M. Vukobratović,
M. Kirćanski

### *Kinematics and Trajectories Synthesis of Manipulation Robots*

1986. X, 267 pp. 66 figs. ISBN 3-540-13071-3

Volume 2
M. Vukobratović,
D. Stokić

### *Control of Manipulation Robots*

*Theory and Application*
1982. XIII, 363 pp. 111 figs. ISBN 3-540-11629-X

Volume 1
M. Vukobratović,
V. Potkonjak

### *Dynamics of Manipulation Robots*

*Theory and Application*
1982. XIII, 303 pp. 149 figs. ISBN 3-540-11628-1

M. Vukobratović

## *Applied Dynamics of Manipulation Robots*

### *Modelling, Analysis and Examples*

1989. XIV, 471 pp. 176 figs.
ISBN 3-540-51468-6


M. Vukobratović, D. Stokić

## *Applied Control of Manipulation Robots*

### *Analysis, Synthesis and Exercises*

1989. XVI, 470 pp. 100 figs.
ISBN 3-540-51469-4


M. Vukobratović

## *Introduction to Robotics*

With contributions by M. Djurović,
D. Hristić, B. Karan, M. Kirćanski,
N. Kirćanski, D. Stokić, D. Vuijić,
M. Vukobratović

1989. XIV, 301 pp. 228 figs.
ISBN 3-540-17452-4


Springer-Verlag Berlin Heidelberg New York London Paris Tokyo Hong Kong